D0876378

Curves and Surfaces
for
Computer Aided Geometric Design
A Practical Guide

Third Edition

This is a volume in
COMPUTER SCIENCE AND SCIENTIFIC COMPUTING

Werner Rheinbolt, editor

Curves and Surfaces
for
Computer Aided
Geometric Design
A Practical Guide

Third Edition

Gerald Farin

Department of Computer Science
Arizona State University
Tempe, Arizona

ACADEMIC PRESS, INC.
Harcourt Brace Jovanovich, Publishers
Boston   San Diego   New York
London   Sydney   Tokyo   Toronto

ACADEMIC PRESS, INC.
1250 Sixth Avenue, San Diego, CA 92101

United Kingdom edition published by
ACADEMIC PRESS LIMITED
24–28 Oval Road, London NW1 7DX

Chapter 1 was written by P. Bézier.
Chapters 11 and 22 were written by W. Boehm.
Plates IX, X, and XI reproduced courtesy of Silicon Graphics, Inc.

Library of Congress Cataloging-in-Publication Data

Farin, Gerald E.
    Curves and surfaces for computer aided geometric design : a
practical guide / Gerald Farin. — 3rd. ed.
        p.   cm. — (Computer science and scientific computing)
    Includes bibliographical references and index.
    ISBN 0-12-249052-5
    1. Computer graphics.   I. Title.   II. Series.
T385.F37   1992   92-26864
006.6—dc20   CIP

Printed in the United States of America
92 93 94 95   MV   9 8 7 6 5 4 3 2 1

To My Parents

Contents

Preface

In the late 1950s, hardware became available that allowed the machining of 3D shapes out of blocks of wood or steel.[1] These shapes could then be used as stamps and dies for products such as the hood of a car. The bottleneck in this production method was soon found to be the lack of adequate software. In order to machine a shape using a computer, it became necessary to produce a computer-compatible description of that shape. The most promising description method was soon identified to be in terms of parametric surfaces. An example for this approach is provided by Plates I and III: Plate I shows the actual hood of a car; Plate III shows how it is represented internally as a collection of parametric surfaces.

The theory of parametric surfaces was well understood in differential geometry. Their potential for the representation of surfaces in a Computer Aided Design (CAD) environment were not known at all, however. The exploration of the use of parametric curves and surfaces can be viewed as the origin of Computer Aided Geometric Design (CAGD).

The major breakthroughs in CAGD were undoubtedly the theory of Bézier surfaces and Coons patches, later combined with B-spline methods. Bézier curves and surfaces were independently developed by P. de Casteljau at Citroën and by P. Bézier at Rénault. De Casteljau's development, slightly earlier than Bézier's, was never published, and so the whole theory of polynomial curves and surfaces in Bernstein form now bears Bézier's name. CAGD became a discipline in its own right after the 1974 conference at the University of Utah (see Barnhill and Riesenfeld [27]).

This book presents a unified treatment of the main ideas in CAGD. During the last years, there has been a trend towards more geometric insight into curve and surface schemes; I have followed this trend by basing most concepts on simple geometric algorithms. For instance, a student will be able

---

[1]A process that is now called CAM for *Computer Aided Manufacturing*.

to construct Bézier curves with hardly any knowledge of the concept of a parametric curve. Later, when parametric curves are discussed in the context of differential geometry, one can apply differential geometry ideas to the concrete curves that were developed before.

The theory of Bézier curves (and rational Bézier curves) plays a central role in this book. They are numerically the most stable among all polynomial bases currently used in CAD systems, as was recently shown by Farouki and Rajan [177]. Thus Bézier curves are the ideal geometric standard for the representation of piecewise polynomial curves. Also, Bézier curves lend themselves easily to a geometric understanding of many CAGD phenomena and may, for instance, be used to derive the theory of rational and nonrational B-spline curves.

While this book offers a comprehensive treatment of the basic methods in curve and surface design, it is not meant to provide solutions to application-oriented problems that arise in practice. In particular, no algorithms are included to handle intersection, rendering, or offset problems. At present, no unified approach exists for these "geometry processing" problems. However, the material presented here should enable the reader to read the advanced literature on these topics; on offsets: [161], [172], [173], [268], [271], [274], [291], [398], [457]; on intersections: [24], [141], [143], [203], [208], [232], [272], [252], [275], [301], [308], [326], [429], [360], [382], [402], [433], [431]; on rendering: [1], [90], [193], [204], [309], [446].

Also, this is not a text on solid modeling. That branch of geometric modeling is concerned with the representation of objects that are enclosed by an assembly of surfaces, mostly very elementary ones such as planes, cylinders, or tori. When solid modeling systems become fully accepted, they will be able to incorporate the free-form curves and surfaces described in this book. The literature includes: [93], [176], [184], [262], [325], [335], [394], [463].

I have taught the material presented here in the form of both conference tutorials and university courses, typically at the intermediate level. The problems are mostly taken from graduate level courses. In general, the material in the text should enable readers to *use* the presented material. The problems are meant as a guide towards deeper understanding. They should be complemented with simpler exercises if this text is used at a lower level. In teaching this material, it is essential that students have access to computing and graphics facilities; practical experience greatly helps the understanding and appreciation of what might otherwise remain dry theory.

When I use this book as a text for a one semester CAGD class at the lower graduate / upper undergraduate level, I typically cover the following chapters: the first half of Chapter 2, Chapters 3, 4, 6, 8, 9, 16, and 17. Material from other chapters is sprinkled in as needed.

For the second edition, a sequence of C programs was added to the text. These programs are my implementations of some (but not all) of the most important methods that are described here. The programs were tested for many examples, but they are not meant to be "industrial strength." In general, no checks have been made for consistency or correctness of input data. Also, modularity was valued higher than efficiency. The programs are in C, but with non-C users in mind – in particular, all modules should be easily translatable into FORTRAN.

More programs have been added to the third edition, and also several data and PostScript files. They have now been transferred to a diskette, and only the program headers are left in the text. More problems were added that build on top of these programs: so instead of, say, having to program cubic spline curves, a student is now expected to modify and expand the provided cubic spline routines. In addition to the programs, several sections have now been included that deal with the new concept of blossoming. This is a powerful technique that unifies many aspects of polynomial and piecewise polynomial curves.

This book would not have been possible without the stimulating environment provided by the CAGD group at Arizona State University (and formerly at the University of Utah), led by Robert E. Barnhill. The book also greatly benefitted from numerous discussions I had with experts such as T. Foley, Q. Fu, H. Hagen, J. Hoschek, S. Kersey, G. Nielson, R. Patterson, and A. Worsey. I would also like to express my appreciation for the funding provided by the National Science Foundation and the Department of Energy.[2] Special thanks go to D. C. Hansford for the numerous helpful suggestions concerning the mathematical side of the material and also to W. Boehm, who was a critical and constructive consultant during all stages of this book.

For the second edition, many improvements over the first versions were suggested by S. Abi-Ezzi, N. Beebe, W. Boehm, R. E. Barnhill, E. Clapp, P. J. Davis, D. Hansford, F. Kimura, S. Mann, G. Nielson, A. Swimmer, K. Voegele, W. Waggenspack, M. Wozny, and Y. Yamaguchi.

For the third edition, special thanks go to T. De Rose, T. Foley, D. Hansford, J. Hoschek, T. Klojcnik, N. Max, K. Opitz, H. Späth, and T. Varady.

Tempe, Ariz. Gerald Farin

[2]Grants DCR-8502858 and DE-FG02-87ER25041 respectively.

Chapter 1

P. Bézier: How a Simple System Was Born

To solve CAD/CAM mathematical problems, many many solutions have been offered, each adapted to specific matters. Most of the systems were invented by mathematicians, but UNISURF, at least at the beginning, was developed by mechanical engineers from the automotive industry. These engineers were familiar with parts mainly described by lines and circles; fillets and other blending auxiliary surfaces were scantly defined, their final shape being left to the skill and experience of patternmakers and die-setters.

About 1960, designers of stamped parts such as car-body panels used French curves and sweeps. The final standard, however, was the "master model," the shape of which, for many and valid reasons, could not coincide with the curves traced on the drawing board. This problem resulted in discussions, arguments, haggling, retouches, expenses, and delay.

Obviously, no significant improvement could be expected as long as a method was not devised that could provide an accurate, complete, and undisputable definition of freeform shapes.

Computing and numerical control (NC) had made great progress at that time, and it was certain that only numbers, transmitted from drawing office to tool drawing office, manufacture, patternshop, and inspection, could provide an answer. Of course, drawings would remain necessary, but they would only be explanatory, their accuracy having no importance. Numbers would be the only and final definition.

Certainly, no system could be devised without the help of mathematics – yet designers, in charge of the operation, had a good knowledge of geometry, especially descriptive geometry, but no basic training in algebra or analysis.

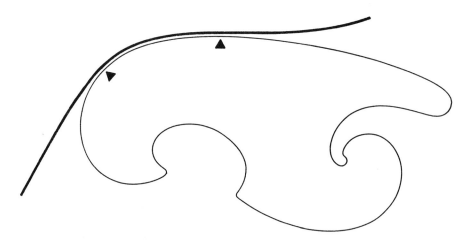

Figure 1.1: An arc of a hand-drawn curve is approximated by a part of a template.

Note that in France very little was known at that time about the work performed in the American aircraft industry. The papers of James Ferguson were not publicized before 1964; Citroen was secretive about the results obtained by Paul de Casteljau, and the famous technical report MAC-TR-41 (by S. A. Coons) did not appear until 1967. The works of W. Gordon and R. Riesenfeld were printed in 1974.

The idea of UNISURF was initially oriented toward geometry rather than analysis, but kept in mind that every datum should be exclusively expressed by numbers. For instance, an arc of a curve could be represented (Figure 1.1) by the coordinates, cartesian of course, of its limit points, i.e., A and B, together with their curvilinear abscissas, related with a grid traced on the edge.

The shape of the middle line of a sweep is a cube, granted that its cross section is constant, its matter is homogeneous, and neglecting the effect of friction on the tracing cloth, but it is difficult to take into account the length between endpoints. Moreover, the curves employed for software for NC machine tools, i.e., 2D milling machines, were lines and circles and, sometimes,

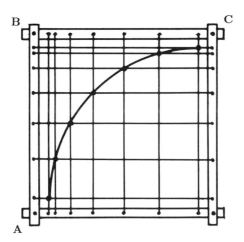

Figure 1.2: A circular arc is obtained by connecting the points in this rectangular grid.

parabolas. Hence, a spline shape should be divided and subdivided into small arcs of circles placed end to end.

To transform an arc of circle into a portion of an ellipse, one could imagine (Figure 1.2) a square frame containing two sets of strings, the intersections of which would be located on an arc of a circle; the frame sides being hinged, the square is transformed into a diamond (Figure 1.3) and the circle becomes an arc of an ellipse, which would be entirely defined as soon as the coordinates of points A, B, and C were known. If the hinged sides of the frame were replaced by pantographs (Figure 1.4), the diamond would become a parallelogram, and the definition of the arc of an ellipse would still result from the coordinates of the three points A, B, and C (Figure 1.5).

Of course, this idea was not realistic, but it was easily replaced by the computation of coordinates of successive points of the curve. Harmonic functions were available with the help of analog computers, which were widely used at that time and gave excellent results.

However, employing only arcs of ellipses limited by conjugate diameters was far too restrictive, and a more flexible definition was required.

Another idea came from the practice of a speaker projecting, with a hand-held torch, a small sign, cross, or arrow onto a screen displaying a figure

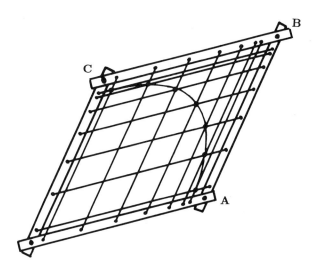

Figure 1.3: If the frame from the previous figure is sheared, an arc of an ellipse is obtained.

printed on a slide. Replacing the arrow with a curve and recording the exact location and orientation of the torch (Figure 1.6) would define the image of the curve projected on the wall of the drawing office. One could even imagine having a variety of slides, each of which would bear a specific curve: circle, parabola, astroid, etc.

Of course, this was not a realistic idea, because the focal plane of the zoom would seldom be square to the axis; an optician's nightmare! But the principle could be translated, via projective geometry and matrix computation, into cartesian coordinates.

At that time, designers defined the shape of a car body by cross sections located one hundred millimeters apart, and sometimes less. The advantage was that, from a drawing, one could derive templates for adjusting a clay model, a master, or a stamping tool. The drawback was that a stylist does not define a shape by cross sections but with so-called "character lines," which seldom are plane curves. Hence, a good system should be capable of manipulating and defining directly "space curves" or "freeform curves." Of course, one could imagine working alternately (Figure 1.7) on two projections of a space curve, but it is very unlikely that a stylist would accept such a solution.

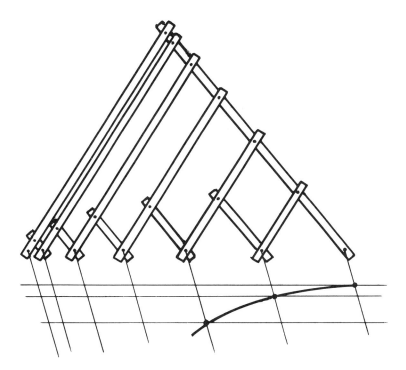

Figure 1.4: Pantograph construction of an arc of an ellipse.

Theoretically, at least, a space curve could be expressed by a sweep having a circular section, constrained by springs or counterweights (Figure 1.8), but this would prove quite impractical.

Would not the best solution be to revert to the basic idea of a frame? Instead of a curve inscribed in a square it would be located in a cube (Figure 1.9) that could become any parallelepiped (Figure 1.10) by a linear transformation that is easy to compute; the first idea was to choose a basic curve that would be the intersection of two circular cylinders. The parallelepiped would be defined (Figure 1.10) by points O, X, Y, and Z, but it is more practical to put the basic vectors end to end so as to obtain a polygon OMNB (Figure 1.10), which defines directly the endpoint B and its tangent NB. Points O, M, N, and B need not be coplanar and can define a space curve.

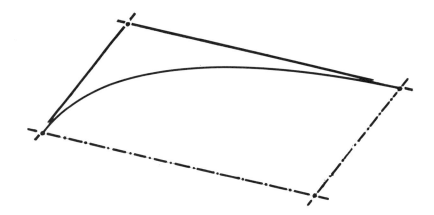

Figure 1.5: A "control polygon" for an arc of an ellipse.

Polygons with three legs can define quite a large variety of curves, but in order to increase it, we can make use of cubes and hypercubes of any order (Figure 1.11) and the relevant polygons (Figure 1.13) (see Figure 3.4 in Section 3.3).

At that moment, it became necessary to do away with harmonic functions and revert to polynomials, which was even more desirable since digital computers were gradually replacing analog computers. The polynomial functions were chosen according to the properties that were considered best: tangency, curvature, etc. Later it was discovered that they could be considered as sums of Bernstein's functions.

When it was suggested that these curves could replace sweeps and French curves, most stylists objected that they had invented their own templates and would not change. The solemn promise was made that their "secret" curves whould be translated in secret listings and buried in the most secret part of the memory of the computer, and that nobody but the stylists could have a key to the vaulted cellar. In fact, the standard curves were flexible enough and secret curves were soon forgotten; designers and draughtsmen easily understood the polygons and their relation to the shape of the corresponding curves.

In the traditional process of body engineering, a set of curves is carved in a 3D model, between which interpolation is left to the experience of highly

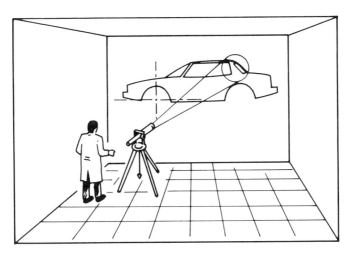

Figure 1.6: A projector producing a "template curve" on the drawing of an object.

skilled patternmakers. However, to obtain a satisfactory numerical definition, the surface had to be totally expressed with numbers.

At that time, around 1960, very little, if anything, had been published about biparametric patches. The basic idea of UNISURF came from a comparison with a process often used in foundries to obtain a core: sand being compacted in a box (Figure 1.12). The shape of the upper surface of the core is obtained by scraping off the surplus with a timber plank cut as a template; of course, a shape obtained by such a method is relatively simple since the shape of the plank is constant and that of the box edges is generally simple. To make the system more flexible, one might change the shape of the template at the same time as it is moved. In fact, this idea takes us back to a very old, and sometimes forgotten, definition of a surface: It is the locus of a curve which is at the same time moved and distorted. About 1970, a Dutch laboratory sculptured blocks of styrofoam with a flexible strip of steel, heated by electricity, the shape of which was controlled by the flexion torque imposed on its extremities.

This process could not produce a large variety of shapes, but the principle could be translated into a mathematical solution: The guiding edges of the box are similar to the curves AB and CD of Figure 1.13, which can be considered

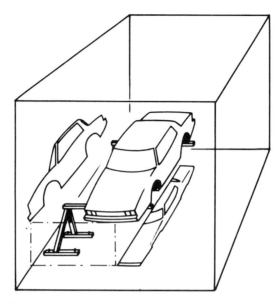

Figure 1.7: Two imaginary projections of a car.

as directrices of a surface, defined by their characteristic polygon. If a curve such as EF is generatrix, defined by its own polygon, the ends of which run along lines AB and CD, and the intermediate vertices of the polygon are on curves GH and JK, the surface ABDC is known as soon as the four polygons are defined; connecting the corresponding vertices of the polygons defines the "characteristic net" of the patch, which plays, regarding the surface, the same part as a polygon a curve. Hence, the cartesian coordinates of the points of the patch are computed according to the values of two parameters.

After expressing this basic idea, a good many problems remained to be solved: choosing adequate functions, blending curves and patches, dealing with degenerate patches, to name only a few. Resolving these was a matter of relatively simple mathematics, the basic principle remaining untouched.

A system has thus been progressively created. If we consider the manner in which an initial idea evolved, we observe that the first solution – parallelogram, pantograph – is the result of an education oriented toward kinematics, the conception of mechanisms. Then geometry and optics appeared, which very likely came from some training in the army, when geometry, cosmography, and topography played an important part. Reflexion was oriented toward

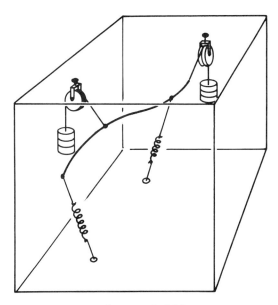

Figure 1.8: A curve held by springs.

analysis, parametric spaces, and, finally, data processing, because a theory, as convenient as it may look, must not impose too heavy a task on the computer and must be easily understood, at least in principle, by the operators.

Note that the various steps of this conception have a point in common: Each idea must be related to the principle of a material system, simple and primitive though it may look, on which a variable solution could be based.

An engineer is a man who defines what is to be done and how it could be done; he not only describes the goal, but he leads the way toward it.

Before looking any deeper into this subject, note that elementary geometry played a major part in its development, and it should not gradually disappear from the courses of a mechanical engineer. Each idea, each hypothesis was expressed by a figure, or a sketch, representing a mechanism. It would have been extremely difficult to build a purely mental image of a somewhat elaborate system without the help of pencil and paper. Let us consider, for instance, Figures 1.9 and 1.11; they are equivalent to Eqs. (4.7) and (16.6) in the subsequent chapters. Evidently, these formulas, conveniently arranged, are best suited to express data given to a computer, but most people would better understand a simple figure than the equivalent algebraic expression.

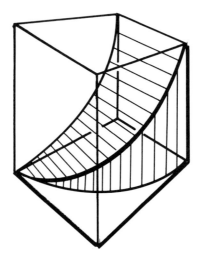

Figure 1.9: A curve defined inside a cube.

Napoleon said: "A short sketch is better than a long report."

Which are the parts played by experience, by theory, and by imagination in the creation of a system? There is no definite answer to such a query. The importance of experience and of theoretical knowledge is not always clearly perceived; imagination seems a gift, a godsend, or the result of a beneficial heredity; but is not imagination, in fact, the result of the maturation of the knowledge gained during education and professional practice? Is it not born from facts apparently forgotten, stored in the dungeon of a distant part of memory, and suddenly remembered when circumstances call them back? Is not imagination based, partly, on the ability to connect notions that, at first sight, look quite unrelated, such as mechanics, electronics, optics, foundry, data processing, etc. Is it not the ability to catch barely seen analogies – as Alice in Wonderland did, to go "through the mirror?"

Will psychologists someday be able to detect in man such a gift that would be applicable to science and technology? Has it a relation with the sense of humor that can detect unexpected relations between facts that look quite unconnected? Shall we learn how to develop it? Will it forever remain a gift, devoted by pure chance to some people while for others carefulness prevails?

It is important that, sometimes, "sensible" men give free rein to imaginative people. "I succeeded," said Henry I. Ford, "because I let some fools try what wise people had advised me not to let them try."

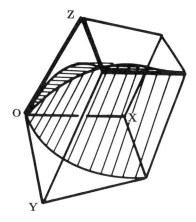

Figure 1.10: A curve defined inside a parallelepiped.

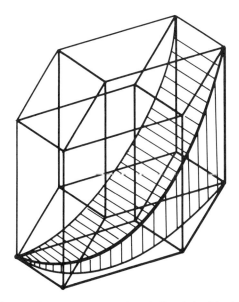

Figure 1.11: Higher order curves can be defined inside higher dimensional cubes.

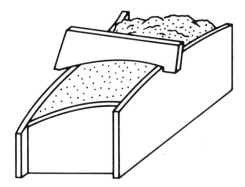

Figure 1.12: A surface is obtained by scraping off excess material with wooden templates.

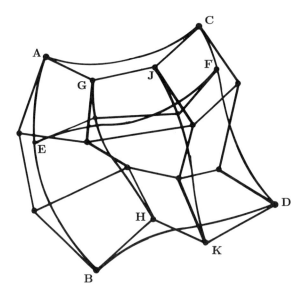

Figure 1.13: The characteristic net of a surface.

Chapter 2

Introductory Material

2.1 Points and Vectors

When a designer or stylist works on an object, he or she does not think of that object in very mathematical terms. A point on the object would not be thought of as a triple of coordinates, but rather in functional terms: as a corner, the midpoint between two other points, and so on. The objective of this book, however, is to discuss objects that *are* defined in mathematical terms, the language that lends itself best to computer implementations. As a first step toward a mathematical description of an object, one therefore defines a *coordinate system* in which it will be described analytically.

The space in which we describe our object does not possess a preferred coordinate system – we have to define one ourselves. Many such systems could be picked (and some will certainly be more practical than others). But whichever one we choose, it should not affect any properties of the object itself. Our interest is in the object and not in its relationship to some arbitrary coordinate system. Therefore, the methods we develop must be independent of a particular choice of a coordinate system. We say that those methods must be *coordinate-free* or *coordinate-independent*.[1]

The concept of coordinate-free methods is stressed throughout this book. It motivates the strict distinction between points and vectors as discussed next. (For more details on this topic, see R. Goldman [207].)

We shall denote *points*, elements of three-dimensional euclidean (or point) space $\mathbb{E}^3$, by lowercase boldface letters such as $\mathbf{a}$, $\mathbf{b}$, etc. (The term "euclidean

---

[1]More mathematically, the geometry of this book is affine geometry. The objects that we will consider "live" in affine spaces, not in linear spaces.

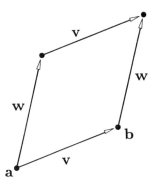

Figure 2.1: Points and vectors: vectors are not affected by translations.

space" is used here because it is a relatively familiar term to most people. More correctly, we should have used the term "affine space.") A point identifies a location, often relative to other objects. Examples are the midpoint of a straight line segment or the center of gravity of a physical object.

The same notation (lowercase boldface) will be used for *vectors*, elements of three-dimensional linear (or vector) space $I\!R^3$. If we represent points or vectors by coordinates relative to some coordinate system, we shall adopt the convention of writing them as coordinate columns.

Although both points and vectors are described by triples of real numbers, we emphasize that there is a clear distinction between them: for any two points **a** and **b**, there is a unique vector **v** that points from **a** to **b**. It is computed by componentwise subtraction:

$$\mathbf{v} = \mathbf{b} - \mathbf{a}; \quad \mathbf{a}, \mathbf{b} \in I\!\!E^3, \quad \mathbf{v} \in I\!R^3.$$

On the other hand, given a vector **v**, there are infinitely many pairs of points **a**, **b** such that $\mathbf{v} = \mathbf{b} - \mathbf{a}$. For if **a**, **b** is one such pair and if **w** is an arbitrary vector, then $\mathbf{a} + \mathbf{w}, \mathbf{b} + \mathbf{w}$ is another such pair since $\mathbf{v} = (\mathbf{b} + \mathbf{w}) - (\mathbf{a} + \mathbf{w})$ also. Figure 2.1 illustrates this fact.

Assigning the point $\mathbf{a} + \mathbf{w}$ to every point $\mathbf{a} \in I\!\!E^3$ is called a *translation*, and the above asserts that vectors are invariant under translations while points are not.

Elements of point space $I\!\!E^3$ can only be *subtracted* – this operation yields a vector. They cannot be *added* – this operation is not defined for points. (It is defined for vectors.) Figure 2.2 gives an example.

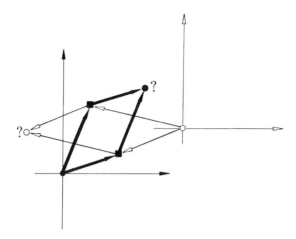

Figure 2.2: Addition of points: this is not a well-defined operation, since different coordinate systems would produce different "solutions." The two points to be "added" are marked by solid squares.

However, addition-like operations are defined for points: they are *barycentric combinations*.[2] These are weighted sums of points where the weights sum to one:

$$\mathbf{b} = \sum_{j=0}^{n} \alpha_j \mathbf{b}_j; \quad \begin{matrix} \mathbf{b}_j \in \mathbb{E}^3 \\ \alpha_0 + \cdots + \alpha_n = 1 \end{matrix} \cdot \tag{2.1}$$

At first glance, this looks like an undefined summation of points, but we can rewrite (2.1) as

$$\mathbf{b} = \mathbf{b}_0 + \sum_{j=1}^{n} \alpha_j (\mathbf{b}_j - \mathbf{b}_0),$$

which is clearly the sum of a point and a vector.

An example of a barycentric combination is the centroid **g** of a triangle with vertices **a**, **b**, **c**, given by

$$\mathbf{g} = \frac{1}{3}\mathbf{a} + \frac{1}{3}\mathbf{b} + \frac{1}{3}\mathbf{c}.$$

The term "barycentric combination" is derived from "barycenter," meaning "center of gravity." The origin of this formulation is in physics: if the $\mathbf{b}_j$

---

[2]They are also called *affine combinations*.

are centers of gravity of objects with masses m_j, then their center of gravity
b is located at $\mathbf{b} = \sum m_j \mathbf{b}_j / \sum m_j$ and has the combined mass $\sum m_j$. (If
some of the m_j are negative, the notion of electric charges may provide a
better analogy; see Coxeter [107], p. 214.) Since a common factor in the m_j is
immaterial for the determination of the center of gravity, we may normalize
them by setting $\sum m_j = 1$.

An important special case of barycentric combinations are the *convex
combinations*. These are barycentric combinations where the coefficients α_j,
in addition to summing to one, are also nonnegative. A convex combination
of points is always "inside" those points, which is an observation that leads
to the definition of the *convex hull* of a point set: this is the set that is formed
by all convex combinations of a point set. Figure 2.3 gives an example; see
also Problems. More intuitively, the convex hull of a set is formed as follows:
for a 2D set, imagine a string that is loosely circumscribed around the set,
with nails driven through the points in the set. Now pull the string tight – it
will become the boundary of the convex hull.

The convex hull of a point set is a *convex set*. Such a set is characterized
by the following: for any two points in the set, the straight line connecting
them is also contained in the set. Examples are ellipses or parallelograms.
It is an easy exercise to verify that affine maps (see next section) preserve
convexity.

Let us return to barycentric combinations, which generate points from
points. If we want to generate a *vector* from a set of points, we may write

$$\mathbf{v} = \sum_{j=0}^{n} \sigma_j \mathbf{P}_j,$$

where we have a new restriction on the coefficients: Now we must demand
that the σ_j sum to zero.

If we are given an equation of the form

$$\mathbf{a} = \sum \beta_j \mathbf{b}_j,$$

and **a** is supposed to be a point, then we must be able to split the sum into
two groups:

$$\mathbf{a} = \sum_{\sum \beta_j = 1} \beta_j \mathbf{b}_j + \sum_{\text{remaining } \beta's} \beta_j \mathbf{b}_j.$$

Then the $\mathbf{b}_j$ in the first sum are points, whereas those in the second one are
vectors.

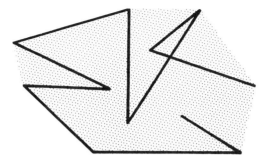

Figure 2.3: Convex hulls: a point set (a polygon) and its convex hull, shown shaded.

The interplay between points and vectors is unusual at first. Later, it will turn out to be of invaluable theoretical and practical help. For example, we can perform quick *type checking* when we derive formulas. If the point coefficients fail to add up to one or zero – depending on the context – we know that something has gone wrong. In a more formal way, T. DeRose has developed the concept of "geometric programming," a graphics language that automatically performs type checks [132], [133]. R. Goldman's article [207] treats the validity of point/vector operations in more detail.

2.2 Affine Maps

Most of the transformations that are used to position or scale an object in a computer graphics or CAD environment are *affine* maps. (More complicated, so-called projective maps are discussed in Chapter 14.) The term "affine map" is due to L. Euler; affine maps were first studied systematically by F. Moebius [337].

The fundamental operation for points is the barycentric combination. We will thus base the definition of an affine map on the notion of barycentric combinations. *A map Φ that maps $\mathbb{E}^3$ into itself is called an affine map if it leaves barycentric combinations invariant.* So if

$$\mathbf{x} = \sum \alpha_j \mathbf{a}_j; \quad \mathbf{x}, \mathbf{a}_j \in \mathbb{E}^3$$

and Φ is an affine map, then also

$$\Phi\mathbf{x} = \sum \alpha_j \Phi\mathbf{a}_j; \quad \Phi\mathbf{x}, \Phi\mathbf{a}_j \in \mathbb{E}^3. \tag{2.2}$$

This definition looks fairly abstract, yet has a simple interpretation. The expression $\mathbf{x} = \sum \alpha_j \mathbf{a}_j$ specifies how we have to weight the points $\mathbf{a}_j$ so that their weighted average is $\mathbf{x}$. This relation is still valid if we apply an affine map to all points $\mathbf{a}_j$ and to $\mathbf{x}$. As an example, the midpoint of a straight line segment will be mapped to the midpoint of the affine image of that straight line segment. Also, the centroid of a number of points will be mapped to the centroid of the image points.

Let us now be more specific. In a given coordinate system, a point $\mathbf{x}$ is represented by a coordinate triple, which we also denote by $\mathbf{x}$. An affine map now takes on the familiar form

$$\Phi\mathbf{x} = A\mathbf{x} + \mathbf{v}, \tag{2.3}$$

where A is a 3×3 matrix and $\mathbf{v}$ is a vector from $I\!R^3$.

A simple computation verifies that (2.3) does in fact describe an affine map, i.e., that barycentric combinations are preserved by maps of that form. For the following, recall that $\sum \alpha_j = 1$:

$$\begin{aligned} \Phi\left(\sum \alpha_j \mathbf{a}_j\right) &= A\left(\sum \alpha_j \mathbf{a}_j\right) + \mathbf{v} \\ &= \sum \alpha_j A\mathbf{a}_j + \sum \alpha_j \mathbf{v} \\ &= \sum \alpha_j \left(A\mathbf{a}_j + \mathbf{v}\right) \\ &= \sum \alpha_j \Phi\mathbf{a}_j, \end{aligned}$$

which concludes our proof. It also shows that the inverse of our initial statement is true as well: Every map of the form (2.3) represents an affine map.

Some examples of affine maps:

The identity. It is given by $\mathbf{v} = \mathbf{0}$, the zero vector, and by $A = I$, the identity matrix.

A translation. It is given by $A = I$, and a *translation vector* $\mathbf{v}$.

A scaling. It is given by $\mathbf{v} = \mathbf{0}$ and by a diagonal matrix A. The diagonal entries define by how much each component of the preimage $\mathbf{x}$ is to be scaled.

A rotation. If we rotate around the z-axis, then $\mathbf{v} = \mathbf{0}$ and

$$A = \begin{bmatrix} \cos\alpha & -\sin\alpha & 0 \\ \sin\alpha & \cos\alpha & 0 \\ 0 & 0 & 1 \end{bmatrix}.$$

Figure 2.4: A shear: this affine map is sometimes used in font design in order to generate slanted fonts. Left: original letter; right: slanted letter.

A shear. An example is given by $\mathbf{v} = \mathbf{0}$ and

$$A = \begin{bmatrix} 1 & a & b \\ 0 & 1 & c \\ 0 & 0 & 1 \end{bmatrix}.$$

This family of shears maps the (x, y)-plane onto itself.

A parallel projection. All of $I\!\!E^3$ is projected onto the x, y-plane if we set

$$A = \begin{bmatrix} 1 & 0 & 0 \\ 0 & 1 & 0 \\ 0 & 0 & 0 \end{bmatrix}.$$

and $\mathbf{v} = \mathbf{0}$. Note that A may also be viewed as a scaling matrix.

We give one example of an affine map that is becoming important in the area of *font design*. A given letter is subjected to a 2D shear and thus transforms into a slanted letter. Figure 2.4 gives an example; see also Section 8.5.

An important special case of affine maps are the *euclidean maps*, also called *rigid body motions*. They are characterized by orthonormal matrices A that are defined by the property $A^{\mathrm{T}}A = I$. Euclidean maps leave lengths and angles unchanged; the most important examples are rotations and translations.

Affine maps can be combined, and a complicated map may be decomposed into a sequence of simpler maps. Every affine map can be composed of translations, rotations, shears, and scalings.

The *rank* of A has an important geometric interpretation: if $\operatorname{rank}(A) = 3$, then the affine map Φ maps three-dimensional objects to three-dimensional objects. If the rank is less than three, Φ is a parallel projection onto a plane $(\operatorname{rank} = 2)$ or even onto a straight line $(\operatorname{rank} = 1)$.

An affine map of $I\!\!E^2$ to $I\!\!E^2$ is uniquely determined by a (nondegenerate) triangle and its image. Thus any two triangles determine an affine map of the plane onto itself. In $I\!\!E^3$, an affine map is uniquely defined by a (nondegenerate) tetrahedron and its image.

More important facts about affine maps are discussed in the following section.

2.3 Linear Interpolation

Let $\mathbf{a}, \mathbf{b}$ be two distinct points in $I\!\!E^3$. The set of all points $\mathbf{x} \in I\!\!E^3$ of the form

$$\mathbf{x} = \mathbf{x}(t) = (1 - t)\mathbf{a} + t\mathbf{b}; \quad t \in I\!\!R \tag{2.4}$$

is called the *straight line* through $\mathbf{a}$ and $\mathbf{b}$. Any three (or more) points on a straight line are said to be *collinear*.

For $t = 0$ the straight line passes through $\mathbf{a}$ and for $t = 1$ it passes through $\mathbf{b}$. For $0 \leq t \leq 1$, the point $\mathbf{x}$ is between $\mathbf{a}$ and $\mathbf{b}$, while for all other values of t it is outside; see Figure 2.5.

Equation (2.4) represents $\mathbf{x}$ as a barycentric combination of two points in $I\!\!E^3$. The same barycentric combination holds for the three points $0, t, 1$ in $I\!\!E^1$: $t = (1 - t) \cdot 0 + t \cdot 1$. So t is related to 0 and 1 by the same barycentric combination that relates $\mathbf{x}$ to $\mathbf{a}$ and $\mathbf{b}$. However, by the definiton of affine maps, the three points $\mathbf{a}, \mathbf{x}, \mathbf{b}$ in three-space are an affine map of the three points $0, t, 1$ in one-space! *Thus linear interpolation is an affine map of the real line onto a straight line in $I\!\!E^3$.*[3]

---

[3]Strictly speaking, we should therefore use the term "affine interpolation" instead of "linear interpolation." We use "linear interpolation" because its use is so widespread.

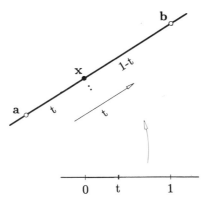

Figure 2.5: Linear interpolation: two points $\mathbf{a}, \mathbf{b}$ define a straight line through them. The point $\mathbf{x}$ divides the straight line segment between $\mathbf{a}$ and $\mathbf{b}$ in the ratio $t : 1 - t$.

It is now almost a tautology when we state: *Linear interpolation is affinely invariant.* Written as a formula: if Φ is an affine map of $I\!\!E^3$ onto itself, and (2.4) holds, then also

$$\Phi\mathbf{x} = \Phi((1-t)\mathbf{a} + t\mathbf{b}) = (1-t)\Phi\mathbf{a} + t\Phi\mathbf{b}. \qquad (2.5)$$

Closely related to linear interpolation is the concept of *barycentric coordinates*, due to Moebius [337]. Let $\mathbf{a}, \mathbf{x}, \mathbf{b}$ be three collinear points in $I\!\!E^3$:

$$\mathbf{x} = \alpha\mathbf{a} + \beta\mathbf{b}; \qquad \alpha + \beta = 1. \qquad (2.6)$$

Then α and β are called *barycentric coordinates* of $\mathbf{x}$ with respect to $\mathbf{a}$ and $\mathbf{b}$. Note that by our previous definitions, $\mathbf{x}$ is a barycentric combination of $\mathbf{a}$ and $\mathbf{b}$.

The connection between barycentric coordinates and linear interpolation is obvious: we have $\alpha = 1 - t$ and $\beta = t$. This shows, by the way, that barycentric coordinates do not always have to be positive: For $t \notin [0, 1]$, either α or β is negative. For any three collinear points $\mathbf{a}, \mathbf{b}, \mathbf{c}$, the barycentric coordinates of $\mathbf{b}$ with respect to $\mathbf{a}$ and $\mathbf{c}$ are given by

$$\alpha = \frac{\mathrm{vol}_1(\mathbf{b}, \mathbf{c})}{\mathrm{vol}_1(\mathbf{a}, \mathbf{c})},$$

$$\beta = \frac{\mathrm{vol}_1(\mathbf{a}, \mathbf{b})}{\mathrm{vol}_1(\mathbf{a}, \mathbf{c})},$$

where vol_1 denotes the one-dimensional volume, which is the signed distance between two points. Barycentric coordinates are not only defined on a straight line, but also on a plane. Section 18.1 has more details.

Another important concept in this context is that of *ratios*. The ratio of three collinear points $\mathbf{a}, \mathbf{b}, \mathbf{c}$ is defined by

$$\text{ratio}(\mathbf{a}, \mathbf{b}, \mathbf{c}) = \frac{\text{vol}_1(\mathbf{a}, \mathbf{b})}{\text{vol}_1(\mathbf{b}, \mathbf{c})}. \tag{2.7}$$

If α and β are barycentric coordinates of $\mathbf{b}$ with respect to $\mathbf{a}$ and $\mathbf{c}$, it follows that

$$\text{ratio}(\mathbf{a}, \mathbf{b}, \mathbf{c}) = \frac{\beta}{\alpha}. \tag{2.8}$$

The barycentric coordinates of a point do not change under affine maps, and neither does their quotient. Thus the ratio of three collinear points is not affected by affine transformations. So if (2.8) holds, then also

$$\text{ratio}(\Phi\mathbf{a}, \Phi\mathbf{b}, \Phi\mathbf{c}) = \frac{\beta}{\alpha}, \tag{2.9}$$

where Φ is an affine map.[4]

The last equation states that *affine maps are ratio preserving*. This property may be used to define affine maps. Every map that takes straight lines to straight lines and is ratio preserving is an affine map.

The concept of ratio preservation may be used to derive another useful property of linear interpolation. We have defined the straight line segment $[\mathbf{a}, \mathbf{b}]$ to be the affine image of the *unit interval* $[0, 1]$, but we can also view that straight line segment as the affine image of any interval $[a, b]$. The interval $[a, b]$ may itself be obtained by an affine map from the interval $[0, 1]$ or vice versa. With $t \in [0, 1]$ and $u \in [a, b]$, that map is given by $t = (u - a)/(b - a)$. The interpolated point on the straight line is now given by both

$$\mathbf{x}(t) = (1 - t)\mathbf{a} + t\mathbf{b}$$

and

$$\mathbf{x}(u) = \frac{b - u}{b - a}\mathbf{a} + \frac{u - a}{b - a}\mathbf{b}. \tag{2.10}$$

[4]This property can be used to *compute* ratios efficiently. Instead of using square roots to compute the distances between points $\mathbf{a}$, $\mathbf{x}$, and $\mathbf{b}$, one would project them onto one of the coordinate axes and then use simple differences. This method works since parallel projection is an affine map!

Since a, u, b and $0, t, 1$ are in the same ratio as the triple $\mathbf{a}, \mathbf{x}, \mathbf{b}$, we have shown that *linear interpolation is invariant under affine domain transformations*. By affine domain transformation, we simply mean an affine map of the real line onto itself. The parameter t is sometimes called a *local parameter* of the interval $[a, b]$.

A concluding remark: we have demonstrated the interplay between the two concepts of linear interpolation and ratios. In this book, we will often describe methods by saying that points have to be collinear and must be in a given ratio. This is the geometric (descriptive) equivalent of the algebraic (algorithmic) statement that one of the three points may be obtained by linear interpolation from the other two.

2.4 Piecewise Linear Interpolation

Let $\mathbf{b}_0, \ldots, \mathbf{b}_n \in I\!\!E^3$ form a *polygon* $\mathbf{B}$. A polygon consists of a sequence of straight line segments, each interpolating to a pair of points $\mathbf{b}_i, \mathbf{b}_{i+1}$. It is therefore also called the *piecewise linear interpolant* $\mathcal{PL}$ to the points $\mathbf{b}_i$. If the points $\mathbf{b}_i$ lie on a curve $\mathbf{c}$, then $\mathbf{B}$ is said to be a piecewise linear interpolant to $\mathbf{c}$, and we write

$$\mathbf{B} = \mathcal{PL}\mathbf{c}. \tag{2.11}$$

One of the important properties of piecewise linear interpolation is *affine invariance*. If the curve $\mathbf{c}$ is mapped onto a curve $\Phi\mathbf{c}$ by an affine map Φ, then the piecewise linear interpolant to $\Phi\mathbf{c}$ is the affine map of the original piecewise linear interpolant:

$$\mathcal{PL}\ \Phi\mathbf{c} = \Phi\ \mathcal{PL}\ \mathbf{c}. \tag{2.12}$$

Another property is the *variation diminishing property*. Consider a continuous curve $\mathbf{c}$, a piecewise linear interpolant $\mathcal{PL}\mathbf{c}$, and an arbitrary plane. Let cross $\mathbf{c}$ be the number of crossings that the curve $\mathbf{c}$ has with this plane, and let cross $\mathcal{PL}\mathbf{c}$ be the number of crossings that the piecewise linear interpolant has with this plane. (Special cases may arise; see Problems.) Then we always have

$$\text{cross}\mathcal{PL}\ \mathbf{c} \leq \text{cross}\ \mathbf{c}. \tag{2.13}$$

This property follows from a simple observation: consider two points $\mathbf{b}_i, \mathbf{b}_{i+1}$. The straight line segment through them can cross a given plane at one point at most, while the curve segment from $\mathbf{c}$ that connects them may cross the

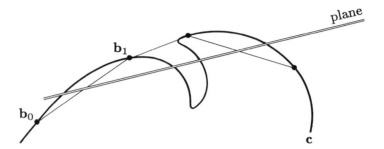

Figure 2.6: The variation diminishing property: a piecewise linear interpolant to a curve has no more intersections with any plane than the curve itself.

same plane in many arbitrary points. The variation diminishing property is illustrated in Figure 2.6.

2.5 Menelaos' Theorem

We use the concept of piecewise linear interpolation to prove one of the most important geometric theorems for the theory of CAGD: Menelaos' theorem. This theorem can be used for the proof of many constructive algorithms, and its importance was already realized by de Casteljau [122].

Referring to Figure 2.7, let

$$
\begin{aligned}
\mathbf{a}_t &= (1-t)\mathbf{p}_1 + t\mathbf{p}_2, \\
\mathbf{a}_s &= (1-s)\mathbf{p}_1 + s\mathbf{p}_2, \\
\mathbf{b}_t &= (1-t)\mathbf{p}_2 + t\mathbf{p}_3, \\
\mathbf{b}_s &= (1-s)\mathbf{p}_2 + s\mathbf{p}_3.
\end{aligned}
$$

Let $\mathbf{c}$ be the intersection of the straight lines $\mathbf{a}_t\mathbf{b}_t$ and $\mathbf{a}_s\mathbf{b}_s$. Then

$$
\text{ratio}(\mathbf{a}_t, \mathbf{c}, \mathbf{b}_t) = \frac{s}{1-s} \quad \text{and} \quad \text{ratio}(\mathbf{a}_s, \mathbf{c}, \mathbf{b}_s) = \frac{t}{1-t}. \tag{2.14}
$$

For a proof, we simply show that $\mathbf{c}$ satisfies the two equations

$$
\mathbf{c} = (1-s)\mathbf{a}_t + s\mathbf{b}_t \quad \text{and} \quad \mathbf{c} = (1-t)\mathbf{a}_s + t\mathbf{b}_s,
$$

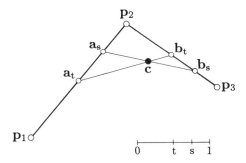

Figure 2.7: Menelaos' theorem: the point **c** may be obtained from linear interpolation at t or at s.

which is straightforward. Notice also that the four collinear points $\mathbf{p}_1, \mathbf{a}_t, \mathbf{a}_s, \mathbf{p}_2$ as well as the four collinear points $\mathbf{p}_2, \mathbf{b}_t, \mathbf{b}_s, \mathbf{p}_3$ are affine maps of the four points $0, t, s, 1$ on the real line.

Equation (2.14) is a "CAGD version" of the original Menelaos' theorem, which may be stated as (see Coxeter [107]):

$$\text{ratio}(\mathbf{b}_s, \mathbf{b}_t, \mathbf{p}_2) \cdot \text{ratio}(\mathbf{p}_2, \mathbf{a}_t, \mathbf{a}_s) \cdot \text{ratio}(\mathbf{a}_s, \mathbf{c}, \mathbf{b}_s) = -1. \tag{2.15}$$

The proof of (2.15) is a direct consequence of (2.14). Note the ordering of points in the second ratio! Menelaos' theorem is closely related to Ceva's, which is given in Section 18.1.

2.6 Function Spaces

This section contains material that will later simplify our work by allowing very concise notation. Although we shall try to develop our material with an emphasis on geometric concepts, it will sometimes simplify our work considerably if we can resort to some elementary topics from functional analysis. Good references are the books by Davis [110] and de Boor [114].

Let $C[a, b]$ be the set of all real-valued continuous functions defined over the interval $[a, b]$ of the real axis. We can define addition and multiplication by a constant for elements $f, g \in C[a, b]$ by setting $(\alpha f + \beta g)(t) = \alpha f(t) + \beta g(t)$ for all $t \in [a, b]$. With these definitions, we can easily show that $C[a, b]$ forms a *linear space* over the reals. The same is true for the sets $C^k[a, b]$, the sets

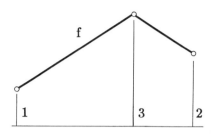

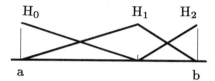

Figure 2.8: Hat functions: the piecewise linear function f can be written as $f = H_0 + 3H_1 + 2H_2$.

of all real-valued functions defined over $[a, b]$ that are k-times continuously differentiable. Furthermore, for every k, C^{k+1} is a *subspace* of C^k.

We say that n functions $f_1, \ldots, f_n \in C[a, b]$ are *linearly independent* if $\sum c_i f_i = 0$ for all $t \in [a, b]$ implies $c_1 = \ldots = c_n = 0$.

We mention some subspaces of $C[a, b]$ that will be of interest later. The spaces $\mathcal{P}^n$ of all *polynomials* of degree n are:

$$p^n(t) = a_0 + a_1 t + a_2 t^2 + \cdots + a_n t^n; \quad t \in [a, b].$$

For fixed n, the dimension of $\mathcal{P}^n$ is $n+1$: each $p^n \in \mathcal{P}^n$ is determined uniquely by the $n + 1$ coefficients $a_0, \ldots, a_n$. These can be interpreted as a vector in $(n + 1)$-dimensional linear space $\mathbb{R}^{n+1}$, which has dimension $n + 1$. We can also name a *basis* for $\mathcal{P}^n$: the *monomials* $1, t, t^2, \ldots, t^n$ are $n + 1$ linearly independent functions and thus form a basis.

Another interesting class of subspaces of $C[a, b]$ is given by piecewise linear functions: let $a = t_0 < t_1 < \cdots < t_n = b$ be a *partition* of the interval $[a, b]$. A continuous function that is linear on each subinterval $[t_i, t_{i+1}]$ is called a *piecewise linear function*. Over a fixed partition of $[a, b]$, the piecewise linear functions form a linear function space. A basis for this space is given by the *hat functions*: a hat function $H_i(t)$ is a piecewise linear function with $H_i(t_i) = 1$ and $H_i(t_j) = 0$ if $i \neq j$. A piecewise linear function f with $f(t_j) = f_j$ can always be written as

$$f(t) = \sum_{j=0}^{n} f_j H_j(t).$$

Figure 2.8 gives an example.

We will also consider *linear operators* that assign a function $\mathcal{A}f$ to a given function f. An operator $\mathcal{A} : C[a, b] \to C[a, b]$ is called *linear* if it leaves linear combinations invariant:

$$\mathcal{A}(\alpha f + \beta g) = \alpha \mathcal{A}f + \beta \mathcal{A}g; \quad \alpha, \beta \in \mathbb{R}.$$

An example is given by the derivative operator that assigns the derivative f' to a given function f: $\mathcal{A}f = f'$.

2.7 Problems

1. We have seen that affine maps leave the ratio of three collinear points constant, i.e., they are ratio-preserving. Show that the converse is also true: every ratio-preserving map is affine.

2. We defined the convex hull of a point set to be the set of all convex combinations formed by the elements of that set. Another definition is the following: the convex hull of a point set is the intersection of all convex sets that contain the given set. Show that both definitions are equivalent.

3. In the definition of the variation diminishing property, we counted the crossings of a polygon with a plane. Discuss the case when the plane contains a whole polygon leg.

4. Show that the $n + 1$ functions $f_i(t) = t^i$; $i = 0, \ldots, n$ are linearly independent.

5. Our definition of barycentric combinations gives the impression that it needs the involved points expressed in terms of some coordinate system. Show that this is not necessary: draw five points on a piece of paper, assign a weight to each one, and *construct* the barycenter of your points

using a ruler (or compass and straightedge, if you are more classically inclined).

Remark: For this construction, it is not necessary for the weights to add to one. This is so because the geometric construction remains the same if we multiplied all weights by a common factor. In fact, one may replace the concept of points (having mass one and requiring barycentric combinations as the basic point operation) by that of *mass points*, having arbitrary weights and yielding their barycenter (with the combined mass of all points) as the basic operation. In such a setting, vectors would also be mass points, but with mass zero.[5]

6. Fix two distinct points **a**, **b** on the x-axis. Let a third point **x** trace out all of the x-axis. For each location of **x**, plot the value of the function ratio(**a**, **x**, **b**), thus obtaining a graph of the ratio function.

---

[5]I was introduced to this concept by A. Swimmer. It was developed by H. Grassmann in 1844.

Chapter 3

The de Casteljau Algorithm

The algorithm described in this chapter is probably the most fundamental one in the field of curve and surface design, yet it is surprisingly simple. Its main attraction is the beautiful interplay between geometry and algebra: a very intuitive geometric construction leads to a powerful theory.

Historically, it is with this algorithm that the work of de Casteljau started in 1959. The only written evidence is in [121] and [122], both of which are technical reports that are not easily accessible. De Casteljau's work went unnoticed until W. Boehm obtained copies of the reports in 1975. From then on, de Casteljau's name gained more popularity.

3.1 Parabolas

We give a simple construction for the generation of a parabola; the straightforward generalization will then lead to Bézier curves. Let $\mathbf{b}_0, \mathbf{b}_1, \mathbf{b}_2$ be any three points in $I\!\!E^3$, and let $t \in I\!\!R$. Construct

$$
\begin{aligned}
\mathbf{b}_0^1(t) &= (1-t)\mathbf{b}_0 + t\mathbf{b}_1, \\
\mathbf{b}_1^1(t) &= (1-t)\mathbf{b}_1 + t\mathbf{b}_2,
\end{aligned}
$$

$$
\mathbf{b}_0^2(t) = (1-t)\mathbf{b}_0^1(t) + t\mathbf{b}_1^1(t).
$$

Inserting the first two equations into the third one, we obtain

$$
\mathbf{b}_0^2(t) = (1-t)^2\mathbf{b}_0 + 2t(1-t)\mathbf{b}_1 + t^2\mathbf{b}_2. \tag{3.1}
$$

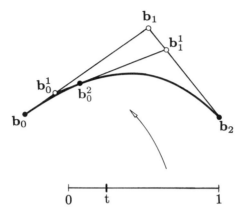

Figure 3.1: Parabolas: construction by repeated linear interpolation.

This is a quadratic expression in t (the superscript denotes the degree), and so $\mathbf{b}_0^2(t)$ traces out a *parabola* as t varies from $-\infty$ to $+\infty$. We denote this parabola by $\mathbf{b}^2$. This construction consists of *repeated linear interpolation*; its geometry is illustrated in Figure 3.1. For t between 0 and 1, $\mathbf{b}^2(t)$ is inside the triangle formed by $\mathbf{b}_0, \mathbf{b}_1, \mathbf{b}_2$; in particular, $\mathbf{b}^2(0) = \mathbf{b}_0$ and $\mathbf{b}^2(1) = \mathbf{b}_2$.

Inspecting the ratios of points in Figure 3.1, we see that

$$\text{ratio}(\mathbf{b}_0, \mathbf{b}_0^1, \mathbf{b}_1) = \text{ratio}(\mathbf{b}_1, \mathbf{b}_1^1, \mathbf{b}_2) = \text{ratio}(\mathbf{b}_0^1, \mathbf{b}_0^2, \mathbf{b}_1^1) = t/(1-t).$$

Thus our construction of a parabola is *affinely invariant* because piecewise linear interpolation is affinely invariant; see Section 2.4.

We also note that a parabola is a plane curve, due to the fact that $\mathbf{b}^2(t)$ is always a barycentric combination of three points, as is clear from inspecting (3.1). A parabola is a special case of *conic sections*, which will be discussed in Chapter 14.

Finally we state a theorem from analytic geometry, closely related to our parabola construction. Let $\mathbf{a}, \mathbf{b}, \mathbf{c}$ be three distinct points on a parabola. Let the tangent at $\mathbf{b}$ intersect the tangents at $\mathbf{a}$ and $\mathbf{c}$ in $\mathbf{e}$ and $\mathbf{f}$, respectively. Let the tangents at $\mathbf{a}$ and $\mathbf{c}$ intersect in $\mathbf{d}$. Then $\text{ratio}(\mathbf{a}, \mathbf{e}, \mathbf{d}) = \text{ratio}(\mathbf{e}, \mathbf{b}, \mathbf{f}) = \text{ratio}(\mathbf{d}, \mathbf{f}, \mathbf{c})$. This *three tangent theorem* describes a property of parabolas; the de Casteljau algorithm can be viewed as the constructive counterpart.

3.2 The de Casteljau Algorithm

Parabolas are plane curves. However, many applications require true space curves.[1] For those purposes, the previous construction for a parabola can be generalized to generate a polynomial curve of arbitrary degree n:

de Casteljau algorithm:

Given: $\mathbf{b}_0, \mathbf{b}_1, \ldots, \mathbf{b}_n \in I\!\!E^3$ and $t \in I\!\!R$,

set

$$\mathbf{b}_i^r(t) = (1-t)\mathbf{b}_i^{r-1}(t) + t\mathbf{b}_{i+1}^{r-1}(t) \qquad \begin{cases} r = 1, \ldots, n \\ i = 0, \ldots, n-r \end{cases} \qquad (3.2)$$

and $\mathbf{b}_i^0(t) = \mathbf{b}_i$. Then $\mathbf{b}_0^n(t)$ is the point with parameter value t on the *Bézier curve* $\mathbf{b}^n$.

The polygon $\mathbf{P}$ formed by $\mathbf{b}_0, \ldots, \mathbf{b}_n$ is called the *Bézier polygon* or *control polygon* of the curve $\mathbf{b}^n$. Similarly, the polygon vertices $\mathbf{b}_i$ are called *control points* or *Bézier points*. Figure 3.2 illustrates the cubic case.

Sometimes we also write $\mathbf{b}^n(t) = \mathcal{B}[\mathbf{b}_0, \ldots, \mathbf{b}_n; t] = \mathcal{B}[\mathbf{P}; t]$ or, shorter, $\mathbf{b}^n = \mathcal{B}[\mathbf{b}_0, \ldots, \mathbf{b}_n] = \mathcal{B}\mathbf{P}$. This notation[2] defines $\mathcal{B}$ to be the (linear) operator that associates the Bézier curve with its control polygon. We say that the curve $\mathcal{B}[\mathbf{b}_0, \ldots, \mathbf{b}_n]$ is the *Bernstein-Bézier approximation* to the control polygon, a terminology borrowed from approximation theory; see also Section 5.10.

The intermediate coefficients $\mathbf{b}_i^r(t)$ are conveniently written into a triangular array of points, the *de Casteljau scheme*. We give the example of the cubic case:

$$\begin{array}{llll} \mathbf{b}_0 & & & \\ \mathbf{b}_1 & \mathbf{b}_0^1 & & \\ \mathbf{b}_2 & \mathbf{b}_1^1 & \mathbf{b}_0^2 & \\ \mathbf{b}_3 & \mathbf{b}_2^1 & \mathbf{b}_1^2 & \mathbf{b}_0^3. \end{array} \qquad (3.3)$$

This triangular array of points seems to suggest the use of a two-dimensional array in writing code for the de Casteljau algorithm. That would be a waste of storage, however: it is sufficient to use the left column only and to overwrite it appropriately.

---

[1] Compare the comments by P. Bézier in Chapter 1!

[2] This notation should not be confused with the blossoming notation used later.

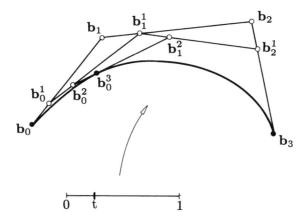

Figure 3.2: The de Casteljau algorithm: the point $\mathbf{b}_0^3(t)$ is obtained from repeated linear interpolation. The cubic case $n = 3$ is shown for $t = 1/4$.

For a numerical example, see Example 3.1. Figure 3.3 shows 50 evaluations of a Bézier curve. The intermediate points $\mathbf{b}_i^r$ are also plotted, and connected.

3.3 Some Properties of Bézier Curves

The de Casteljau algorithm allows us to infer several important properties of Bézier curves. We will infer these properties from the geometry underlying the algorithm. In the next chapter, we will show how they can also be derived analytically.

Affine invariance. Affine maps were discussed in Section 2.2. They are in the tool kit of every CAD system: objects must be repositioned, scaled, and so on. An important property of Bézier curves is that they are invariant under affine maps, which means that the following two procedures yield the same result: (1) first, compute the point $\mathbf{b}^n(t)$ and then apply an affine map to it; (2) first, apply an affine map to the control polygon and then evaluate the mapped polygon at parameter value t.

Affine invariance is, of course, a direct consequence of the de Casteljau algorithm: the algorithm is composed of a sequence of linear interpolations (or, equivalently, of a sequence of affine maps). These are themselves affinely invariant, and so is a finite sequence of them.

A de Casteljau scheme for a planar cubic and for $t = \frac{1}{2}$:

$$\begin{bmatrix} 0 \\ 0 \end{bmatrix}$$
$$\begin{bmatrix} 0 \\ 2 \end{bmatrix} \quad \begin{bmatrix} 0 \\ 1 \end{bmatrix}$$
$$\begin{bmatrix} 8 \\ 2 \end{bmatrix} \quad \begin{bmatrix} 4 \\ 2 \end{bmatrix} \quad \begin{bmatrix} 2 \\ \frac{3}{2} \end{bmatrix}$$
$$\begin{bmatrix} 4 \\ 0 \end{bmatrix} \quad \begin{bmatrix} 6 \\ 1 \end{bmatrix} \quad \begin{bmatrix} 5 \\ \frac{3}{2} \end{bmatrix} \quad \begin{bmatrix} \frac{7}{2} \\ \frac{3}{2} \end{bmatrix}$$

Example 3.1: Computing a point on a Bézier curve with the Casteljau algorithm.

Let us discuss a practical aspect of affine invariance. Suppose we plot a cubic curve $\mathbf{b}^3$ by evaluating at 100 points and then plotting the resulting point array. Suppose now that we would like to plot the curve after a rotation has been applied to it. We can take the hundred computed points, apply the rotation to each of them, and plot. Or, we can apply the rotation to the four control points, then evaluate one hundred times and plot. The first method needs one hundred applications of the rotation, while the second needs only four!

Affine invariance may not seem to be a very exceptional property for a useful curve scheme; in fact, it is not straightforward to think of a curve scheme that does not have it (exercise!). It is perhaps worth noting that Bézier curves do *not* enjoy another, also very important, property: they are not *projectively invariant*. Projective maps are used in computer graphics when an object is to be rendered realistically. So if we try to make life easy and simplify a perspective map of a Bézier curve by mapping the control polygon and then computing the curve, we have actually cheated: that curve is not the perspective image of the original curve! More details on perspective maps can be found in Chapter 14.

Invariance under affine parameter transformations. Very often, one thinks of a Bézier curve as being defined over the interval $[0, 1]$. This is

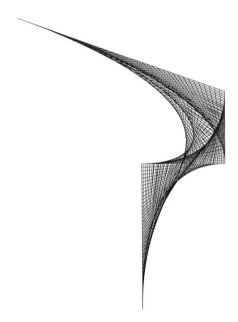

Figure 3.3: The de Casteljau algorithm: 50 points are computed on a quartic curve, and the intermediate points $\mathbf{b}_i^r$ are connected.

done because it is convenient, not because it is necessary: the de Casteljau algorithm is "blind" to the actual interval that the curve is defined over because it uses ratios only. One may therefore think of the curve as being defined over any arbitrary interval $a \leq u \leq b$ of the real line – after the introduction of local coordinates $t = (u - a)/(b - a)$, the algorithm proceeds as usual. This property is inherited from linear interpolation, see Eq. (2.10). The corresponding generalized de Casteljau algorithm is of the form:

$$\mathbf{b}_i^r(u) = \frac{b - u}{b - a}\mathbf{b}^{r-1}(u) + \frac{u - a}{b - a}\mathbf{b}_{i+1}^{r-1}. \qquad (3.4)$$

The transition from the interval $[0, 1]$ to the interval $[a, b]$ is an *affine map*. Therefore, we can say that Bézier curves are invariant under affine parameter transformations. Sometimes, one sees the term *linear parameter transformation* in this context, but this terminology is not quite correct: the transformation of the interval $[0, 1]$ to $[a, b]$ typically includes a translation, which is not a linear map.

Convex hull property. For $t \in [0, 1]$, $\mathbf{b}^n(t)$ lies in the convex hull (see Figure 2.3) of the control polygon. This follows since every intermediate $\mathbf{b}_i^r$ is obtained as a convex barycentric combination of previous $\mathbf{b}_j^{r-1}$ – at no step of the de Casteljau algorithm do we produce points outside the convex hull of the $\mathbf{b}_i$.

A simple consequence of the convex hull property is that a planar control polygon always generates a planar curve.

The importance of the convex hull property lies in what is known as *interference checking*. Suppose we want to know if two Bézier curves intersect each other – for example, each might represent the path of a robot arm, and our aim is to make sure that the two paths do not intersect, thus avoiding expensive collisions of the robots. Instead of actually computing a possible intersection, we can perform a much cheaper test: circumscribe the smallest possible box around the control polygon of each curve such that it has its edges parallel to some coordinate system. Such boxes are called *minmax boxes*, since their faces are created by the minimal and maximal coordinates of the control polygons. Clearly each box contains its control polygon, and, by the convex hull property, also the corresponding Bézier curve. If we can verify that the two boxes do not overlap (a trivial test), we are assured that the two curves do not intersect. If the boxes do overlap, we would have to perform more checks on the curves. The possibility for a quick decision of no interference is extremely important, since in practice one often has to check one object against thousands of others, most of which can be labeled as "no interference" by the minmax box test.[3]

Endpoint interpolation. The Bézier curve passes through $\mathbf{b}_0$ and $\mathbf{b}_n$: we have $\mathbf{b}^n(0) = \mathbf{b}_0, \mathbf{b}^n(1) = \mathbf{b}_n$. This is easily verified by writing down the scheme of Eq. (3.3) for the cases $t = 0$ and $t = 1$. In a design situation, the endpoints of a curve are certainly two very important points. It is therefore essential to have direct control over them, which is assured by endpoint interpolation.

---

[3]It is possible to create volumes (or areas, in the 2D case) that hug the given curve closer than the minmax box does. See Sederberg *et al.* [436].

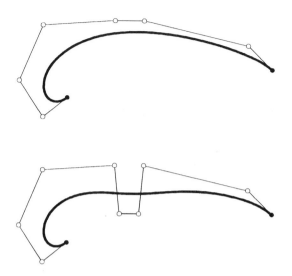

Figure 3.4: Bézier curves: some examples.

Designing with Bézier curves. Figure 3.4 shows two Bézier curves. From the inspection of these examples, one gets the impression that in some sense the Bézier curve "mimicks" the Bézier polygon – this statement will be made more precise later. It is the reason why Bézier curves provide such a handy tool for the *design* of curves: To reproduce the shape of a hand-drawn curve, it is sufficient to specify a control polygon that somehow "exaggerates" the shape of the curve. One lets the computer draw the Bézier curve defined by the polygon, and, if necessary, adjusts the location (possibly also the number) of the polygon vertices. Typically, an experienced person will reproduce a given curve after two to three iterations of this *interactive* procedure.

3.4 The Blossom

In recent years, a new way to look at Bézier curves has been developed; it is called the principle of *blossoming*. This principle was independently developed by de Casteljau [123] and Ramshaw [386][388]. Other literature includes Seidel [437], [440], [441]; DeRose and Goldman [137]; Boehm [67]; and Lee [305].

We introduce blossoms as a generalization of the de Casteljau algorithm. Written in a scheme as in (3.3), we have to compute n columns. Our generalization is as follows: in column r, do not again perform a de Casteljau step for parameter value t, but use a new value t_r. Restricting ourselves to the cubic case, we obtain:

$$
\begin{aligned}
&\mathbf{b}_0 \\
&\mathbf{b}_1 \quad \mathbf{b}_0^1[t_1] \\
&\mathbf{b}_2 \quad \mathbf{b}_1^1[t_1] \quad \mathbf{b}_0^2[t_1, t_2] \\
&\mathbf{b}_3 \quad \mathbf{b}_2^1[t_1] \quad \mathbf{b}_1^2[t_1, t_2] \quad \mathbf{b}_0^3[t_1, t_2, t_3].
\end{aligned}
\tag{3.5}
$$

The resulting point $\mathbf{b}_0^3[t_1, t_2, t_3]$ is now a function of three independent variables; thus it no longer traces out a curve, but a region of $I\!\!E^3$. This trivariate function $\mathbf{b}[\cdot, \cdot, \cdot]$ is called the *blossom* of the curve $\mathbf{b}^3(t)$, after L. Ramshaw [386]. The original curve is recovered if we set all three arguments equal: $t = t_1 = t_2 = t_3$.

To understand the blossom better, we now evaluate it for several special arguments. We already know, of course, that $\mathbf{b}[0, 0, 0] = \mathbf{b}_0$ and $\mathbf{b}[1, 1, 1] = \mathbf{b}_3$. Let us start with $[t_1, t_2, t_3] = [0, 0, 1]$. The scheme (3.5) reduces to:

$$
\begin{aligned}
&\mathbf{b}_0 \\
&\mathbf{b}_1 \quad \mathbf{b}_0 \\
&\mathbf{b}_2 \quad \mathbf{b}_1 \quad \mathbf{b}_0 \\
&\mathbf{b}_3 \quad \mathbf{b}_2 \quad \mathbf{b}_1 \quad \mathbf{b}_1 = \mathbf{b}[0, 0, 1].
\end{aligned}
\tag{3.6}
$$

Similarly, we can show that $\mathbf{b}[0, 1, 1] = \mathbf{b}_2$. Thus the original Bézier points can be found by evaluating the curve's blossom at arguments consisting only of 0's and 1's.

But the remaining entries in (3.3) may also be written as values of the blossom for special arguments. For instance, setting $[t_1, t_2, t_3] = [0, 0, t]$, we have the scheme

$$
\begin{aligned}
&\mathbf{b}_0 \\
&\mathbf{b}_1 \quad \mathbf{b}_0 \\
&\mathbf{b}_2 \quad \mathbf{b}_1 \quad \mathbf{b}_0 \\
&\mathbf{b}_3 \quad \mathbf{b}_2 \quad \mathbf{b}_1 \quad \mathbf{b}_0^1 = \mathbf{b}[0, 0, t].
\end{aligned}
\tag{3.7}
$$

Continuing in the same manner, we may write the complete scheme (3.3) as:

$$
\begin{aligned}
\mathbf{b}_0 &= \mathbf{b}[0,0,0] \\
\mathbf{b}_1 &= \mathbf{b}[0,0,1] \quad \mathbf{b}[0,0,t] \\
\mathbf{b}_2 &= \mathbf{b}[0,1,1] \quad \mathbf{b}[0,t,1] \quad \mathbf{b}[0,t,t] \\
\mathbf{b}_3 &= \mathbf{b}[1,1,1] \quad \mathbf{b}[t,1,1] \quad \mathbf{b}[t,t,1] \quad \mathbf{b}[t,t,t].
\end{aligned}
\tag{3.8}
$$

For arbitrary degrees, one can also express the Bézier points as blossom values:

$$
\mathbf{b}_i = \mathbf{b}[0^{<n-i>}, 1^{<i>}],
\tag{3.9}
$$

where $t^{<r>}$ means that t appears r times as an argument. For example, $\mathbf{b}[0^{<1>}, t^{<2>}, 1^{<0>}] = \mathbf{b}[0,t,t]$. A proof of (3.9) is given with (4.9) in Section 4.1.

The de Casteljau recursion (3.2) can now be expressed in terms of the blossom $\mathbf{b}[]$:

$$
\begin{aligned}
\mathbf{b}[0^{<n-r-i>}, t^{<r>}, 1^{<i>}] &= (1-t)\mathbf{b}[0^{<n-r-i+1>}, t^{<r-1>}, 1^{<i>}] \\
&+ t\mathbf{b}[0^{<n-r-i>}, t^{<r-1>}, 1^{<i+1>}].
\end{aligned}
\tag{3.10}
$$

The point on the curve is given by $\mathbf{b}[t^{<n>}]$.

We next note that it does not matter in which order we use the t_i for the blossom's evaluation. So we have, again for the cubic case, that $\mathbf{b}[t_1, t_2, t_3] = \mathbf{b}[t_2, t_3, t_1]$, etc. A proof of this statement is obtained using Figure 2.7: point $\mathbf{c}$ in that figure may be written as the value of a quadratic blossom: $\mathbf{c} = \mathbf{b}[t, s] = \mathbf{b}[s, t]$. The general result follows from this special instance.

Functions whose values do not depend on the order of their arguments are called *symmetric*; thus a blossom is a symmetric polynomial function of n variables. Every polynomial curve has a unique blossom associated with it – it is a symmetric polynomial of n variables, mapping $I\!R^n$ into $I\!E^3$.

The blossom has yet another important property. If the first argument of the blossom is a barycentric combination of two (or more) numbers, we may compute the blossom values for each argument and then form their barycentric combination:

$$
\mathbf{b}[\alpha r + \beta s, t_2, \ldots, t_n] = \alpha \mathbf{b}[r, t_2, \ldots, t_n] + \beta \mathbf{b}[s, t_2, \ldots, t_n]; \quad \alpha + \beta = 1.
\tag{3.11}
$$

Equation (3.11) states that the blossom $\mathbf{b}$ is affine with respect to its first argument, but it is affine for any of the remaining arguments as well. This is the reason why the blossom is called *multiaffine*. Blossoms are multiaffine since they can be obtained by repeated steps of the de Casteljau algorithm. Each of these steps consists of linear interpolation, an affine map itself [Eq. (2.5)].

We may also consider the blossom of a Bézier curve that is not defined over $[0,1]$ but over the more general interval $[a,b]$. Proceeding exactly as above – but now utilizing (3.4) – we find that the Bézier points $\mathbf{b}_i$ are found as the blossom values

$$\mathbf{b}_i = \mathbf{b}[a^{<n-i>}, b^{<i>}]. \tag{3.12}$$

Thus a cubic over $u \in [a,b]$ has Bézier points $\mathbf{b}[a,a,a], \mathbf{b}[a,a,b], \mathbf{b}[a,b,b], \mathbf{b}[b,b,b]$.

3.5 Implementation

The header of the de Casteljau algorithm program is:

```
float decas(degree,coeff,t)
/*   uses  de Casteljau to compute one coordinate
     value of a  Bezier curve. Has to be called
     for each coordinate  (x,y, and/or z) of a control polygon.
Input:   degree: degree of curve.
         coeff:  array with coefficients of curve.
         t:      parameter value.
Output: coordinate value.
*/
```

This procedure invites several comments. First, we see that it requires the use of an auxiliary array `coeffa`. Moreover, this auxiliary array has to be filled for each function call! So on top of the already high computational cost of the de Casteljau algorithm we add another burden to the routine, keeping it from being very efficient. A faster evaluation method is given at the end of the next chapter.

To plot a Bézier curve, we would then call the routine several times:

```
void bez_to_points(degree,npoints,coeff,points)
/*     Converts Bezier curve into point sequence. Works on
       one coordinate only.
 Input:   degree: degree of curve.
          npoints: # of coordinates to be generated. (counting
                   from 0!)
          coeff:   coordinates of control polygon.
```

Output: points: coordinates of points on curve.

       Remark: For a 2D curve, this routine needs to be called twice,
          once for the x-coordinates and once for y.
*/

    The last subroutine has to be called once for each coordinate, i.e., two or three times. The main program **decasmain.c** on the enclosed disk gives an example of how to use it and how to generate postscript output.

3.6   Problems

1. Use the notation $\mathbf{b}^n(t) = \mathcal{B}[\mathbf{b}_0, \ldots, \mathbf{b}_n; t]$ to formulate the de Casteljau algorithm. Show that the computation count is exponential (in terms of the degree) if you implement such a recursive algorithm in a language like C.

2. Suppose a planar Bézier curve has a control polygon that is symmetric with respect to the y-axis. Is the curve also symmetric with respect to the y-axis? Be sure to consider the control polygon $(-1, 0), (0, 1), (1, 1),$ $(0, 2), (0, 1), (-1, 1), (0, 2), (0, 1), (1, 0)$. Generalize to other symmetry properties.

3. Show that every nonplanar cubic in $I\!\!E^3$ can be obtained as an affine map of the *standard cubic* (see Boehm [63])

$$\mathbf{x}(t) = \begin{bmatrix} t \\ t^2 \\ t^3 \end{bmatrix}.$$

4. Use the de Casteljau algorithm to design a curve of degree four that has its middle control point on the curve. More specifically, try to achieve

$$\mathbf{b}_2 = \mathbf{b}_0^4(\frac{1}{2}).$$

5. Write an experimental program that replaces $(1-t)$ and t in the recursion (3.2) by $[1 - f(t)]$ and $f(t)$, where f is some "interesting" function. Change the routine **decas** and comment on your results.

6. Rewrite the routine **decas** to handle blossoms. Evaluate and plot for some "interesting" arguments.

Chapter 4

The Bernstein Form of a Bézier Curve

Bézier curves can be defined by a recursive algorithm, which is how de Casteljau first developed them. It is also necessary, however, to have an *explicit* representation for them, i.e., to express a Bézier curve in terms of a nonrecursive formula rather than in terms of an algorithm. This will facilitate further theoretical development considerably.

4.1 Bernstein Polynomials

We will express Bézier curves in terms of *Bernstein polynomials*, defined explicitly by

$$B_i^n(t) = \binom{n}{i} t^i (1-t)^{n-i}, \tag{4.1}$$

where the binomial coefficients are given by

$$\binom{n}{i} = \begin{cases} \frac{n!}{i!(n-i)!} & \text{if} \quad 0 \le i \le n \\ 0 & \text{else.} \end{cases}$$

There is a fair amount of literature on these polynomials. We cite just a few: Bernstein [47], Lorentz [319], Davis [110], and Korovkin [293]. An extensive bibliography is given in Gonska and Meier [213].

Before we explore the importance of Bernstein polynomials to Bézier curves, let us first examine them more closely. One of their important properties is that they satisfy the following recursion:

$$B_i^n(t) = (1-t)B_i^{n-1}(t) + tB_{i-1}^{n-1}(t) \tag{4.2}$$

41

with

$$B_0^0(t) \equiv 1 \tag{4.3}$$

and

$$B_j^n(t) \equiv 0 \quad \text{for } j \notin \{0, \dots, n\}. \tag{4.4}$$

The proof is simple:

$$
\begin{aligned}
B_i^n(t) &= \binom{n}{i} t^i (1-t)^{n-i} \\
&= \binom{n-1}{i} t^i (1-t)^{n-i} + \binom{n-1}{i-1} t^i (1-t)^{n-i} \\
&= (1-t) B_i^{n-1}(t) + t B_{i-1}^{n-1}(t).
\end{aligned}
$$

Another important property is that Bernstein polynomials form a *partition of unity*:

$$\sum_{j=0}^{n} B_j^n(t) \equiv 1. \tag{4.5}$$

This fact is proved with the help of the binomial theorem:

$$1 = [t + (1-t)]^n = \sum_{j=0}^{n} \binom{n}{j} t^j (1-t)^{n-j} = \sum_{j=0}^{n} B_j^n(t).$$

Figure 4.1 shows the family of the five quartic Bernstein polynomials. Note that the B_i^n are nonnegative over the interval $[0,1]$.

We are now ready to see why Bernstein polynomials are important for the development of Bézier curves. The intermediate de Casteljau points $\mathbf{b}_i^r$ can be expressed in terms of Bernstein polynomials of degree r:

$$\mathbf{b}_i^r(t) = \sum_{j=0}^{r} \mathbf{b}_{i+j} B_j^r(t) \quad \begin{array}{l} \in \{0, \dots, n\} \\ i \in \{0, \dots, n-r\}. \end{array} \tag{4.6}$$

This equation shows exactly how the intermediate point $\mathbf{b}_i^r$ depends on the given Bézier points $\mathbf{b}_i$. Figure 3.3 shows how these intermediate points form Bézier curves themselves. The main importance of (4.6) lies, of course, in the case $r = n$. The corresponding de Casteljau point is the point on the curve and is given by

$$\mathbf{b}^n(t) = \mathbf{b}_0^n(t) = \sum_{j=0}^{n} \mathbf{b}_j B_j^n(t). \tag{4.7}$$

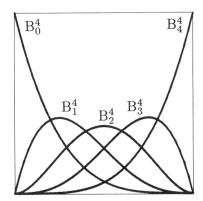

Figure 4.1: Bernstein polynomials: the quartic case.

We still have to prove (4.6). To that end, we use the recursive definition of the $\mathbf{b}_i^r$ [Eq. (3.2)] and the recursion for the Bernstein polynomials (4.2) and (4.4) in an inductive proof:

$$
\begin{aligned}
\mathbf{b}_i^r(t) &= (1-t)\mathbf{b}_i^{r-1}(t) + t\mathbf{b}_{i+1}^{r-1}(t) \\
&= (1-t)\sum_{j=i}^{i+r-1}\mathbf{b}_j B_{j-i}^{r-1}(t) + t\sum_{j=i+1}^{i+r}\mathbf{b}_j B_{j-i-1}^{r-1}(t).
\end{aligned}
$$

Reindexing and invoking (4.4), we can rewrite this as

$$
\begin{aligned}
\mathbf{b}_i^r(t) &= (1-t)\sum_{j=i}^{i+r}\mathbf{b}_j B_{j-i}^{r-1}(t) + t\sum_{j=i}^{i+r}\mathbf{b}_j B_{j-i-1}^{r-1}(t) \\
&= \sum_{j=i}^{i+r}\mathbf{b}_j[(1-t)B_{j-i}^{r-1}(t) + tB_{j-i-1}^{r-1}(t)].
\end{aligned}
$$

Application of (4.2) then completes the proof. Note that (4.2) also defines B_0^n and B_n^n, since $B_{-1}^{n-1} = B_n^{n-1} = 0$ by (4.4).

With the intermediate points $\mathbf{b}_i^r$ at hand, we can write a Bézier curve in the form

$$
\mathbf{b}^n(t) = \sum_{i=0}^{r}\mathbf{b}_i^{n-r}(t)B_i^r(t). \tag{4.8}
$$

This is to be interpreted as follows: first, compute $n - r$ levels of the de Casteljau algorithm with respect to t. Then, take the resulting points $\mathbf{b}_i^{n-r}(t)$ as control points of an rth degree Bézier curve and evaluate it at t.

There is a link between the last statement and the blossom of a Bézier curve, as explained in Section 3.4. Let us reformulate our statement: "first, compute $n - r$ levels of the de Casteljau algorithm with respect to 0. Then, take the resulting points $\mathbf{b}_i^{n-r}(0)$ as control points of an rth degree Bézier curve and evaluate it at 1." This leads to

$$\sum_{i=0}^{r} \mathbf{b}_i^{n-r}(0) B_i^r(1) = \mathbf{b}[0^{<n-r>}, 1^{<r>}] = \mathbf{b}_r, \tag{4.9}$$

thus proving (3.9).

4.2 Properties of Bézier Curves

Many of the properties in this section have already appeared in the previous chapter. They were derived using geometric arguments. We shall now rederive several of them, using algebraic arguments. If the same heading is used here as in Chapter 3, the reader should look there for a complete description of the property in question.

Affine invariance. Barycentric combinations are invariant under affine maps. Therefore, (4.5) gives the algebraic verification of this property. We note again that this does not imply invariance under perspective maps!

Invariance under affine parameter transformations. Algebraically, this property reads

$$\sum_{i=0}^{n} \mathbf{b}_i B_i^n(t) = \sum_{i=0}^{n} \mathbf{b}_i B_i^n \left(\frac{u-a}{b-a} \right). \tag{4.10}$$

Convex hull property. This follows, since for $t \in [0, 1]$, the Bernstein polynomials are nonnegative. They sum to one as shown in (4.5).

Endpoint interpolation. This is a consequence of the identities

$$\begin{aligned} B_i^n(0) &= \delta_{i,0} \\ B_i^n(1) &= \delta_{i,n} \end{aligned} \tag{4.11}$$

and (4.5). Here, $\delta_{i,j}$ is the Kronecker delta function: it equals one when its arguments agree, and zero otherwise.

Symmetry. Looking at the examples in Figure 3.4, it is clear that it does not matter if the Bézier points are labeled $\mathbf{b}_0, \mathbf{b}_1, \ldots, \mathbf{b}_n$ or $\mathbf{b}_n, \mathbf{b}_{n-1}, \ldots, \mathbf{b}_0$. The curves that correspond to the two different orderings look the same; they differ only in the direction in which they are traversed. Written as a formula:

$$\sum_{j=0}^{n} \mathbf{b}_j B_j^n(t) = \sum_{j=0}^{n} \mathbf{b}_{n-j} B_j^n(1-t). \tag{4.12}$$

This follows from the identity

$$B_j^n(t) = B_{n-j}^n(1-t), \tag{4.13}$$

which follows from inspection of (4.1). We say that Bernstein polynomials are *symmetric* with respect to t and $1-t$.

Invariance under barycentric combinations. The process of forming the Bézier curve from the Bézier polygon leaves barycentric combinations invariant. For $\alpha + \beta = 1$, we obtain

$$\sum_{j=0}^{n}(\alpha \mathbf{b}_j + \beta \mathbf{c}_j)B_j^n(t) = \alpha \sum_{j=0}^{n} \mathbf{b}_j B_j^n(t) + \beta \sum_{j=0}^{n} \mathbf{c}_j B_j^n(t). \tag{4.14}$$

In words: we can construct the weighted average of two Bézier curves either by taking the weighted average of corresponding points on the curves, or by taking the weighted average of corresponding control vertices and then computing the curve.

This linearity property is essential for many theoretical purposes, the most important one being the definition of tensor product surfaces in Chapter 16.

Linear precision. The following is a useful identity:

$$\sum_{j=0}^{n} \frac{j}{n} B_j^n(t) = t, \tag{4.15}$$

which has the following application: suppose the polygon vertices $\mathbf{b}_j$ are uniformly distributed on a straight line joining two points $\mathbf{p}$ and $\mathbf{q}$:

$$\mathbf{b}_j = (1 - \frac{j}{n})\mathbf{p} + \frac{j}{n}\mathbf{q}; \quad j = 0, \ldots, n.$$

The curve that is generated by this polygon is the straight line between **p** and **q**, i.e., the initial straight line is reproduced. This property is called *linear precision*.[1]

Pseudo-local control. The Bernstein polynomial B_i^n has only one maximum and attains it at $t = i/n$. This has a design application: if we move only one of the control polygon vertices, say, $\mathbf{b}_i$, then the curve is mostly affected by this change in the region of the curve around the parameter value i/n. This makes the effect of the change reasonably predictable, although the change does affect the whole curve. As a rule of thumb (mentioned to me by P. Bézier), the maximum of each B_i^n is roughly $\frac{1}{3}$; thus a change of $\mathbf{b}_i$ by three units will change the curve by one unit.

4.3 The Derivative of a Bézier Curve

The derivative of a Bernstein polynomial B_i^n is obtained as

$$
\begin{aligned}
\frac{\mathrm{d}}{\mathrm{d}t} B_i^n(t) &= \frac{\mathrm{d}}{\mathrm{d}t} \binom{n}{i} t^i (1-t)^{n-i} \\
&= \frac{i\, n!}{i!(n-i)!} t^{i-1}(1-t)^{n-i} - \frac{(n-i)n!}{i!(n-i)!} t^i (1-t)^{n-i-1} \\
&= \frac{n(n-1)!}{(i-1)!(n-i)!} t^{i-1}(1-t)^{n-i} - \frac{n(n-1)!}{i!(n-i-1)!} t^i (1-t)^{n-i-1} \\
&= n\left[B_{i-1}^{n-1}(t) - B_i^{n-1}(t) \right].
\end{aligned}
$$

Thus

$$
\frac{\mathrm{d}}{\mathrm{d}t} B_i^n(t) = n\left[B_{i-1}^{n-1}(t) - B_i^{n-1}(t) \right]. \tag{4.16}
$$

We can now determine the derivative of a Bézier curve $\mathbf{b}^n$:

$$
\frac{\mathrm{d}}{\mathrm{d}t} \mathbf{b}^n(t) = n \sum_{j=0}^{n} \left[B_{j-1}^{n-1}(t) - B_j^{n-1}(t) \right] \mathbf{b}_j.
$$

---

[1]If the points are not uniformly spaced, we will also recapture the straight line segment. However, it will not be linearly parametrized.

Because of (4.4), this can be simplified to

$$\frac{d}{dt}\mathbf{b}^n(t) = n\sum_{j=1}^{n} B_{j-1}^{n-1}(t)\mathbf{b}_j - n\sum_{j=0}^{n-1} B_j^{n-1}(t)\mathbf{b}_j,$$

and now an index transformation of the first sum yields

$$\frac{d}{dt}\mathbf{b}^n(t) = n\sum_{j=0}^{n-1} B_j^{n-1}(t)\mathbf{b}_{j+1} - n\sum_{j=0}^{n-1} B_j^{n-1}(t)\mathbf{b}_j,$$

and finally

$$\frac{d}{dt}\mathbf{b}^n(t) = n\sum_{j=0}^{n-1} (\mathbf{b}_{j+1} - \mathbf{b}_j)B_j^{n-1}(t).$$

The last formula can be simplified somewhat by the introduction of the *forward difference operator* Δ:

$$\Delta\mathbf{b}_j = \mathbf{b}_{j+1} - \mathbf{b}_j. \tag{4.17}$$

We now have for the derivative of a Bézier curve:

$$\frac{d}{dt}\mathbf{b}^n(t) = n\sum_{j=0}^{n-1} \Delta\mathbf{b}_j B_j^{n-1}(t); \qquad \Delta\mathbf{b}_j \in I\!\!R^3. \tag{4.18}$$

The derivative of a Bézier curve is thus another Bézier curve, obtained by differencing the original control polygon. However, this derivative Bézier curve does not "live" in $I\!\!E^3$ anymore! Its coefficients are differences of points, i.e., *vectors*, which are elements of $I\!\!R^3$. To visualize the derivative curve and polygon in $I\!\!E^3$, we can construct a polygon in $I\!\!E^3$ that consists of the points $\mathbf{a} + \Delta\mathbf{b}_0, \ldots, \mathbf{a} + \Delta\mathbf{b}_{n-1}$. Here $\mathbf{a}$ is arbitrary; one reasonable choice is $\mathbf{a} = \mathbf{0}$. Figure 4.2 illustrates a Bézier curve and its derivative curve (with the choice $\mathbf{a} = \mathbf{0}$). This derivative curve is sometimes called a *hodograph*. For more information on hodographs, see Forrest [191], Bézier [53], or Sederberg and Wang [435].

4.4 Higher Order Derivatives

To compute higher derivatives, we first generalize the forward difference operator (4.17): the *iterated forward difference operator* Δ^r is defined by

$$\Delta^r\mathbf{b}_j = \Delta^{r-1}\mathbf{b}_{j+1} - \Delta^{r-1}\mathbf{b}_j. \tag{4.19}$$

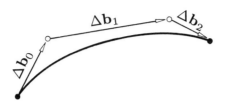

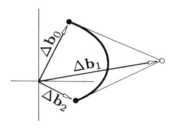

Figure 4.2: Derivatives: a Bézier curve and its first derivative curve (scaled down by a factor of three). Note that this derivative curve does not change if a translation is applied to the original curve.

We list a few examples:

$$
\begin{aligned}
\Delta^0 \mathbf{b}_i &= \mathbf{b}_i \\
\Delta^1 \mathbf{b}_i &= \mathbf{b}_{i+1} - \mathbf{b}_i \\
\Delta^2 \mathbf{b}_i &= \mathbf{b}_{i+2} - 2\mathbf{b}_{i+1} + \mathbf{b}_i \\
\Delta^3 \mathbf{b}_i &= \mathbf{b}_{i+3} - 3\mathbf{b}_{i+2} + 3\mathbf{b}_{i+1} - \mathbf{b}_i.
\end{aligned}
$$

The factors on the right-hand sides are binomial coefficients, forming a Pascal-like triangle. This pattern holds in general:

$$
\Delta^r \mathbf{b}_i = \sum_{j=0}^{r} \binom{r}{j} (-1)^{r-j} \mathbf{b}_{i+j}. \tag{4.20}
$$

We are now in a position to give the formula for the rth derivative of a Bézier curve:

$$
\frac{d^r}{dt^r} \mathbf{b}^n(t) = \frac{n!}{(n-r)!} \sum_{j=0}^{n-r} \Delta^r \mathbf{b}_j B_j^{n-r}(t). \tag{4.21}
$$

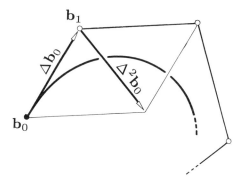

Figure 4.3: Endpoint derivatives: the first and second derivative vectors at $t = 0$ are multiples of the first and second difference vectors at $\mathbf{b}_0$.

The proof of (4.21) is by repeated application of (4.18).

Two important special cases of (4.21) are given by $t = 0$ and $t = 1$. Because of (4.11) we obtain

$$\frac{\mathrm{d}^r}{\mathrm{d}t^r}\mathbf{b}^n(0) = \frac{n!}{(n-r)!}\Delta^r\mathbf{b}_0 \qquad (4.22)$$

and

$$\frac{\mathrm{d}^r}{\mathrm{d}t^r}\mathbf{b}^n(1) = \frac{n!}{(n-r)!}\Delta^r\mathbf{b}_{n-r}. \qquad (4.23)$$

Thus the rth derivative of a Bézier curve at an endpoint depends only on the $r + 1$ Bézier points near (and including) that endpoint. For $r = 0$, we get the already established property of endpoint interpolation. The case $r = 1$ states that $\mathbf{b}_0$ and $\mathbf{b}_1$ define the tangent at $t = 0$, provided they are distinct.[2] Similarly, $\mathbf{b}_{n-1}$ and $\mathbf{b}_n$ determine the tangent at $t = 1$. The cases $r = 1, r = 2$ are illustrated in Figure 4.3.

If one knows all derivatives of a function at one point, corresponding to $t = 0$, say, one can generate its Taylor series. The Taylor series of a polynomial is just that polynomial itself, in the *monomial form*:

$$\mathbf{x}(t) = \sum_{j=0}^{n} \frac{1}{j!}\mathbf{x}^{(j)}(0)t^j.$$

---

[2]In general, the tangent at $\mathbf{b}_0$ is determined by $\mathbf{b}_0$ and the first $\mathbf{b}_i$ that is distinct from $\mathbf{b}_0$. Thus the tangent may be defined even if the tangent vector is the zero vector.

Utilizing (4.22), we have

$$\mathbf{b}^n(t) = \sum_{j=0}^{n} \frac{j!}{(n-j)!} \Delta^j \mathbf{b}_0 \, t^j. \tag{4.24}$$

4.5   Derivatives and the de Casteljau Algorithm

Derivatives of a Bézier curve can be expressed in terms of the intermediate points generated by the de Casteljau algorithm:

$$\frac{d^r}{dt^r} \mathbf{b}^n(t) = \frac{n!}{(n-r)!} \Delta^r \mathbf{b}_0^{n-r}(t). \tag{4.25}$$

This follows since summation and taking differences commute:

$$\sum_{j=0}^{n-1} \Delta \mathbf{b}_j = \sum_{j=1}^{n} \mathbf{b}_j - \sum_{j=0}^{n-1} \mathbf{b}_j = \Delta \sum_{j=0}^{n-1} \mathbf{b}_j. \tag{4.26}$$

Using this, we have

$$\frac{d^r}{dt^r} \mathbf{b}^n(t) = \frac{n!}{(n-r)!} \sum_{j=0}^{n-r} \Delta^r \mathbf{b}_j B_j^{n-r}(t) \tag{4.27}$$

$$= \frac{n!}{(n-r)!} \Delta^r \sum_{j=0}^{n-r} \mathbf{b}_j B_j^{n-r}(t) \tag{4.28}$$

$$= \frac{n!}{(n-r)!} \Delta^r \mathbf{b}_0^{n-r}(t). \tag{4.29}$$

The first and the last of these three equations suggest two different ways of computing the rth derivative of a Bézier curve: for the first method (4.27), compute all r^{th} forward differences of the control points, then interpret them as a new Bézier polygon of degree $n - r$ and evaluate it at t.

The second method, using (4.29), computes the rth derivative as a "by-product" of the de Casteljau algorithm. If we compute a point on a Bézier curve using a triangular arrangement as in (3.3), then for any $n - r$, the corresponding $\mathbf{b}_i^{n-r}$ form a column (with r entries) in that scheme. To obtain the rth derivative at t, we simply take the rth difference of these points and then multiply by the constant $n!/(n-r)!$ In some applications (curve/plane intersection, for example), one needs not only a point on the curve, but its

To compute the derivative of the Bézier curve from Example 3.1, we could form the first differences of the control points and evaluate the corresponding quadratic curve at $t = \frac{1}{2}$:

$$\begin{bmatrix} 0 \\ 2 \end{bmatrix}$$

$$\begin{bmatrix} 8 \\ 0 \end{bmatrix} \quad \begin{bmatrix} 4 \\ 1 \end{bmatrix}$$

$$\begin{bmatrix} -4 \\ -2 \end{bmatrix} \quad \begin{bmatrix} 2 \\ -1 \end{bmatrix} \quad \begin{bmatrix} 3 \\ 0 \end{bmatrix}$$

Alternatively, we could compute the difference $\mathbf{b}_1^2 - \mathbf{b}_0^2$:

$$\begin{bmatrix} 5 \\ \frac{3}{2} \end{bmatrix} - \begin{bmatrix} 2 \\ \frac{3}{2} \end{bmatrix} = \begin{bmatrix} 3 \\ 0 \end{bmatrix}.$$

In both cases, the result needs to be multiplied by a factor of 3.

Example 4.1: Two ways to compute derivatives.

first and/or second derivative at the same time. The de Casteljau algorithm offers a quick solution to this problem.

A summary of both methods: to compute the rth derivative of a Bézier curve, perform r difference steps and $n-r$ evaluation steps. It does not matter in which order we perform these two steps.

The case $r = 1$ is important enough to warrant special attention:

$$\frac{\mathrm{d}}{\mathrm{d}t}\mathbf{b}^n(t) = n[\mathbf{b}_1^{n-1}(t) - \mathbf{b}_0^{n-1}(t)]. \tag{4.30}$$

The intermediate points $\mathbf{b}_0^{n-1}$ and $\mathbf{b}_1^{n-1}$ thus determine the *tangent vector* at $\mathbf{b}^n(t)$, which is illustrated in Figures 3.1 and 3.2.

Two ways to compute the tangent vector of a Bézier curve are demonstrated in Example 4.1.

4.6 Subdivision

A Bézier curve $\mathbf{b}^n$ is usually defined over the interval (the domain) $[0, 1]$, but it can also be defined over any interval $[0, c]$. The part of the curve that

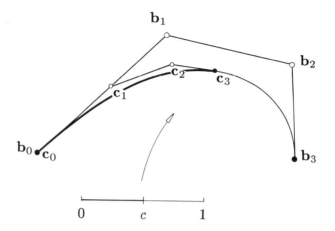

Figure 4.4: Subdivision: two Bézier polygons describing the same curve: one (the $\mathbf{b}_i$) is associated with the parameter interval $[0,1]$, the other (the $\mathbf{c}_i$) with $[0,c]$.

corresponds to $[0,c]$ can also be defined by a Bézier polygon, as illustrated in Figure 4.4. Finding this Bézier polygon is referred to as *subdivision* of the Bézier curve.

The unknown Bézier points $\mathbf{c}_i$ are found without much work if we use the blossoming principle from Section 3.4. Equation (3.12) gave us the Bézier points of a polynomial curve that is defined over an arbitrary interval $[a,b]$. We are currently interested in the interval $[0,c]$, and so our Bézier points are:

$$\mathbf{c}_i = \mathbf{b}[0^{<n-i>}, c^{<i>}].$$

Thus each $\mathbf{c}_i$ is obtained by carrying out i de Casteljau steps with respect to c, in nonblossom notation:

$$\mathbf{c}_j = \mathbf{b}_0^j(c). \tag{4.31}$$

This formula is called the *subdivision formula* for Bézier curves.

Thus it turns out that the de Casteljau algorithm not only computes the point $\mathbf{b}^n(c)$, but also provides the control vertices of the Bézier curve corresponding to the interval $[0,c]$. Because of the symmetry property (4.12), it follows that the control vertices of the part corresponding to $[c,1]$ are given by the $\mathbf{b}_j^{n-j}$. Thus, in Figures 3.1 and 3.2, we see the two subpolygons defining the arcs from $\mathbf{b}^n(0)$ to $\mathbf{b}^n(c)$ and from $\mathbf{b}^n(c)$ to $\mathbf{b}^n(1)$.

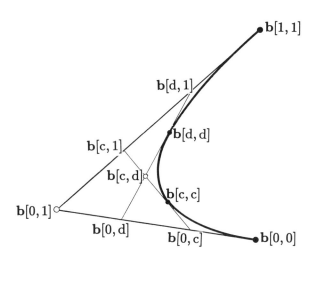

Figure 4.5: Generalized subdivision: evaluation of a quadratic at two parameter values c and d subdivides it into three segments. Its Bézier points are shown in blossom notation.

Figure 4.5 shows the blossom notation if we subdivide at *two* parameter values c and d simultaneously.

Instead of subdividing a Bézier curve, we may also *extrapolate* it: in that case, we might be interested in the Bézier points $\mathbf{d}_i$ corresponding to an interval $[1, d]$. They are given by

$$\mathbf{d}_j = \mathbf{b}[1^{<n-j>}, d^{<j>}] = \mathbf{b}_{n-j}^j(d).$$

Subdivision for Bézier curves, although mentioned by de Casteljau [122], was rigorously proved by E. Staerk [451]. Our blossom development is due to Ramshaw [386] and de Casteljau [123].

Subdivision may be repeated: we may subdivide a curve at $t = 1/2$, then split the two resulting curves at $t = 1/2$ of their respective parameters, and

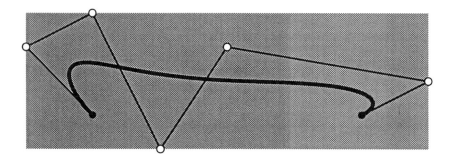

Figure 4.6: The minmax box of a Bézier curve: the smallest rectangle that contains the curve's control polygon.

so on. After k levels of subdivisions, we end up with 2^k Bézier polygons, each describing a small arc of the original curve. These polygons converge to the curve if we keep increasing k, as was shown by Lane and Riesenfeld [298]. We will prove a more general statement in Section 10.7.

Convergence of this repeated subdivision process is very fast (see Cohen and Schumaker [100] and Dahmen [108],) and thus it has many practical applications. We shall discuss here the process of intersecting a straight line with a Bézier curve: Suppose we are given a planar Bézier curve and we wish to find intersection points with a given straight line $\mathbf{L}$, if they exist.

If the curve and $\mathbf{L}$ are far apart, we would like to be able to flag such configurations as quickly as possible, and then abandon any further attempts to find intersection points. To do this, we create the *minmax box* of the control polygon: this is the smallest rectangle, with sides parallel to the coordinate axes, that contains the polygon. It is found very quickly, and by the convex hull property of Bézier curves, we know that it also contains the curve. Figure 4.6 gives an example.

Having found the minmax box, it is trivial to determine if it interferes with $\mathbf{L}$; if not, we know we will not have any intersections. This quick test is called *trivial reject*.

Now suppose the minmax box *does* interfere with $\mathbf{L}$. Then there may be an intersection. We now subdivide the curve at $t = 1/2$ and carry out our trivial reject test for both subpolygons.[3] If the outcome is still inconclusive, we repeat. Eventually the size of the involved minmax boxes will be so small that we can simply take their centers as the desired intersection points.

The routine `intersect` employs this idea, and a little more: as we keep subdividing the curve, zooming in toward the intersection points, the generated subpolygons become simpler and simpler in shape. If the control points

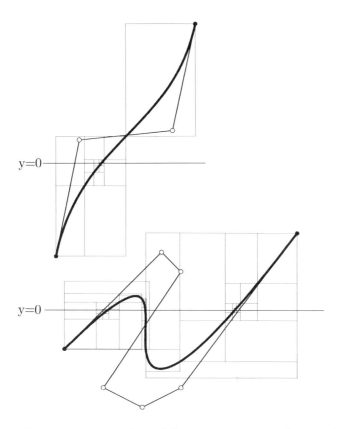

Figure 4.7: Curve intersection by subdivision: two examples are shown. Intersection is with the x-axis in both cases. Note the clustering of minmax boxes near the intersection points.

---

[3]The choice $t = 1/2$ is arbitrary, but works well. One might try to find better places to subdivide, but it is most likely cheaper to just perform a few more subdivisions instead.

of a polygon are almost collinear, we may replace them by a straight line. We could then intersect this straight line with **L** in order to find an intersection point. The extra work here lies in determining if a control polygon is "linear" or not. In our case, this is done by the routine `checkflat`. Figure 4.7 gives two examples. Note how the subdivision process finds *all* intersection points in the bottom example. These points will not, however, be recorded by increasing values of t.

4.7   Blossom and Polar

After the first de Casteljau step with respect to a parameter value t_1, the resulting $\mathbf{b}_0^1(t_1), \ldots, \mathbf{b}_{n-1}^1(t_1)$ may be interpreted as a control polygon of a curve $\mathbf{p}_1(t)$ of degree $n-1$. In the blossoming terminology from Section 3.4, we can write:

$$\mathbf{p}_1(t) = \mathbf{b}[t_1, t^{<n-1>}].$$

Invoking our knowledge about derivatives, we have:

$$
\begin{aligned}
\mathbf{p}_1(t) &= \sum_{i=0}^{n-1} [(1 - t_1)\mathbf{b}_i + t_1\mathbf{b}_{i+1}]B_i^{n-1}(t) \\
&= \sum_{i=0}^{n-1} [(1 - t_1)\mathbf{b}_i + t_1\mathbf{b}_{i+1} - \mathbf{b}_i^1(t)]B_i^{n-1}(t) + \sum_{i=0}^{n-1} \mathbf{b}_i^1(t)B_i^{n-1}(t) \\
&= (t_1 - t)\sum_{i=0}^{n-1} [\mathbf{b}_{i+1} - \mathbf{b}_i]B_i^{n-1}(t) + \sum_{i=0}^{n-1} \mathbf{b}_i^1(t)B_i^{n-1}(t).
\end{aligned}
$$

Therefore,

$$\mathbf{p}_1(t) = \mathbf{b}(t) + \frac{t_1 - t}{n}\frac{\mathrm{d}}{\mathrm{d}t}\mathbf{b}(t). \tag{4.32}$$

The polynomial $\mathbf{p}_1$ is called *first polar* of $\mathbf{b}(t)$ with respect to t_1. Figure 4.8 illustrates the geometric significance of (4.32): the tangent at any point $\mathbf{b}(t)$ intersects the polar $\mathbf{p}_1(t)$ at $\mathbf{p}_1(t)$. Keep in mind that this is not restricted to planar curves, but is equally valid for space curves!

For the special case of a (nonplanar) cubic, we may then conclude the following: the polar $\mathbf{p}_1$ lies in the osculating plane (see Section 11.2) of the cubic at $\mathbf{b}(t_1)$. If we intersect all tangents to the cubic with this osculating plane, we will trace out the polar. We can also conclude that for three different

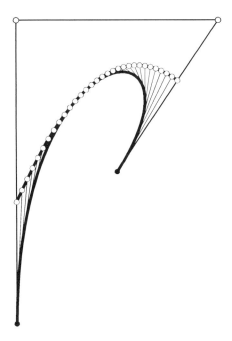

Figure 4.8: Polars: the polar $\mathbf{p}_1(t)$ with respect to $t_1 = 0.4$ is intersected by the tangents of the given curve $\mathbf{b}(t)$.

parameters t_1, t_2, t_3, the blossom value $\mathbf{b}[t_1, t_2, t_3]$ is the intersection of the corresponding osculating planes.

Another special case is given by $\mathbf{b}[0, t^{<n-1>}]$: this is the polynomial defined by $\mathbf{b}_0, \ldots, \mathbf{b}_{n-1}$. Similarly, $\mathbf{b}[1, t^{<n-1>}]$ is defined by $\mathbf{b}_1, \ldots, \mathbf{b}_n$. This observation may be used for a proof of (3.9).

Returning to the general case, we may repeat the process of forming polars, thus obtaining a second polar $\mathbf{p}_{1,2}(t) = \mathbf{b}[t_1, t_2, t^{<n-2>}]$, etc. We finally arrive at the nth polar, which we have already encountered as the blossom $\mathbf{b}[t_1, \ldots, t_n]$ of $\mathbf{b}(t)$. The relationship between blossoms and polars was observed by Ramshaw in [388]. The above geometric arguments are due to S. Jolles, who developed a geometric theory of blossoming as early as 1886 in [280].[4]

---

[4]W. Boehm first noted the relevance of Jolles's work to the theory of blossoming.

Section 3.4 provided a way to generate the blossom of a curve recursively. We may also find explicit formulas for it; here is the case of a cubic:

$$
\begin{aligned}
&\mathbf{b}[t_1, t_2, t_3] \\
={} &(1 - t_1)\mathbf{b}[0, t_2, t_3] + t_1\mathbf{b}[1, t_2, t_3] \\
={} &(1 - t_1)[(1 - t_2)\mathbf{b}[0, 0, t_3] + t_2\mathbf{b}[0, 1, t_3]] + t_1[(1 - t_2)\mathbf{b}[0, 1, t_3] \\
&+ t_2\mathbf{b}[1, 1, t_3]] = \mathbf{b}[0, 0, 0](1 - t_1)(1 - t_2)(1 - t_3) \\
&+ \mathbf{b}[0, 0, 1][(1 - t_1)(1 - t_2)t_3 + (1 - t_1)t_2(1 - t_3) + t_1(1 - t_2)(1 - t_3)] \\
&+ \mathbf{b}[0, 1, 1][t_1 t_2(1 - t_3) + t_1(1 - t_2)t_3 + (1 - t_1)t_2 t_3] \\
&+ \mathbf{b}[1, 1, 1]t_1 t_2 t_3.
\end{aligned}
$$

For each step, we have exploited the fact that blossoms are multiaffine.

Note how we recover the cubic Bernstein polynomials for $t_1 = t_2 = t_3$. The above development would hold for parameter intervals other than $[0, 1]$ equally well, because of the invariance under affine parameter transformations.

We should add that not every multivariate polynomial function can be interpreted as the blossom of a Bézier curve. To qualify as a blossom, the function must be both symmetric and multiaffine.

4.8   The Matrix Form of a Bézier Curve

Some authors (Faux and Pratt [180], Mortenson [339], Chang [87]) prefer to write Bézier curves and other polynomial curves in matrix form. A curve of the form

$$
\mathbf{x}(t) = \sum_{j=0}^{n} \mathbf{c}_i C_i(t)
$$

can be interpreted as a dot product:

$$
\mathbf{x}(t) = \begin{bmatrix} \mathbf{c}_0 & \cdots & \mathbf{c}_n \end{bmatrix} \begin{bmatrix} C_0(t) \\ \vdots \\ C_n(t) \end{bmatrix}.
$$

One can take this a step further and write

$$
\begin{bmatrix} C_0(t) \\ \vdots \\ C_n(t) \end{bmatrix} = \begin{bmatrix} m_{00} & \cdots & m_{0n} \\ \vdots & & \vdots \\ m_{n0} & \cdots & m_{nn} \end{bmatrix} \begin{bmatrix} t^0 \\ \vdots \\ t^n \end{bmatrix}. \tag{4.33}
$$

The matrix $M = \{m_{ij}\}$ describes the basis transformation between the basis polynomials $C_i(t)$ and the *monomial basis* t^i.

If the C_i are Bernstein polynomials, $C_i = B_i^n$, the matrix M has elements

$$
m_{ij} = (-1)^{j-i} \binom{n}{j} \binom{j}{i}. \tag{4.34}
$$

We list the cubic case explicitly:

$$
M = \begin{bmatrix} 1 & -3 & 3 & -1 \\ 0 & 3 & -6 & 3 \\ 0 & 0 & 3 & -3 \\ 0 & 0 & 0 & 1 \end{bmatrix}.
$$

Why the matrix form? Mathematically, it is equivalent to other curve formulations. When it comes to computer implementations, however, the matrix form may be advantageous if matrix multiplication is hard-wired.

A warning is appropriate: the monomial form is *numerically unstable* and should be avoided where accuracy in computation is of any importance. See the discussion in Section 24.3 for more details.

4.9   Implementation

First, we provide a routine that evaluates a Bézier curve more efficiently than **decas** from the last chapter. It will have the flavor of Horner's scheme for the evaluation of a polynomial in monomial form. To give an example of Horner's scheme, also called *nested multiplication*, we list the cubic case:

$$
\mathbf{c}_0 + t\mathbf{c}_1 + t^2\mathbf{c}_2 + t^3\mathbf{c}_3 = \mathbf{c}_0 + t[\mathbf{c}_1 + t(\mathbf{c}_2 + t\mathbf{c}_3)].
$$

A similar nested form can be devised for Bézier curves; again, the cubic case:

$$
\mathbf{b}^3(t) = \left\{ \left[\binom{3}{0} s\mathbf{b}_0 + \binom{3}{1} t\mathbf{b}_1 \right] s + \binom{3}{2} t^2 \mathbf{b}_2 \right\} s + \binom{3}{3} t^3 \mathbf{b}_3,
$$

where $s = 1 - t$. Recalling the identity

$$\binom{n}{i} = \frac{n-i+1}{i}\binom{n}{i-1}; \quad i > 0,$$

we arrive at the following program (for the general case):

```
float hornbez(degree,coeff,t)
/*   uses   a Horner-like  scheme to compute one coordinate
     value of a  Bezier curve. Has to be called
     for each coordinate  (x,y, and/or z) of a control polygon.
Input:    degree: degree of curve.
          coeff:  array with coefficients of curve.
          t:      parameter value.
Output:           coordinate value.
*/
```

To use this routine for plotting a Bézier curve, we would replace the call to decas in bez_to_points by an identical call to hornbez. Replacing decas with hornbez results in a significant savings of time: we do not have to save the control polygon in an auxiliary array; also, hornbez is of order n, whereas decas is of order n^2.

This is not to say, however, that we have produced super-efficient code for plotting points on a Bézier curve. For instance, we have to call hornbez once for each coordinate, and thus have to generate the binomial coefficients n_choose_i twice. This could be improved by writing a routine that combines the two calls. A further improvement could be to compute the sequence of binomial coefficents only once, and not over and over for each new value of t. All these (and possibly more) improvements would speed up the program, but would be less modular and thus less understandable. For the code in this book, modularity is placed above efficiency (in most cases).

We also include the programs to convert from the Bézier form to the monomial form:

```
void bezier_to_power(degree,bez,coeff)
/*Converts Bezier form to power (monomial) form. Works on
one coordinate only.

    Input:    degree:    degree of curve.
              bez:       coefficients of Bezier form
    Output:   coeff:     coefficients of power form.
```

Remark: For a 2D curve, this routine needs to be called twice,
once for the x-coordinates and once for y.
*/

The conversion program internally calls iterated forward differences:

```
void differences(degree,coeff,diffs)
/*
Computes all forward differences Delta^i(b_0).
Has to be called for each coordinate (x,y, and/or z) of a control polygon.
    Input:    degree: length (from 0) of coeff.
              coeff:  array of coefficients.
      Output: diffs:  diffs[i]= Delta^i(coeff[0]).
*/
```

Once the power form is found, it may be evaluated using Horner's scheme:

```
float horner(degree,coeff,t)
/*
    uses  Horner's scheme to compute one coordinate
    value of a curve in power form. Has to be called
    for each coordinate  (x,y, and/or z) of a control polygon.
    Input:    degree: degree of curve.
              coeff:  array with coefficients of curve.
              t:      parameter value.
      Output: coordinate value.
*/
```

The subdivision routine:

```
void subdiv(degree,coeff,weight,t,bleft,bright,wleft,wright)
/*
        subdivides ratbez curve at parameter value t.
  Input: degree:    degree of Bezier curve
         coeff:     Bezier points (one coordinate only)
         weight:    weights for rational case
         t:         where to subdivide
  Output:
         bleft,bright: left and reight subpolygons
         wleft,wright: their weights


  Note:  1. For the polynomial case, set all entries in weight to 1.
         2. Ordering of right polygon bright is reversed.
*/
```

Actually, this routine computes a more general case than is described in this chapter; namely, it computes subdivison for a *rational* Bézier curve. This will be discussed later; if the entries in `weight` are all unity, then `wleft` and `wright` will also be unity and can be safely ignored in the context of this chapter.

Now the routine to intersect a Bézier curve with a straight line (the straight line is assumed to be the y-axis):

```
void intersect(bx,by,w,degree,tol)
/* Intersects Bezier curve with x-axis  by  adaptive subdivision.
   Subdivision is controlled by tolerance tol. There is
   no check for stack depth! Intersection points are not found  in
   'natural' order.  Results are written into file outfile.
   Input: bx,by,w:    rational Bezier curve
          degree:     its degree
          tol:        accuracy for results
   Output:            intersection points, written into a file
   */
```

This routine (again covering the rational case as well) uses a routine to check if a control polygon is flat:

```
int check_flat(bx,by,degree,tol)
/* Checks if a polygon is flat. If all points
   are closer  than tol to the connection of the
   two endpoints, then it is flat. Crashes if the endpoints
   are identical.

   Input:    bx,by, degree: the Bezier curve
             tol:               tolerance
   Output:   1 if flat, 0 else.
   */
```

4.10   Problems

1. Show that the Bernstein polynomials B_i^n form a basis for the linear space of all polynomials of degree n.

2. Show that the Bernstein polynomial B_i^n attains its maximum at $t = i/n$. Find the maximum value. What happens for large n?

3. A cusp is a point on a curve where the first derivative vector vanishes. Show that a nonplanar cubic Bézier curve cannot have a cusp.

4. Compare the runtimes of `decas` and `hornbez` for curves of various degrees.

5. The blossom of a parabola is a map $I\!E^2 \to I\!E^2$. Mark the area in $I\!E^2$ that is obtainable as an image of the blossom. Then consider 2D and 3D cubics.

6. Use subdivision to create *smooth fractals*. Start with a degree four Bézier curve. Subdivide it into two curves and then perturb the middle control point b_2 for each of the two subpolygons. Continue for several levels. Try to perturb the middle control point by a random displacement and then by a controlled displacement. Literature on fractals: [28], [322].

7. Use subdivision to approximate a high-order $(n > 2)$ Bézier curve by a collection of quadratic Bézier curves. You will have to write a routine that determines if a given Bézier curve may be replaced by a quadratic one within a given tolerance. Literature on approximating higher order curves by lower order ones: [270], [274].

Chapter 5

Bézier Curve Topics

5.1 Degree Elevation

Suppose we were designing with Bézier curves as described in Section 3.3, trying to use a Bézier curve of degree n. After modifying the polygon a few times, it may turn out that a degree n curve does not possess sufficient flexibility to model the desired shape. One way to proceed in such a situation is to increase the flexibility of the polygon by adding another vertex to it. As a first step, one might want to add another vertex yet leave the shape of the curve unchanged – this corresponds to raising the degree of the Bézier curve by one. The new vertices $\mathbf{b}_j^{(1)}$ must satisfy

$$\sum_{j=0}^{n} \mathbf{b}_j \binom{n}{j} t^j (1-t)^{n-j} = \sum_{j=0}^{n+1} \mathbf{b}_j^{(1)} \binom{n+1}{j} t^j (1-t)^{n+1-j}. \qquad (5.1)$$

We multiply the left-hand side by $[t + (1-t)]$ to obtain

$$\sum_{j=0}^{n} \mathbf{b}_j \binom{n}{j} \left[t^j (1-t)^{n+1-j} + t^{j+1} (1-t)^{n-j} \right] = \sum_{j=0}^{n+1} \mathbf{b}_j^{(1)} \binom{n+1}{j} t^j (1-t)^{n+1-j}.$$

We now compare coefficients of $t^j (1-t)^{n+1-j}$ on both sides and obtain

$$\mathbf{b}_j^{(1)} \binom{n+1}{j} = \mathbf{b}_j \binom{n}{j} + \mathbf{b}_{j-1} \binom{n}{j-1},$$

which is equivalent to

$$\mathbf{b}_j^{(1)} = \frac{j}{n+1} \mathbf{b}_{j-1} + (1 - \frac{j}{n+1}) \mathbf{b}_j; \quad j = 0, \ldots, n+1. \qquad (5.2)$$

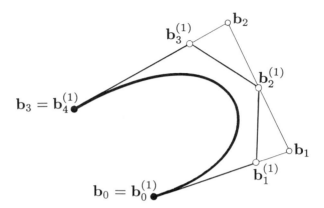

Figure 5.1: Degree elevation: both polygons define the same (degree three) curve.

Thus the new vertices $\mathbf{b}_j^{(1)}$ are obtained from the old polygon by piecewise linear interpolation at the parameter values $j/(n+1)$. It follows that the new polygon $\mathcal{E}\mathbf{P}$ lies in the convex hull of the old one. Figure 5.1 gives an example. Note how $\mathcal{E}\mathbf{P}$ is "closer" to the curve $\mathcal{B}\mathbf{P}$ than the original polygon $\mathbf{P}$.

Degree elevation has important application in surface design: for several algorithms that produce surfaces from curve input, it is necessary that these curves be of the same degree. Using degree elevation, we may achieve this by raising the degree of all input curves to the one of the highest degree. Another application lies in the area of *data transfer* between different CAD/CAM or graphics systems: Suppose you have generated a parabola (i.e., a degree two Bézier curve), and you want to feed it into a system that only knows about cubics. All you have to do is degree elevate your parabola.

5.2   Repeated Degree Elevation

The process of degree elevation assigns a polygon $\mathcal{E}\mathbf{P}$ to an original polygon $\mathbf{P}$. We may repeat this process and obtain a sequence of polygons

$\mathbf{P}, \mathcal{E}\mathbf{P}, \mathcal{E}^2\mathbf{P}$, etc. After r degree elevations, the polygon $\mathcal{E}^r\mathbf{P}$ has the vertices $\mathbf{b}_0^{(r)}, \ldots, \mathbf{b}_{n+r}^{(r)}$, and each $\mathbf{b}_i^{(r)}$ is explicitly given by

$$\mathbf{b}_i^{(r)} = \sum_{j=0}^{n} \mathbf{b}_j \binom{n}{j} \frac{\binom{r}{i-j}}{\binom{n+r}{i}}. \tag{5.3}$$

This formula is easily proved by induction (see Problems).

Let us now investigate what happens if we repeat the process of degree elevation again and again. As we shall see, the polygons $\mathcal{E}^r\mathbf{P}$ converge to the curve that all of them define:

$$\lim_{r\to\infty} \mathcal{E}^r\mathbf{P} = \mathcal{B}\mathbf{P}. \tag{5.4}$$

To prove this result, fix some parameter value t. For each r, find the index i such that $i/(n+r)$ is closest to t. We can think of $i/(n+r)$ as a parameter on the polygon $\mathcal{E}^r\mathbf{P}$, and as $r \to \infty$, this ratio tends to t. One can now show (using Stirling's formula) that

$$\lim_{i/(n+r)\to t} \frac{\binom{r}{i-j}}{\binom{r+n}{i}} = t^j(1-t)^{n-j},$$

and therefore

$$\lim_{i/(n+r)\to t} \mathbf{b}_i^{(r)} = \sum_{j=0}^{n} \mathbf{b}_j B_j^n(t) = [\mathcal{B}\mathbf{P}](t).$$

Figure 5.2 shows an example of the limit behavior of the polygons $\mathcal{E}^r\mathbf{P}$.

The polygons $\mathcal{E}^r\mathbf{P}$ approach the curve very slowly; this convergence result has no practical consequences. However, it helps in the investigation of some theoretical properties, as is seen in the next section.

The convergence of the polygons $\mathcal{E}^r\mathbf{P}$ to the curve was conjectured by R. Forrest [191] and proved in Farin [150]. The above proof follows an approach taken by J. Zhou [480]. Degree elevation may be generalized to "corner-cutting"; for a brief description, see Section 10.7.

5.3 The Variation Diminishing Property

We can now show that Bézier curves enjoy the *variation diminishing property*[1]: the curve $\mathcal{B}\mathbf{P}$ has no more intersections with any plane other than

---

[1] The variation diminishing property was first investigated by I. Schoenberg [421] in the context of B-spline approximation.

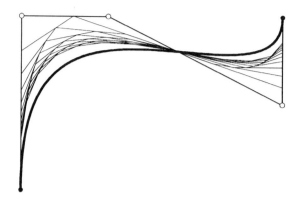

Figure 5.2: Degree elevation: a sequence of polygons approaching the curve that is defined by each of them.

the polygon $\mathbf{P}$. Degree elevation is an instance of piecewise linear interpolation, and we know that operation is variation diminishing (see Section 2.4). Thus each $\mathcal{E}^r\mathbf{P}$ has fewer intersections with a given plane than has its predecessor $\mathcal{E}^{(r-1)}\mathbf{P}$. Since the curve is the limit of these polygons, we have proved our statement. For high-degree Bézier curves, variation diminution may become so strong that the control polygon no longer resembles the curve.

A special case is obtained for *convex* polygons: a planar polygon (or curve) is said to be convex if it has no more than two intersections with any plane. The variation diminishing property thus asserts that a convex polygon generates a convex curve. Note that the inverse statement is not true: convex curves exist that have a nonconvex control polygon!

5.4   Degree Reduction

Degree elevation can be viewed as a process that introduces redundancy: a curve is described by more information than is actually necessary. The inverse process might seem more interesting: can we *reduce* possible redundancy in a curve representation? More specifically, can we write a given curve of degree n as one of degree $n-1$? We shall call this process *degree reduction*.

In general, exact degree reduction is not possible. For example, a cubic with a point of inflection cannot possibly be written as a quadratic. Degree reduction, therefore, can be viewed only as a method to *approximate* a given curve by one of lower degree. Our problem can now be stated as follows: given a Bézier curve with control vertices $\mathbf{b}_i; i = 0, \ldots, n$, can we find a Bézier curve with control vertices $\hat{\mathbf{b}}_i; i = 0, \ldots, n-1$ that approximates the first curve?

Let us now pretend that the $\mathbf{b}_i$ were obtained from the $\hat{\mathbf{b}}_i$ by the process of degree elevation (this is not true, in general, but makes a good working assumption). Then they would be related by

$$\mathbf{b}_i = \frac{i}{n}\hat{\mathbf{b}}_{i-1} + \frac{n-i}{n}\hat{\mathbf{b}}_i; \quad i = 0, 1, \ldots, n. \tag{5.5}$$

This equation can be used to derive two recursive formulas for the generation of the $\hat{\mathbf{b}}_i$ from the $\mathbf{b}_i$:

$$\hat{\mathbf{b}}_i = \frac{n\mathbf{b}_i - i\hat{\mathbf{b}}_{i-1}}{n-i}; \quad i = 0, 1, \ldots, n-1 \tag{5.6}$$

and

$$\hat{\mathbf{b}}_{i-1} = \frac{n\mathbf{b}_i - (n-i)\hat{\mathbf{b}}_i}{i}; \quad i = n, n-1, \ldots, 1. \tag{5.7}$$

Figure 5.3 illustrates the first of the two recursive formulas: the polygon of the $\mathbf{b}_i$ is given, and the degree $n-1$ approximation to it is constructed by "unraveling" the degree elevation process from left to right. If the given n^{th} degree curve had actually been of degree $n-1$, Eq. (5.6) would have produced the lower degree polygon. Since in general this is not true, we only obtain an approximation – quite a bad one in most cases. The reason is that both (5.6) and (5.7) are *extrapolation* formulas, which tend to be numerically unstable.

One observes that (5.6) tends to produce reasonable approximations near $\mathbf{b}_0$ and that (5.7) behaves decently near $\mathbf{b}_n$. We may take advantage of this and combine both approximations; for example, we could take the left half of the polygon from (5.6) and the right half of the polygon from (5.7) and merge them to obtain the final polygon of degree $n-1$. (Caution: there is a case distinction depending on n being even or odd!) An example is shown in Figure 5.4.

The first appearance of degree reduction is in Forrest [191]. It has been used for curve rendering by F. Little. A detailed treatment is in Watkins and Worsey [470].

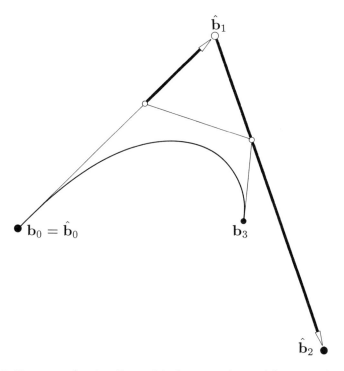

Figure 5.3: Degree reduction I: a cubic is approximated by a quadratic. The approximation is very poor.

5.5   Nonparametric Curves

We have so far considered three-dimensional parametric curves $\mathbf{b}(t)$. Now we shall restrict ourselves to *functional curves* of the form $y = f(x)$, where f denotes a polynomial. These (planar) curves can be written in parametric form:

$$\mathbf{b}(t) = \left[\begin{array}{c} x(t) \\ y(t) \end{array} \right] = \left[\begin{array}{c} t \\ f(t) \end{array} \right].$$

We are interested in functions f that are expressed in terms of the Bernstein basis:

$$f(t) = b_0 B_0^n(t) + \cdots + b_n B_n^n(t).$$

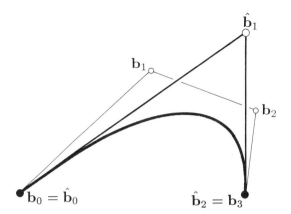

Figure 5.4: Degree reduction II: combining the two degree reduction methods (solving from left to right and from right to left), a reasonable approximation is obtained. Here, $\hat{\mathbf{b}}_1$ is the midpoint of the two corresponding points from the two methods.

Note that now the coefficients b_j are real numbers, not points. The b_j therefore do not form a polygon, yet functional curves are a subset of parametric curves and therefore must possess a control polygon. To find it, we recall the linear precision property of Bézier curves, as defined by (4.15). We can now write our functional curve as

$$\mathbf{b}(t) = \sum_{j=0}^{n} \begin{bmatrix} j/n \\ b_j \end{bmatrix} B_j^n(t). \tag{5.8}$$

Thus the control polygon of the function $f(t) = \sum b_j B_j^n$ is given by the points $(j/n, b_j); j = 0, \ldots, n$. If we want to distinguish clearly between the parametric and the nonparametric cases, we call $f(t)$ a *Bézier function*. Figure 5.5 illustrates the cubic case. We also emphasize that the b_i are real numbers, not points; we call the b_i *Bézier ordinates*.

Because Bézier curves are invariant under affine reparametrizations, we may consider any interval $[a, b]$ instead of the special interval $[0, 1]$. Then the abscissa values are $a + i(b - a)/n; \quad i = 0, \ldots, n$.

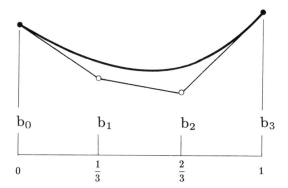

Figure 5.5: Functional curves: the control polygon of a cubic polynomial has abscissa values of 0,1/3,2/3,1.

5.6   Cross Plots

Parametric Bézier curves are composed of coordinate functions: each component is a Bézier function. For two-dimensional curves, this can be used to construct the *cross plot* of a curve. Figure 5.6 shows the decomposition of a Bézier curve into its two coordinate functions. A cross plot can be a very helpful tool for the investigation not only of Bézier curves, but of general two-dimensional curves. We will use it for the analysis of Bézier and B-spline curves. It can be generalized to more than two dimensions, but is not as useful then.

5.7   Integrals

As we have seen, the Bézier polygon $\mathbf{P}$ of a Bézier function is formed by points $(j/n, b_j)$. Let us assign an area $\mathcal{A}\mathbf{P}$ to $\mathbf{P}$ by

$$\mathcal{A}\mathbf{P} = \frac{1}{n+1} \sum_{j=0}^{n} b_j. \tag{5.9}$$

An example for this area is shown in Figure 5.7; it corresponds to approximating the area under the polygon by a particular Riemann sum (of the polygon).

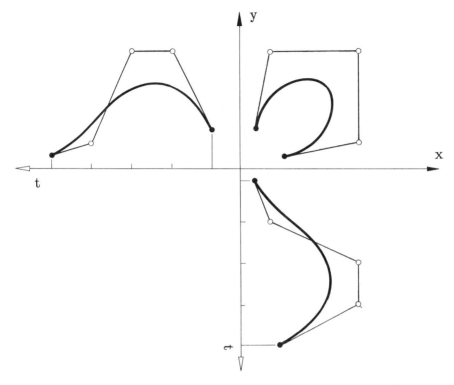

Figure 5.6: Cross plots: a two-dimensional Bézier curve together with its two coordinate functions.

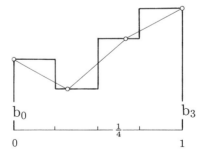

Figure 5.7: Integrals: an approximation to the area under **P**.

It is now easy to show that this "approximation area" is the same for the polygon $\mathcal{E}\mathbf{P}$, obtained from degree elevation (Section 5.1):

$$\begin{aligned}
\mathcal{AEP} &= \frac{1}{n+2}\sum_{j=0}^{n+1}\frac{j}{n+1}b_{j-1} + \left(1 - \frac{j}{n+1}\right)b_j \\
&= \frac{1}{n+2}\sum_{j=0}^{n}\frac{n+2}{n+1}b_j \\
&= \mathcal{AP}.
\end{aligned}$$

If we repeat the process of degree elevation, we know that the polygons $\mathcal{E}^r\mathbf{P}$ converge to the function $\mathcal{B}\mathbf{P}$. Their area $\mathcal{AE}^r\mathbf{P}$ stays the same, and in the limit is equal to the Riemann sum of the function, which converges to the integral:

$$\int_0^1 \sum_{j=0}^n b_j B_j^n(x)\mathrm{d}x = \frac{1}{n+1}\sum_{j=0}^n b_j. \tag{5.10}$$

The special case $b_i = \delta_{i,j}$ gives

$$\int_0^1 B_i^n(x)\mathrm{d}x = \frac{1}{n+1}, \tag{5.11}$$

i.e., all basis functions B_i^n (for a fixed n) have the same integral.

5.8 The Bézier Form of a Bézier Curve

In his work ([50], [51], [52], [53], [54], [55], [57], see also Vernet [464]), Bézier did not use the Bernstein polynomials as basis functions. He wrote the curve $\mathbf{b}^n$ as a linear combination of functions F_i^n:

$$\mathbf{b}^n(t) = \sum_{j=0}^n \mathbf{c}_j F_j^n(t), \tag{5.12}$$

where the F_j^n are polynomials that obey the following recursion:

$$F_i^n(t) = (1-t)F_i^{n-1}(t) + tF_{i-1}^{n-1}(t) \tag{5.13}$$

with

$$F_0^0(t) = 1, \quad F_{r+1}^r(t) = 0, \quad F_{-1}^r(t) = 1. \tag{5.14}$$

Note that the third condition in the last equation is the only instance where the definition of the F_i^n differs from that of the B_i^n! An explicit expression for the F_i^n is given by

$$F_i^n = \sum_{j=i}^{n} B_j^n. \tag{5.15}$$

A consequence of (5.14) is that $F_0^n \equiv 1$ for all n. Since $F_j^n(t) \geq 0$ for $t \in [0,1]$, it follows that (5.12) is not a barycentric combination of the c_j. In fact, c_0 is a point while the other c_j are vectors. The following relations hold:

$$c_0 = b_0, \tag{5.16}$$
$$c_j = \Delta b_{j-1}; \quad j > 0. \tag{5.17}$$

This undesirable distinction between points and vectors was abandoned soon after R. Forrest's discovery that the Bézier form (5.12) of a Bézier curve could be written in terms of Bernstein polynomials (see the appendix in [53]).

5.9 The Barycentric Form of a Bézier Curve

In this section, we present different notation for Bézier curves that will be useful later. Let p_1 and p_2 be two distinct points on the real line. Then, as described in Section 2.3, we can write any point p on the straight line in terms of barycentric coordinates of p_1 and p_2: $p = up_1 + vp_2$, thus identifying p with $u = (u,v)$ and $u+v = 1$. In particular, $p_1 = (1,0)$ and $p_2 = (0,1)$. The real line can be mapped into E^3, where it defines a polynomial curve $b(u)$; namely,

$$b(u) = \sum_{\substack{i+j=n \\ i,j \geq 0}} \binom{n}{i,j} u^i v^j b_{i,j} = \sum_{i+j=n} B_{i,j}^n(u) b_{i,j}, \tag{5.18}$$

where

$$\binom{n}{i,j} = \frac{n!}{i!j!}.$$

Note that, although (5.18) *looks* bivariate, it really isn't: the condition $u+v = 1$ ensures that we still define a curve, not a surface. The connection with the standard Bézier form is established by setting $t = v, b_j = b_{i,j}$.

The barycentric form demonstrates nicely two important properties of Bézier curves: invariance under affine parameter transformations and, as a consequence, symmetry, as discussed in Section 3.3. The location of the two points $\mathbf{p}_1$ and $\mathbf{p}_2$ becomes completely irrelevant – all that matters is the relative location of $\mathbf{p}$ with respect to them, described by u and v.

Here is what the de Casteljau algorithm becomes in barycentric notation:

$$\mathbf{b}_{i,j}^r(\mathbf{u}) = u\mathbf{b}_{i+1,j}^{r-1}(\mathbf{u}) + v\mathbf{b}_{i,j+1}^{r-1}(\mathbf{u}) \quad \begin{cases} r = 1,\ldots,n \\ i+j = n-r \end{cases}. \tag{5.19}$$

The point on the curve is then given by $\mathbf{b}_{0,0}^n(\mathbf{u})$.

We can also define derivatives in terms of the barycentric form. Derivatives produce tangent vectors, and these have a sense of direction, which we abandoned for the sake of symmetry. We may reintroduce a direction into our calculations by relating $\mathbf{u}$ to the "standard" parameter t:

$$\mathbf{u} = \mathbf{u}(t) = (1 - t, t).$$

We obtain

$$\frac{d}{dt}\mathbf{b}[\mathbf{u}(t)] = \frac{\partial}{\partial u}\mathbf{b} \cdot \frac{du}{dt} + \frac{\partial}{\partial v}\mathbf{b} \cdot \frac{dv}{dt}.$$

Inserting the known values for $\frac{du}{dt}$ and $\frac{dv}{dt}$, we have

$$\frac{d}{dt}\mathbf{b}[\mathbf{u}(t)] = \frac{\partial}{\partial v}\mathbf{b} - \frac{\partial}{\partial u}\mathbf{b}. \tag{5.20}$$

If we define a vector $\mathbf{d}$ by $\mathbf{d} = \mathbf{p}_2 - \mathbf{p}_1 = (-1,1)$, this equation may be written as a directional derivate with respect to $\mathbf{d}$:

$$\frac{d}{dt}\mathbf{b}[\mathbf{u}(t)] = D_{\mathbf{d}}\mathbf{b}(\mathbf{u}). \tag{5.21}$$

We shall now see how the de Casteljau algorithm ties in with these directional derivatives.

Instead of evaluating at a *point* $\mathbf{u}$ with $u + v = 1$, let us evaluate at the *vector* $\mathbf{d} = (-1,1)$. The de Casteljau algorithm (5.19) becomes

$$\mathbf{b}_{i,j}^r(\mathbf{d}) = -\mathbf{b}_{i+1,j}^{r-1}(\mathbf{d}) + \mathbf{b}_{i,j+1}^{r-1}(\mathbf{d}).$$

Thus a *vector* argument for the de Casteljau algorithm produces forward differences! In other words,

$$\mathbf{b}_{i,j}^r(\mathbf{d}) = \Delta^r\mathbf{b}_j,$$

where the term on the right-hand side is in standard, nonbarycentric notation.

We thus have, for the first derivative,

$$D_{\mathbf{d}}\mathbf{b}[\mathbf{u}(t)] = n \sum_{j=0}^{n-1} \Delta \mathbf{b}_j B_j^{n-1}(t) = n \sum_{i+j=n-1} \mathbf{b}_{i,j}^1(\mathbf{d}) B_{i,j}^{n-1}(\mathbf{u}). \qquad (5.22)$$

The last part of this equation asserts that our directional derivative is obtained by taking one de Casteljau step with respect to $\mathbf{d}$ and $n-1$ steps with respect to $\mathbf{u}$. This calls for the blossom notation!

The Bézier points of a curve can be expressed as blossom values of the arguments $\mathbf{p}_1$ and $\mathbf{p}_2$; we thus have three possible ways to label Bézier points, using the standard, the barycentric, and the blossom notation:

$$\mathbf{b}_j = \mathbf{b}_{i,j} = \mathbf{b}[\mathbf{p}_1^{<i>}, \mathbf{p}_2^{<j>}]; \quad i+j = n.$$

The intermediate points in the de Casteljau algorithm can now be written as

$$\mathbf{b}_{i,j}^r = \mathbf{b}[\mathbf{p}_1^{<i>}, \mathbf{p}_2^{<j>}, \mathbf{u}^{<r>}]; \quad i+j+r = n,$$

and the point on the curve is given by $\mathbf{b}[\mathbf{u}^{<n>}]$.

Returning to (5.22), we get

$$D_{\mathbf{d}}\mathbf{b}(\mathbf{u}) = D_{\mathbf{d}}\mathbf{b}[\mathbf{u}^{<n>}] = n\mathbf{b}[\mathbf{u}^{<n-1>}, \mathbf{d}].$$

The above arguments easily generalize this to

$$D_{\mathbf{d}}^r\mathbf{b}(\mathbf{u}) = D_{\mathbf{d}}^r\mathbf{b}[\mathbf{u}^{<n>}] = \frac{n!}{(n-r)!}\mathbf{b}[\mathbf{u}^{<n-r>}, \mathbf{d}^{<r>}]. \qquad (5.23)$$

Thus the rth derivative of a curve involves r vector steps and $n-r$ point steps of the de Casteljau algorithm. Of course, it is immaterial in which order these steps are performed. Figure 5.8 illustrates the quadratic case.

We finish this section with an identity that is due to L. Euler. We may formally replace $\mathbf{d}$ by $\mathbf{u}$ in Eq. (5.23):

$$D_{\mathbf{u}}^r\mathbf{b}(\mathbf{u}) = \frac{n!}{(n-r)!}\mathbf{b}[\mathbf{u}^{<n>}].$$

This shows how closely related the processes of differentiating and evaluating are when combining the barycentric notation and blossoms.

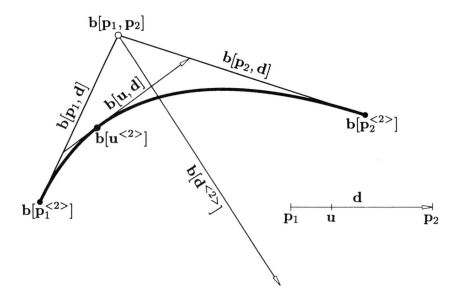

Figure 5.8: Blossoms and derivatives in barycentric form: a point on a quadratic, together with its derivatives. The constant second derivative is given by $\mathbf{b}[\mathbf{d}^{<2>}]$.

5.10   The Weierstrass Approximation Theorem

One of the most important results in approximation theory is the Weierstrass approximation theorem. S. Bernstein invented the polynomials that now bear his name in order to formulate a constructive proof of this theorem. The interested reader is referred to Davis [110] or to Korovkin [293].

    We will give a "customized" version of the theorem, namely, we state it in the context of parametric curves. So let $\mathbf{c}$ be a continuous curve that is defined over $[0, 1]$. For some fixed n, we can sample $\mathbf{c}$ at parameter values i/n. The points $\mathbf{c}(i/n)$ can now be interpreted as the Bézier polygon of a polynomial curve $\mathbf{x}_n$:

$$\mathbf{x}_n(t) = \sum_{i=0}^{n} \mathbf{c}(\frac{i}{n}) B_i^n(t).$$

We say that $\mathbf{x}_n$ is the nth degree Bernstein-Bézier approximation to $\mathbf{c}$.

We are next going to increase the density of our samples, i.e., we increase n. This generates a sequence of approximations $\mathbf{x}_n, \mathbf{x}_{n+1}, \ldots$. The Weierstrass approximation theorem states that this sequence of polynomials converges to the curve $\mathbf{c}$:

$$\lim_{n \to \infty} \mathbf{x}_n(t) = \mathbf{c}(t).$$

At first sight, this looks like a handy way to approximate a given curve by polynomials: we just have to pick a degree n that is sufficiently large, and we are as close to the curve as we like. This is only theoretically true, however. In practice, one would have to choose values of n in the thousands or even millions in order to obtain a reasonable closeness of fit (see Korovkin [293] for more details).

The value of the theorem is therefore more of a theoretical nature. It shows that every curve may be approximated arbitrarily closely by a polynomial curve.

5.11 Formulas for Bernstein Polynomials

This section is a collection of formulas; some appeared in the text, some did not. Credit for some of these goes to R. Goldman, R. Farouki, and V. Rajan [178].

A Bernstein polynomial is defined by

$$B_i^n(t) = \begin{cases} \binom{n}{i} t^i (1-t)^{n-i} & \text{if } i \in [0, n], \\ 0 & \text{else.} \end{cases}$$

The power basis $\{t^i\}$ and the Bernstein basis $\{B_i^n\}$ are related by

$$t^i = \sum_{j=i}^{n} \frac{\binom{j}{i}}{\binom{n}{i}} B_j^n(t) \tag{5.24}$$

and

$$B_i^n(t) = \sum_{j=i}^{n} (-1)^{j-i} \binom{n}{j} \binom{j}{i} t^j. \tag{5.25}$$

Recursion:

$$B_i^n(t) = (1-t) B_i^{n-1}(t) + t B_{i-1}^{n-1}(t).$$

Subdivision:

$$B_i^n(ct) = \sum_{j=0}^{n} B_i^j(c) B_j^n(t). \tag{5.26}$$

Derivative:

$$\frac{\mathrm{d}}{\mathrm{d}t} B_i^n(t) = n[B_{i-1}^{n-1}(t) - B_i^{n-1}(t)].$$

Integral:

$$\int_0^t B_i^n(x)\mathrm{d}x = \frac{1}{n+1} \sum_{j=i+1}^{n+1} B_j^{n+1}(t), \tag{5.27}$$

$$\int_0^1 B_i^n(x)\mathrm{d}x = \frac{1}{n+1}.$$

Three degree elevation formulas:

$$(1-t)B_i^n(t) = \frac{n+1-i}{n+1} B_i^{n+1}(t), \tag{5.28}$$

$$tB_i^n(t) = \frac{i+1}{n+1} B_{i+1}^{n+1}(t), \tag{5.29}$$

$$B_i^n(t) = \frac{n+1-i}{n+1} B_i^{n+1}(t) + \frac{i+1}{n+1} B_{i+1}^{n+1}(t). \tag{5.30}$$

Product:

$$B_i^m(u)B_j^n(u) = \frac{\binom{m}{i}\binom{n}{j}}{\binom{m+n}{i+j}} B_{i+j}^{m+n}(u). \tag{5.31}$$

5.12 Implementation

A C routine for degree elevation follows. Note that we have to treat the cases $i = 0$ and $i = n+1$ separately; the program would not like the corresponding nonexisting array elements. The program actually handles the rational case, which will be covered later. For the polynomial case, fill **wb** with 1's and ignore **wc**.

```
void degree_elevate(bx,by,wb,degree,cx,cy,wc)
/* input: two-d Bezier polygon in bx, by and with weights
          in wb. Degree is degree.
       Output:degree elevated curve in cx,cy and with weights in wc.
Note: for nonrational (polynomial) case, fill wc with 1's.
*/
```

5.13 Problems

1. Prove Eq. (5.15).

2. Prove the relationship between the "Bézier" and the Bernstein form for a Bézier curve (5.17).

3. Prove that

$$\int_0^t b^n(x)\mathrm{d}x = \frac{t}{n+1}\sum_{j=0}^n b_0^j(t).$$

4. With the result from the previous problem, prove

$$F_i^n(t) = n\int_0^t B_i^{n-1}(x)\mathrm{d}x.$$

5. Degree reduction: let $\hat{\mathbf{b}}_{n-1}$ be the polygon endpoint obtained from (5.6). Show that

$$\hat{\mathbf{b}}_{n-1} - \mathbf{b}_n = \Delta^n\mathbf{b}_0.$$

Hint: find an explicit form for the $\hat{\mathbf{b}}_i$ first.

6. Prove Eq. (5.3). The coefficients in that formula will look familiar to those with a background in probability; they represent the hypergeometric distribution.

7. Generalize the concept of cross plots to 3D curves. Write a program to visualize your concepts.

Chapter 6

Polynomial Interpolation

Polynomial interpolation is the most fundamental of all interpolation concepts; the earliest method is probably attributable to I. Newton. Nowadays, polynomial interpolation is mostly of theoretical value; faster and more accurate methods have been developed. Those methods are *piecewise polynomial*; thus they intrinsically rely on the polynomial methods that are presented in this chapter.

6.1 Aitken's Algorithm

A common problem in curve design is *point data interpolation*: from data points $\mathbf{p}_i$ with corresponding parameter values t_i, find a curve that passes through the $\mathbf{p}_i$.[1] One of the oldest techniques to solve this problem is to find an *interpolating polynomial* through the given points. That polynomial must satisfy the interpolatory constraints

$$\mathbf{p}(t_i) = \mathbf{p}_i; \quad i = 0, \ldots, n.$$

Several algorithms exist for this problem – any textbook on numerical analysis will discuss several of them. In this section we shall present a recursive technique that is from A. Aitken.

We have already solved the linear case, $n = 1$, in Section 2.3. The Aitken recursion computes a point on the interpolating polynomial through a sequence of repeated linear interpolations, starting with

$$\mathbf{p}_i^1(t) = \frac{t_{i+1} - t}{t_{i+1} - t_i}\mathbf{p}_i + \frac{t - t_i}{t_{i+1} - t_i}\mathbf{p}_{i+1}; \quad i = 0, \ldots, n - 1.$$

---

[1]The shape of the curve depends heavily on the parameter values t_i. Methods for their determination will be discussed later in the context of spline interpolation; see Section 9.4.

83

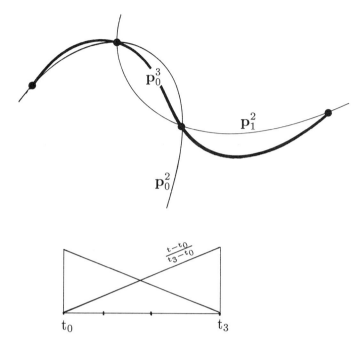

Figure 6.1: Polynomial interpolation: a cubic interpolating polynomial may be obtained as a "blend" of two quadratic interpolants.

Let us now suppose (as one does in recursive techniques) that we have already solved the problem for the case $n - 1$. To be more precise, assume that we have found a polynomial $\mathbf{p}_0^{n-1}$ that interpolates to the n first data points $\mathbf{p}_0, \ldots, \mathbf{p}_{n-1}$, and also a polynomial $\mathbf{p}_1^{n-1}$ that interpolates to the n last data points $\mathbf{p}_1, \ldots, \mathbf{p}_n$. Under these assumptions, it is easy to write down the form of the final interpolant, now called $\mathbf{p}_0^n$:

$$\mathbf{p}_0^n(t) = \frac{t_n - t}{t_n - t_0}\mathbf{p}_0^{n-1}(t) + \frac{t - t_0}{t_n - t_0}\mathbf{p}_1^{n-1}(t). \qquad (6.1)$$

Figure 6.1 illustrates this form for the cubic case.

Let us verify that (6.1) does in fact interpolate to all given data points $\mathbf{p}_i$: for $t = t_0$,

$$\mathbf{p}_0^n(t_0) = 1 * \mathbf{p}_0^{n-1}(t_0) + 0 * \mathbf{p}_1^{n-1}(t_0) = \mathbf{p}_0.$$

A similar result is derived for $t = t_n$. Under our assumption, we have $\mathbf{p}_0^{n-1}(t_i) = \mathbf{p}_1^{n-1}(t_i) = \mathbf{p}_i$ for all other values of i.

Since the weights in (6.1) sum to one identically, we get the desired $\mathbf{p}_0^n(t_i) = \mathbf{p}_i$.

We can now generalize (6.1) to solve the polynomial interpolation problem: starting with the given parameter values t_i and the data points $\mathbf{p}_i = \mathbf{p}_i^0$, we set

$$\mathbf{p}_i^r(t) = \frac{t_{i+r} - t}{t_{i+r} - t_i}\mathbf{p}_i^{r-1}(t) + \frac{t - t_i}{t_{i+r} - t_i}\mathbf{p}_{i+1}^{r-1}(t); \begin{cases} r = 1, \ldots, n; \\ i = 0, \ldots, n - r \end{cases} \tag{6.2}$$

It is clear from the above consideration that $\mathbf{p}_0^n(t)$ is indeed a point on the interpolating polynomial. The recursive evaluation (6.2) is called *Aitken's algorithm*.[2]

It has the following geometric interpretation: to find $\mathbf{p}_i^r$, map the interval $[t_i, t_{i+r}]$ onto the straight line segment through $\mathbf{p}_i^{r-1}, \mathbf{p}_{i+1}^{r-1}$. That affine map takes t to $\mathbf{p}_i^r$. The geometry of Aitken's algorithm is illustrated in Figure 6.2 for the cubic case.

It is convenient to write the intermediate $\mathbf{p}_i^r$ in a triangular array; the cubic case would look like

$$\begin{array}{llll} \mathbf{p}_0 & & & \\ \mathbf{p}_1 & \mathbf{p}_0^1 & & \\ \mathbf{p}_2 & \mathbf{p}_1^1 & \mathbf{p}_0^2 & \\ \mathbf{p}_3 & \mathbf{p}_2^1 & \mathbf{p}_1^2 & \mathbf{p}_0^3. \end{array} \tag{6.3}$$

We can infer several properties of the interpolating polynomial from Aitken's algorithm:

- *Affine invariance:* This follows since the Aitken algorithm uses only barycentric combinations.

- *Linear precision:* If all $\mathbf{p}_i$ are uniformly distributed[3] on a straight line segment, all intermediate $\mathbf{p}_i^r(t)$ are identical for $r > 0$. Thus the straight line segment is reproduced.

---

[2]The particular organization of the algorithm as presented here is from Neville.

[3]If the points are on a straight line, but distributed unevenly, we will still recapture the graph of the straight line, but it will not be parametrized linearly.

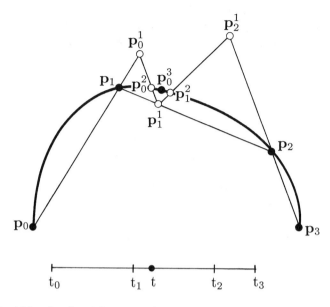

Figure 6.2: Aitken's algorithm: a point on an interpolating polynomial may be found from repeated linear interpolation.

- *No convex hull property:* The parameter t in (6.2) does not have to lie between t_i and t_{i+r}. Therefore, Aitken's algorithm does not use convex combinations only: $\mathbf{p}_0^n(t)$ is not guaranteed to lie within the convex hull of the $\mathbf{p}_i$. We should note, however, that no smooth curve interpolation scheme exists that has the convex hull property.

- *No variation diminishing property:* By the same reasoning, we do not get the variation diminishing property. Again, no "decent" interpolation scheme has this property. However, interpolating polynomials can be variation augmenting to an extent that renders them useless for practical problems.

6.2   Lagrange Polynomials

Aitken's algorithm allows us to compute a point $\mathbf{p}^n(t)$ on the interpolating polynomial through $n+1$ data points. It does not provide an answer to the following questions: (1) Is the interpolating polynomial unique? (2) What is a closed form for it? Both questions are resolved by the use of the *Lagrange polynomials* L_i^n.

The explicit form of the interpolating polynomial $\mathbf{p}$ is given by

$$\mathbf{p}(t) = \sum_{i=0}^{n} \mathbf{p}_i L_i^n(t), \tag{6.4}$$

where the L_i^n are *Lagrange polynomials*

$$L_i^n(t) = \frac{\prod_{\substack{j=0 \\ j \neq i}}^{n} (t - t_j)}{\prod_{\substack{j=0 \\ j \neq i}}^{n} (t_i - t_j)}. \tag{6.5}$$

Before we proceed further, we should note that the L_i^n must sum to one in order for (6.4) to be a barycentric combination and thus be geometrically meaningful; we will return to this topic later.

We verify (6.4) by observing that the Lagrange polynomials are *cardinal*: they satisfy

$$L_i^n(t_j) = \delta_{i,j}, \tag{6.6}$$

with $\delta_{i,j}$ being the Kronecker delta. In other words, the ith Lagrange polynomial vanishes at all knots except at the ith one, where it assumes the value 1. Because of this property of Lagrange polynomials, (6.4) is called the *cardinal* form of the interpolating polynomial $\mathbf{p}$. The polynomial $\mathbf{p}$ has many other representations, of course (we can rewrite it in monomial form, for example), but (6.4) is the only form in which the data points appear explicitly.

We have thus justified our use of the term *the* interpolating polynomial. In fact, the polynomial interpolation problem always has a solution, and it always has a *unique* solution. The reason is that, because of (6.6), the L_i^n form a basis of all polynomials of degree n. Thus, (6.4) is the unique representation of the polynomial $\mathbf{p}$ in this basis. This is why one sometimes refers to all polynomial interpolation schemes as *Lagrange interpolation*.[4]

We can now be sure that Aitken's algorithm yields the same point as does (6.4). Based on that knowlegde, we can conclude a property of Lagrange

---

[4]More precisely, we refer to all those schemes that interpolate to a given set of data points. Other forms of polynomial interpolation exist and are discussed later.

polynomials that was already mentioned right after (6.5), namely, that they sum to one:

$$\sum_{i=0}^{n} L_i^n(t) \equiv 1.$$

This is a simple consequence of the affine invariance of polynomial interpolation, as shown for Aitken's algorithm.

6.3   The Vandermonde Approach

Suppose we want the interpolating polynomial $\mathbf{p}^n$ in the monomial basis:

$$\mathbf{p}^n(t) = \sum_{j=0}^{n} \mathbf{a}_j t^j. \tag{6.7}$$

The standard approach to finding the unknown coefficients from the known data is simply to write down everything one knows about the problem:

$$
\begin{aligned}
\mathbf{p}^n(t_0) = \mathbf{p}_0 &= \mathbf{a}_0 + \mathbf{a}_1 t_0 + \ldots + \mathbf{a}_n t_0^n, \\
\mathbf{p}^n(t_1) = \mathbf{p}_1 &= \mathbf{a}_0 + \mathbf{a}_1 t_1 + \ldots + \mathbf{a}_n t_1^n, \\
&\vdots \\
\mathbf{p}^n(t_n) = \mathbf{p}_n &= \mathbf{a}_0 + \mathbf{a}_1 t_n + \ldots + \mathbf{a}_n t_n^n.
\end{aligned}
$$

In matrix form:

$$
\begin{bmatrix} \mathbf{p}_0 \\ \mathbf{p}_1 \\ \vdots \\ \mathbf{p}_n \end{bmatrix}
=
\begin{bmatrix}
1 & t_0 & t_0^2 & \cdots & t_0^n \\
1 & t_1 & t_1^2 & \cdots & t_1^n \\
\vdots & \vdots & \vdots & & \vdots \\
1 & t_n & t_n^2 & \cdots & t_n^n
\end{bmatrix}
\begin{bmatrix} \mathbf{a}_0 \\ \mathbf{a}_1 \\ \vdots \\ \mathbf{a}_n \end{bmatrix}. \tag{6.8}
$$

We can shorten this to

$$\mathbf{p} = T\mathbf{a}. \tag{6.9}$$

We already know that a solution $\mathbf{a}$ to this linear system exists, but one can show independently that the determinant $\det T$ is nonzero (for distinct parameter values t_i). This determinant is known as the *Vandermonde* of the interpolation problem. The solution, i.e., the vector $\mathbf{a}$ containing the coefficients $\mathbf{a}_i$, can be found from

$$\mathbf{a} = T^{-1}\mathbf{p}. \tag{6.10}$$

This should be taken only as a shorthand notation for the solution – not as an algorithm! Note that the linear system (6.9) really consists of *three* linear systems with the same coefficient matrix, one system for each coordinate. It is known from numerical analysis that in such cases the LU decomposition of T is a more economical way to obtain the solution $\mathbf{a}$. This will be even more important when we discuss tensor product surface interpolation in Section 17.4.

The interpolation problem can also be solved if we use basis functions other than the monomials. Let $\{F_i^n\}_{i=0}^n$ be such a basis. We then seek an interpolating polynomial of the form

$$\mathbf{p}^n(t) = \sum_{j=0}^n \mathbf{c}_j F_i^n(t). \qquad (6.11)$$

The above reasoning again leads to a linear system (three linear systems, to be more precise) for the coefficients $\mathbf{c}_j$, this time with the *generalized Vandermonde F*:

$$F = \begin{bmatrix} F_0^n(t_0) & F_1^n(t_0) & \ldots & F_n^n(t_0) \\ F_0^n(t_1) & F_1^n(t_1) & \ldots & F_n^n(t_1) \\ \vdots & \vdots & & \vdots \\ F_0^n(t_n) & F_1^n(t_n) & \ldots & F_n^n(t_n) \end{bmatrix}. \qquad (6.12)$$

Since the F_i^n form a basis for all polynomials of degree n, it follows that the generalized Vandermonde $\det F$ is nonzero.

Thus, for instance, we are able to find the Bézier curve that passes through a given set of data points: the F_i^n would then be the Bernstein polynomials B_i^n.

6.4 Limits of Lagrange Interpolation

We have seen that polynomial interpolation is simple, unique, and has a nice geometric interpretation. One might therefore expect this interpolation scheme to be used frequently; however, nobody uses it in a design situation. The main reason is illustrated in Figure 6.3: polynomial interpolants *oscillate*. For quite reasonable data points and parameter values, the polynomial interpolant exhibits wild wiggles that are not inherent in the data. One may say that polynomial interpolation is not *shape preserving*.

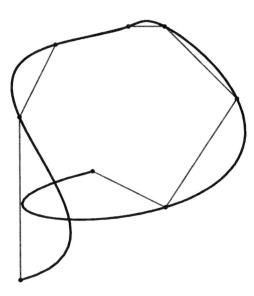

Figure 6.3: Lagrange interpolation: while the data points suggest a convex interpolant, the Lagrange interpolant exhibits extraneous wiggles.

This phenomenon is not due to numerical effects; it is actually inherent in the polynomial interpolation process. Suppose we are given a finite arc of a smooth curve **c**. We can then sample the curve at parameter values t_i and pass the interpolating polynomial through those points. If we increase the number of points on the curve, thus producing interpolants of higher and higher degree, one would expect the corresponding interpolants to converge to the sampled curve **c**. This, however, is not generally true: smooth curves exist for which this sequence of interpolants diverges. This fact is dealt with in numerical analysis, where it is known by the name of its discoverer: it is called the "Runge phenomenon" [400].

As a second consideration, let us examine the cost of polynomial interpolation, i.e., the number of operations necessary to construct and then evaluate the interpolant. Solving the Vandermonde system (6.8) requires roughly n^3 operations; subsequent computation of a point on the curve requires n operations. The operation count for the construction of the interpolant is much smaller for other schemes, as is the cost of evaluations (here piecewise schemes are far superior). This latter cost is the more important one, of course:

construction of the interpolant happens once, but it may have to be evaluated thousands of times!

6.5 Cubic Hermite Interpolation

Polynomial interpolation is not restricted to interpolation to point data; one can also interpolate to other information, such as derivative data. This leads to an interpolation scheme that is more useful than Lagrange interpolation: it is called *Hermite interpolation*. We treat the cubic case first, in which one is given two points $\mathbf{p}_0, \mathbf{p}_1$ and two tangent vectors $\mathbf{m}_0, \mathbf{m}_1$. The objective is to find a cubic polynomial curve $\mathbf{p}$ that interpolates to these data:

$$\begin{aligned}
\mathbf{p}(0) &= \mathbf{p}_0, \\
\dot{\mathbf{p}}(0) &= \mathbf{m}_0, \\
\dot{\mathbf{p}}(1) &= \mathbf{m}_1, \\
\mathbf{p}(1) &= \mathbf{p}_1,
\end{aligned}$$

where the dot denotes differentiation.

We will write $\mathbf{p}$ in cubic Bézier form, and therefore must determine four Bézier points $\mathbf{b}_0, \ldots, \mathbf{b}_3$. Two of them are quickly determined:

$$\mathbf{b}_0 = \mathbf{p}_0, \qquad \mathbf{b}_3 = \mathbf{p}_1.$$

For the remaining two, we recall (from Section 4.3) the endpoint derivative for Bézier curves:

$$\dot{\mathbf{p}}(0) = 3\Delta\mathbf{b}_0, \qquad \dot{\mathbf{p}}(1) = 3\Delta\mathbf{b}_2.$$

We can easily solve for $\mathbf{b}_1$ and $\mathbf{b}_2$:

$$\mathbf{b}_1 = \mathbf{p}_0 + \frac{1}{3}\mathbf{m}_0, \qquad \mathbf{b}_2 = \mathbf{p}_1 - \frac{1}{3}\mathbf{m}_1.$$

This situation is shown in Figure 6.4.

Having solved the interpolation problem, we now attempt to write it in *cardinal form*; we would like to have the given data appear *explicitly* in the equation for the interpolant. So far, our interpolant is in Bézier form:

$$\mathbf{p}(t) = \mathbf{p}_0 B_0^3(t) + (\mathbf{p}_0 + \frac{1}{3}\mathbf{m}_0)B_1^3(t) + (\mathbf{p}_1 - \frac{1}{3}\mathbf{m}_1)B_2^3(t) + \mathbf{p}_1 B_3^3(t).$$

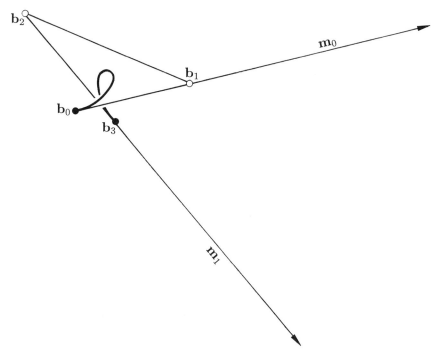

Figure 6.4: Cubic Hermite interpolation: the given data – points and tangent vectors – together with the interpolating cubic in Bézier form.

To obtain the cardinal form, we simply rearrange:

$$\mathbf{p}(t) = \mathbf{p}_0 H_0^3(t) + \mathbf{m}_0 H_1^3(t) + \mathbf{m}_1 H_2^3(t) + \mathbf{p}_1 H_3^3(t), \qquad (6.13)$$

where we have set[5]

$$H_0^3(t) = B_0^3(t) + B_1^3(t),$$

$$H_1^3(t) = \frac{1}{3} B_1^3(t),$$

$$H_2^3(t) = -\frac{1}{3} B_2^3(t), \qquad\qquad (6.14)$$

$$H_3^3(t) = B_2^3(t) + B_3^3(t).$$

---

[5]This is a deviation from standard notation. Standard notation groups by orders of derivatives, i.e., first the two positions, then the two derivatives. The form of Eq. (6.13) was chosen since it groups coefficients according to their geometry.

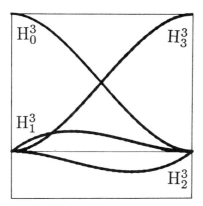

Figure 6.5: Cubic Hermite polynomials: the four H_i^3 are shown over the interval $[0, 1]$.

The H_i^3 are called "cubic Hermite polynomials" and are shown in Figure 6.5.

What are the properties necessary to make the H_i^3 cardinal functions for the cubic Hermite interpolation problem? They must be cardinal with respect to evaluation and differentiation at $t = 0$ and $t = 1$, i.e., each of the H_i^3 equals 1 for one of these four operations and is zero for the remaining three:

$$H_0^3(0) = 1, \quad \frac{d}{dt}H_0^3(0) = 0, \quad \frac{d}{dt}H_0^3(1) = 0, \quad H_0^3(1) = 0,$$

$$H_1^3(0) = 0, \quad \frac{d}{dt}H_1^3(0) = 1, \quad \frac{d}{dt}H_1^3(1) = 0, \quad H_1^3(1) = 0,$$

$$H_2^3(0) = 0, \quad \frac{d}{dt}H_2^3(0) = 0, \quad \frac{d}{dt}H_2^3(1) = 1, \quad H_2^3(1) = 0,$$

$$H_3^3(0) = 0, \quad \frac{d}{dt}H_3^3(0) = 0, \quad \frac{d}{dt}H_3^3(1) = 0, \quad H_3^3(1) = 1.$$

Another important property of the H_i^3 follows from the geometry of the interpolation problem; (6.13) contains combinations of points and vectors. We know that the point coefficients must sum to one if (6.13) is to be geometrically meaningful:

$$H_0^3(t) + H_3^3(t) \equiv 1.$$

This is of course also verified by inspection of (6.14).

Cubic Hermite interpolation has one annoying peculiarity: it is not invariant under affine domain transformations. Let a cubic Hermite interpolant be given as in (6.13), i.e., having the interval $[0, 1]$ as its domain. Now apply an affine domain transformation to it by changing t to $\hat{t} = (1 - t)a + tb$, thereby changing $[0, 1]$ to some $[a, b]$. The interpolant (6.13) becomes

$$\hat{\mathbf{p}}(\hat{t}) = \mathbf{p}_0 \hat{H}_0^3(\hat{t}) + \mathbf{m}_0 \hat{H}_1^3(\hat{t}) + \mathbf{m}_1 \hat{H}_2^3(\hat{t}) + \mathbf{p}_1 \hat{H}_3^3(\hat{t}), \qquad (6.15)$$

where the $\hat{H}_i^3(\hat{t})$ are defined through their cardinal properties:

$$\hat{H}_0^3(a) = 1, \quad \frac{\mathrm{d}}{\mathrm{d}t}\hat{H}_0^3(a) = 0, \quad \frac{\mathrm{d}}{\mathrm{d}t}\hat{H}_0^3(b) = 0, \quad \hat{H}_0^3(b) = 0,$$

$$\hat{H}_1^3(a) = 0, \quad \frac{\mathrm{d}}{\mathrm{d}t}\hat{H}_1^3(a) = 1, \quad \frac{\mathrm{d}}{\mathrm{d}t}\hat{H}_1^3(b) = 0, \quad \hat{H}_1^3(b) = 0,$$

$$\hat{H}_2^3(a) = 0, \quad \frac{\mathrm{d}}{\mathrm{d}t}\hat{H}_2^3(a) = 0, \quad \frac{\mathrm{d}}{\mathrm{d}t}H_2^3(b) = 1, \quad H_2^3(b) = 0,$$

$$\hat{H}_3^3(a) = 0, \quad \frac{\mathrm{d}}{\mathrm{d}t}\hat{H}_3^3(a) = 0, \quad \frac{\mathrm{d}}{\mathrm{d}t}\hat{H}_3^3(b) = 0, \quad \hat{H}_3^3(b) = 1.$$

To satisfy these requirements, the new $\hat{H}_i^3$ must differ from the original H_i^3. We obtain

$$\hat{H}_0^3(\hat{t}) = H_0^3(t),$$
$$\hat{H}_1^3(\hat{t}) = (b - a)H_1^3(t),$$
$$\hat{H}_2^3(\hat{t}) = (b - a)H_2^3(t), \qquad (6.16)$$
$$\hat{H}_3^3(\hat{t}) = H_3^3(t),$$

where $t \in [0, 1]$ is the local parameter of the interval $[a, b]$.

Evaluation of (6.15) at $\hat{t} = a$ and $\hat{t} = b$ yields $\hat{\mathbf{p}}(a) = \mathbf{p}_0, \hat{\mathbf{p}}(b) = \mathbf{p}_1$. The derivatives have changed, however. Invoking the chain rule, we find that $\mathrm{d}\hat{\mathbf{p}}(a)/\mathrm{d}t = (b - a)\mathbf{m}_0$ and, similarly, $\mathrm{d}\hat{\mathbf{p}}(b)/\mathrm{d}t = (b - a)\mathbf{m}_1$.

Thus an affine domain transformation changes the curve unless the defining tangent vectors are changed accordingly – a result quite unlike the Bézier curve case.

To maintain the same curve after a domain transformation, we must change the length of the tangent vectors: if the length of the domain interval is changed by a factor α, we must replace $\mathbf{m}_0$ and $\mathbf{m}_1$ by $\mathbf{m}_0/\alpha$ and $\mathbf{m}_1/\alpha$, respectively. There is an intuitive argument for this: interpreting the parameter as time, we assume we had one time unit to traverse the curve.

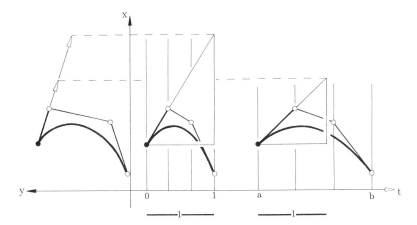

Figure 6.6: Lengths of tangent vectors and domain intervals: the longer the domain interval (right cubic function), the shorter the tangent vector of the parametric curve.

After changing the interval length by a factor of 10, for example, we have 10 time units to traverse the same curve, resulting in a much smaller speed of traversal. Since the magnitude of the derivative equals that speed, it must also shrink by a factor of 10.

   An illustration is given in Figure 6.6. It shows – using a parametric cubic and the x-portion of its cross plot – how a stretching of the domain interval "flattens" the x-component function. This results in a shorter tangent vector of the parametric curve. In this figure, we have made use of the fact that the slope of a function may be expressed as the height of a right triangle with base length one.

   We also note that the Hermite form is not symmetric: if we replace t by $1-t$ (assuming again the interval $[0,1]$ as the domain), the curve coefficients cannot simply be renumbered (as in the case of Bézier curves). Rather, the tangent vectors must be *reversed*. This follows from the above by applying the affine map to the $[0,1]$ that maps that interval to $[1,0]$, thus reversing its direction.

   The dependence of the cubic Hermite form on the domain interval is rather unpleasant – it is often not accounted for and can be blamed for countless programming errors by both students and professionals. We will use the Bézier form whenever possible.

6.6   Quintic Hermite Interpolation

Instead of prescribing only position and first derivative information at two points, one might add that information for second-order derivatives. Then our data are $\mathbf{p}_0, \mathbf{m}_0, \mathbf{s}_0$ and $\mathbf{p}_1, \mathbf{m}_1, \mathbf{s}_1$, "$s''$" denoting second derivative. The lowest order polynomial to interpolate to these data is of degree five. Its Bézier points are easily obtained following the preceding approach. If we rearrange the Bézier form to obtain a cardinal form of the interpolant $\mathbf{p}$, we find

$$\mathbf{p}(t) = \mathbf{p}_0 H_0^5(t) + \mathbf{m}_0 H_1^5(t) + \mathbf{s}_0 H_2^5(t) + \mathbf{s}_1 H_3^5(t) + \mathbf{m}_1 H_4^5(t) + \mathbf{p}_1 H_5^5(t), \quad (6.17)$$

where

$$
\begin{aligned}
H_0^5 &= B_0^5 + B_1^5 + B_2^5, \\
H_1^5 &= \frac{1}{5}[B_1^5 + 2B_2^5], \\
H_2^5 &= \frac{1}{20}B_2^5, \\
H_3^5 &= \frac{1}{20}B_3^5, \\
H_4^5 &= -\frac{1}{5}[2B_3^5 + B_4^5], \\
H_5^5 &= B_3^5 + B_4^5 + B_5^5.
\end{aligned}
$$

It is easy to verify the cardinal properties of the H_i^5: they are the straightforward generalization of the cardinal properties for cubic Hermite polynomials. If used in the context of composite curves, the quintic Hermite polynomials guarantee C^2 continuity. For most applications, one will have to estimate the second derivatives that are needed as input. This estimation is a very sensitive procedure – so unless the quintic form is mandated by a particular problem, the simpler C^2 cubic splines from Chapter 9 are recommended.

6.7   The Newton Form and Forward Differencing

We finish this chapter with a Lagrange-type polynomial interpolant of which a spinoff, called *forward differencing*, has established itself in the computer graphics community. We will assume that the underlying parametrization is uniform, i.e., $t_{j+1} = t_j + h$ for some interval length h. It can be shown that a form of the nth degree interpolating polynomial $\mathbf{p}^n(t)$ is given by

$$\mathbf{p}^n(t) = \mathbf{p}_0 + \frac{1}{h}\Delta\mathbf{p}_0(t - t_0) + \ldots + \frac{1}{n!h^n}\Delta^n\mathbf{p}_0(t - t_0)\cdots(t - t_{n-1}). \quad (6.18)$$

The derivation of this *Newton form* is in any standard text on numerical analysis.

The differences $\Delta^i\mathbf{p}_j$ are defined as

$$\Delta^i\mathbf{p}_j = \Delta^{i-1}\mathbf{p}_{j+1} - \Delta^{i-1}\mathbf{p}_j. \quad (6.19)$$

The coefficients in (6.18) are conveniently written in a table such as the following (setting $g = 1/h$):

$$
\begin{array}{llll}
\mathbf{p}_0 & & & \\
\mathbf{p}_1 & g\Delta\mathbf{p}_0 & & \\
\mathbf{p}_2 & g\Delta\mathbf{p}_1 & g^2\Delta^2\mathbf{p}_0 & \\
\mathbf{p}_3 & g\Delta\mathbf{p}_2 & g^2\Delta^2\mathbf{p}_1 & g^3\Delta^3\mathbf{p}_0.
\end{array}
$$

The diagonal contains the coefficients of the Newton form. We have only considered the cubic case here, and shall continue to do so; the general case is a very straightforward generalization.

We could now evaluate $\mathbf{p}^3$ at any parameter value t by simply evaluating (6.18) there. For the case in which our evaluation points t_j are equally spaced, again with $t_{j+1} = t_j + h$, a much faster way exists. Suppose we had computed $\mathbf{p}_j = \mathbf{p}^3(t_j)$, etc., from (6.18). Then we could compute all entries in the following table:

$$
\begin{array}{lllll}
\mathbf{p}_0 & & & & \\
\mathbf{p}_1 & y\Delta\mathbf{p}_0 & & & \\
\mathbf{p}_2 & g\Delta\mathbf{p}_1 & g^2\Delta^2\mathbf{p}_0 & & \\
\mathbf{p}_3 & g\Delta\mathbf{p}_2 & g^2\Delta^2\mathbf{p}_1 & g^3\Delta^3\mathbf{p}_0 & \\
\mathbf{p}_4 & g\Delta\mathbf{p}_3 & g^2\Delta^2\mathbf{p}_2 & g^3\Delta^3\mathbf{p}_1 & \\
\mathbf{p}_5 & g\Delta\mathbf{p}_4 & g^2\Delta^2\mathbf{p}_3 & g^3\Delta^3\mathbf{p}_2 & \\
\vdots & \vdots & \vdots & \vdots &
\end{array}
\qquad (6.20)
$$

Now consider the last column of this table, containing terms of the form $g^3\Delta^3\mathbf{p}_j$. All these terms are equal! This is so because the nth derivative of an nth degree polynomial is constant, and because the nth derivative of (6.18) is given by $g^n\Delta^n\mathbf{p}_0 = g^n\Delta^n\mathbf{p}_1 = \ldots$.

We thus have a new way of constructing the table (6.20) from *right to left:* instead of computing the entry $\mathbf{p}_4$ from (6.18), first compute $g^2\Delta^2\mathbf{p}_2$ from Eq. (6.19):

$$g^2\Delta^2\mathbf{p}_2 = g^3\Delta^3\mathbf{p}_1 + g^2\Delta^2\mathbf{p}_1,$$

then compute $g\Delta\mathbf{p}_3$ from

$$g\Delta\mathbf{p}_3 = g^2\Delta\mathbf{p}_2 + g\Delta\mathbf{p}_2,$$

and finally

$$\mathbf{p}_4 = g\Delta\mathbf{p}_3 + \mathbf{p}_3.$$

Then compute $\mathbf{p}_5$ in the same manner, and so on. The general formula is, with $\mathbf{q}_j^i = g^i\Delta^i\mathbf{p}_j$:

$$\mathbf{q}_j^i = \mathbf{q}_{j-1}^{i+1} + \mathbf{q}_{j-1}^i; \quad i = 2, 1, 0. \tag{6.21}$$

It yields the points $\mathbf{p}_j = \mathbf{q}_j^0$.[6]

This way of computing the $\mathbf{p}_j$ does not involve a single multiplication after the startup phase! It is therefore extremely fast and has been implemented in many graphics systems, known under the name *forward differencing.* Given four initial points $\mathbf{p}_0, \mathbf{p}_1, \mathbf{p}_2, \mathbf{p}_3$ and a stepsize h, it generates a sequence of points on the cubic polynomial through the initial four points. Typically the polynomial will be given in Bézier form, so those four points have to be computed as a startup operation.

In a graphics environment, it is desirable to adjust the stepsize h such that each pixel along the curve is hit. One way of doing this is to adjust the stepsize while marching along the curve. This is called *adaptive forward differencing* and is described by Lien, Shantz, and Pratt [312] and by Chang, Shantz, and Rochetti [90].

Although fast, forward differencing is not foolproof: As we compute more and more points on the curve, they begin to be affected by roundoff. So while we intend to march along our curve, we may instead leave its path, deviating from it more and more as we continue. For more literature on this method, see Abi-Ezzi [1], Bartels *et al.* [42], or Shantz and Chang [446].

---

[6]It also holds for any degree n if we replace $i = 2, 1, 0$ by $i = n - 1, n - 2, \ldots, 0$.

6.8 Implementation

The code for Aitken's algorithm is very similar to that for the de Casteljau algorithm. Here is its header:

```
float aitken(degree,coeff,t)
/*   uses  Aitken to compute one coordinate
     value of a  Lagrange interpolating polynomial. Has to be called
     for each coordinate  (x,y, and/or z) of data points.
Input:   degree: degree of curve.
         coeff:  array with coordinates to be interpolated.
         t:      parameter value.
Output:  coordinate value.

       Note: we  assume a uniform knot sequence!
*/
```

6.9 Problems

1. The de Casteljau algorithm for Bézier curves has as its "counterpart" the recursion formula (4.2) for Bernstein polynomials. Deduce a recursion formula for Lagrange polynomials from Aitken's algorithm.

2. Aitken's algorithm looks very similar to the de Casteljau algorithm. Use both to define a whole class of algorithms, of which each would be a special case (see [165]). Write a program that uses as input a parameter specifying if the output curve should be "more Bézier" or "more Lagrange."

3. The Hermite form is not invariant under affine domain transformations, while the Bézier form is. What about the Lagrange and monomial forms? What are the general conditions for a curve scheme to be invariant under affine domain transformations?

4. In Lagrange interpolation, each $\mathbf{p}_i$ is assigned a corresponding parameter value t_i. Experiment (graphically) by interchanging two parameter values t_i and t_j without interchanging $\mathbf{p}_i$ and $\mathbf{p}_j$. Explain your results.

5. Show that the cubic and quintic Hermite polynomials are linearly independent.

6. Generalize the above approaches to cubic and quintic Hermite interpolation to degrees 7, 9, etc.

7. The function that was used by Runge to demonstrate the effect that now bears his name is given by

$$f(x) = \frac{1}{1+x^2}; \quad x \in [-1, 1].$$

Use the routine `aitken` to interpolate at equidistant parameter intervals. Keep increasing the degree of the interpolating polynomial until you notice "bad" behavior on the part of the interpolant.

Chapter 7

Spline Curves in Bézier Form

Bézier curves provide a powerful tool in curve design, but they have some limitations: if the curve to be modeled has a complex shape, then its Bézier representation will have a prohibitively high degree (for practical purposes, degrees exceeding 10 are prohibitive). Such complex curves can, however, be modeled using *composite Bézier curves*. We shall also use the name *spline curves* for such *piecewise polynomial curves*. This chapter describes the main properties of cubic and quadratic spline curves. More general spline curves will be presented in Chapter 10.

7.1 Global and Local Parameters

Before we start to develop a theory for piecewise curves, let us establish the main definitions that we will use. When we considered single Bézier curves, we assumed that they were the map of the interval $0 \leq t \leq 1$. We could make this assumption because of the invariance of Bézier curves under affine parameter transformations; see Section 3.3. Life is not quite that easy with piecewise curves: while we can assume that each individual segment of a spline curve $\mathbf{s}$ is the map of the interval $[0, 1]$, the curve as a whole is the map of a collection of intervals, and their relative lengths play an important role.

A *spline curve* $\mathbf{s}$ is the continuous map of a collection of intervals $u_0 < \ldots < u_L$ into E^3, where each interval $[u_i, u_{i+1}]$ is mapped onto a polynomial curve segment. Each real number u_i is called a *breakpoint* or a *knot*. The

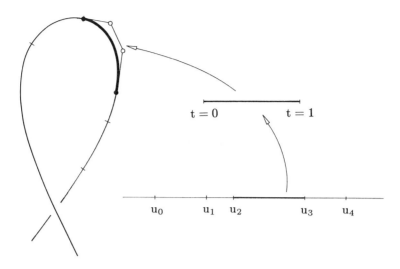

Figure 7.1: Local coordinates: the interval $[u_2, u_3]$ has been endowed with a local coordinate t. The third segment of the spline curve is shown with its Bézier polygon.

collection of all u_i is called the *knot sequence*. For every parameter value u, we thus have a corresponding point $\mathbf{s}(u)$ on the curve $\mathbf{s}$. Let this value u be from an interval $[u_i, u_{i+1}]$. We can introduce a *local coordinate* (or local parameter) t for the interval $[u_i, u_{i+1}]$ by setting

$$t = \frac{u - u_i}{u_{i+1} - u_i} = \frac{u - u_i}{\Delta_i}. \qquad (7.1)$$

One checks that t varies from 0 to 1 as u varies from u_i to u_{i+1}.

When we talk about the whole curve $\mathbf{s}$, it will be more convenient to do so in terms of the global parameter u. (An example of such a property is the concept of differentiability.) The individual segments of $\mathbf{s}$ may be written as Bézier curves, and it is often easier to describe each one of them in terms of local coordinates. We adopt the definition $\mathbf{s}_i$ for the ith segment of $\mathbf{s}$, and we write $\mathbf{s}(u) = \mathbf{s}_i(t)$ to denote a point on it. Figure 7.1 illustrates the interplay between local and global coordinates.

The introduction of local coordinates has some ramifications concerning the use of derivatives. For $u \in [u_i, u_{i+1}]$, the chain rule gives

$$\frac{d\mathbf{s}(u)}{du} = \frac{d\mathbf{s}_i(t)}{dt} \frac{dt}{du} \tag{7.2}$$

$$= \frac{1}{\Delta_i} \frac{d\mathbf{s}_i(t)}{dt}. \tag{7.3}$$

Two more definitions: the points $\mathbf{s}(u_i) = \mathbf{s}_i(0) = \mathbf{s}_{i-1}(1)$ are called *junction points* or *joints*. The collection of the Bézier polygons for all curve segments itself forms a polygon; it is called the *piecewise Bézier polygon* of $\mathbf{s}$.

7.2   Smoothness Conditions

Suppose we are given two Bézier curves $\mathbf{s}_0$ and $\mathbf{s}_1$, with polygons $\mathbf{b}_0, \ldots, \mathbf{b}_n$ and $\mathbf{b}_n, \ldots, \mathbf{b}_{2n}$, respectively. We may think of each curve as existing by itself, defined over the interval $t \in [0, 1]$ or some other interval. We may also think of the two curves as two segments of one *composite* curve, defined as the map of the interval $[u_0, u_2]$ into $I\!E^3$. The "left" segment $\mathbf{s}_0$ is defined over an interval $[u_0, u_1]$, while the "right" segment $\mathbf{s}_1$ is defined over $[u_1, u_2]$ (see Section 7.1).

Let us pretend for a moment that both curves are arcs of one global polynomial curve $\mathbf{b}^n(u)$, defined over the interval $[u_0, u_2]$. Section 4.6 tells us that the two polygons $\mathbf{b}_0, \ldots, \mathbf{b}_n$ and $\mathbf{b}_n, \ldots, \mathbf{b}_{2n}$ must be the result of a subdivision process. Then their control vertices must be related by

$$\mathbf{b}_{n+i} = \mathbf{b}_{n-i}^i(t); \quad i = 0, \ldots, n, \tag{7.4}$$

where $t = (u_2 - u_0)/(u_1 - u_0)$ is the local coordinate of u_2 with respect to the interval $[u_0, u_1]$.

Now suppose we arbitrarily change $\mathbf{b}_{2n}$; the two curves then no longer describe the same global polynomial. However, they still agree in all derivatives of order $0, \ldots, n-1$ at $u = u_1$! This is simply because $\mathbf{b}_{2n}$ has no influence on derivatives of order less than n at $u = u_1$. Similarly, we may change $\mathbf{b}_{2n-r}$ and still maintain continuity of all derivatives of order $0, \ldots, n-r-1$.

We therefore have the C^r condition for Bézier curves: the two Bézier curves defined over $u_0 \le u \le u_1$ and $u_1 \le u \le u_2$, by the polygons $\mathbf{b}_0, \ldots, \mathbf{b}_n$ and $\mathbf{b}_n, \ldots, \mathbf{b}_{2n}$, respectively, are r times continuously differentiable at $u = u_1$ if and only if

$$\mathbf{b}_{n+i} = \mathbf{b}_{n-i}^i(t); \quad i = 0, \ldots, r, \tag{7.5}$$

Suppose the curve from Example 3.1 is defined over [0,1]. What are the Bézier points of a second Bézier curve, defined over [1,3] and with a C^2 join to the first curve? We have to evaluate at $t = 3$:

$$\begin{bmatrix} 0 \\ 0 \end{bmatrix}$$

$$\begin{bmatrix} 0 \\ 2 \end{bmatrix} \quad \begin{bmatrix} 0 \\ 6 \end{bmatrix}$$

$$\begin{bmatrix} 8 \\ 2 \end{bmatrix} \quad \begin{bmatrix} 24 \\ 2 \end{bmatrix} \quad \begin{bmatrix} 72 \\ -6 \end{bmatrix}$$

$$\begin{bmatrix} \mathbf{4} \\ \mathbf{0} \end{bmatrix} \quad \begin{bmatrix} \mathbf{-4} \\ \mathbf{-4} \end{bmatrix} \quad \begin{bmatrix} \mathbf{-60} \\ \mathbf{-18} \end{bmatrix}.$$

The boldface points are the desired ones. If they are the first three Bézier points of the "right" curve, both curves will be C^2 over the interval [0,3].

Example 7.1: Computing the C^2 extension of a Bézier curve.

where $t = (u_2 - u_0)/(u_1 - u_0)$ is the local coordinate of u_2 with respect to the interval $[u_0, u_1]$. See Example 7.1 for a specific case.

Thus the de Casteljau algorithm also governs the continuity conditions between adjacent Bézier curves. Note that (7.5) is a theoretical tool; it should not be used to *construct* C^r curves – this would lead to numerical problems because of the extrapolations that are used in (7.5).

Another condition for C^r continuity should also be mentioned here. By equating derivatives using (4.21) and applying the chain rule,[1] we obtain

$$(\frac{1}{\Delta_0})^i \Delta^i \mathbf{b}_{n-i} = (\frac{1}{\Delta_1})^i \Delta^i \mathbf{b}_n \quad i = 0, \ldots, r. \tag{7.6}$$

Conditions for continuity of higher derivatives of Bézier curves were first derived by E. Staerk [451] in 1976. The cases $r = 1$ and $r = 2$ are probably the ones of most practical relevance, and we shall discuss them in more detail next.

---

[1]Equation (4.21) is with respect to the local parameter of an interval. We are interested in differentiability with respect to the global parameter.

7.3 C^1 Continuity

We know that only the three Bézier points $\mathbf{b}_{n-1}, \mathbf{b}_n, \mathbf{b}_{n+1}$ influence the first derivatives at the junction point $\mathbf{b}_n$. According to (7.5), $\mathbf{b}_{n+1}$ is obtained by linear interpolation of the two points $\mathbf{b}_{n-1}, \mathbf{b}_n$. These three points must therefore be collinear and must also be in the ratio $(u_1 - u_0) : (u_2 - u_1) = \Delta_0 : \Delta_1$. This is illustrated in Figure 7.2.

We can also obtain this result in a different way. Let s and t be local parameters of the intervals $[u_0, u_1]$ and $[u_1, u_2]$, respectively. Let the C^1 curve consisting of the two polynomial segments be called $\mathbf{s}$, having the global parameter u. Then from Section 7.1,

$$\frac{\mathrm{d}}{\mathrm{d}u}\mathbf{s}(u) = \frac{1}{\Delta_0}\frac{\mathrm{d}}{\mathrm{d}s}\mathbf{s}_0(s) = \frac{1}{\Delta_1}\frac{\mathrm{d}}{\mathrm{d}t}\mathbf{s}_1(t).$$

Since

$$\frac{\mathrm{d}}{\mathrm{d}s}\mathbf{s}_0(1) = n\Delta\mathbf{b}_{n-1}$$
$$\frac{\mathrm{d}}{\mathrm{d}t}\mathbf{s}_1(0) = n\Delta\mathbf{b}_n,$$

we get

$$\Delta_1\Delta\mathbf{b}_{n-1} = \Delta_0\Delta\mathbf{b}_n, \tag{7.7}$$

which confirms the above result.

It is important to note that collinearity of three distinct control points $\mathbf{b}_{n-1}, \mathbf{b}_n, \mathbf{b}_{n+1}$ is not sufficient to guarantee C^1 continuity! This is because the notion of C^1 continuity is based on the interplay between domain and range

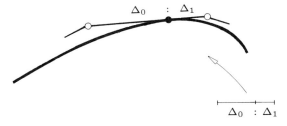

Figure 7.2: C^1 condition: the three shown Bézier points must be collinear with ratio $\Delta_0 : \Delta_1$.

configurations. Collinearity of three points is purely a range phenomenon. Without additional information on the domain of the curve under consideration, we cannot make any statements concerning differentiability. However, collinearity of three distinct control points $\mathbf{b}_{n-1}, \mathbf{b}_n, \mathbf{b}_{n+1}$ does guarantee a continuously varying tangent.

A special situation arises if $\Delta\mathbf{b}_{n-1} = \Delta\mathbf{b}_n = \mathbf{0}$, i.e., if all three points $\mathbf{b}_{n-1}, \mathbf{b}_n, \mathbf{b}_{n+1}$ coincide. In this case, the composite curve s has a zero tangent vector at the junction point $\mathbf{b}_n$, and is differentiable regardless of the interval lengths Δ_0, Δ_1. Zero tangent vectors may give rise to corners or cusps in curves, a fact that intuitively contradicts the concept of differentiability.

Smoothness and differentiability only agree for functional curves – the connection between them is lost in the parametric case. Differentiable curves may not be smooth (see cusps above) and smooth curves may not be differentiable (see Figures 7.6 and 7.7).

7.4 C^2 Continuity

As above, let s be a spline curve that consists of two segments $\mathbf{s}_0$ and $\mathbf{s}_1$, defined over $[u_0, u_1]$ and $[u_1, u_2]$, respectively. Let us assume that s is C^1, so that (7.7) is met. The additional C^2 condition, with $r = 2$ in (7.5), states that the two quadratic polynomials with control polygons $\mathbf{b}_{n-2}, \mathbf{b}_{n-1}, \mathbf{b}_n$ and $\mathbf{b}_n, \mathbf{b}_{n+1}, \mathbf{b}_{n+2}$, defined over $[u_0, u_1]$ and $[u_1, u_2]$, describe the same global quadratic polynomial. Therefore, a polygon $\mathbf{b}_{n-2}, \mathbf{d}, \mathbf{b}_{n+2}$ must exist that describes that polynomial over the interval $[u_0, u_2]$. The two subpolygons are then obtained from it by subdivision at the parameter value u_1.

The C^2 condition for a C^1 curve s at u_1 is thus the existence of a point $\mathbf{d}$ such that

$$\mathbf{b}_{n-1} = (1 - t_1)\mathbf{b}_{n-2} + t_1\mathbf{d}, \qquad (7.8)$$
$$\mathbf{b}_{n+1} = (1 - t_1)\mathbf{d} + t_1\mathbf{b}_{n+2}, \qquad (7.9)$$

where $t_1 = \Delta_0/(u_2 - u_0)$ is the local parameter of u_1 with respect to the interval $[u_0, u_2]$. Figure 7.3 gives an example.

This condition provides us with an easy test to see if a curve is C^2 at a given breakpoint u_i: we simply construct auxiliary points $\mathbf{d}_-, \mathbf{d}_+$ from both the right and the left and check for equality. Figure 7.4 shows two curve segments that fail the C^2 test.

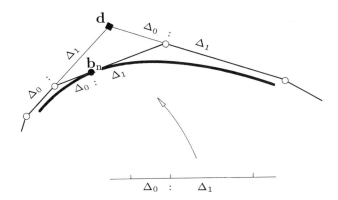

Figure 7.3: C^2 condition: two Bézier curves are twice differentiable at the junction point $\mathbf{b}_n$ if the auxiliary point $\mathbf{d}$ exists uniquely.

Another derivation of the C^2 condition would be to compute the left and right second derivatives at the junction point $\mathbf{b}_n$ and to equate them. The second derivatives at a junction point are essentially second differences of nearby Bézier points. For the simpler case of uniform parameter spacing, $\Delta_0 = \Delta_1$, Figure 7.5 shows how this approach leads to the same C^2 condition as earlier.

While both ways of checking C^2 continuity are mathematically equivalent, the first one is more practical: it compares *points* ($\mathbf{d}_-$ and $\mathbf{d}_+$), while the second one compares *vectors* (left and right second derivatives). One usually has a point tolerance[2] present in an application, but it would be hard to define a tolerance according to which two second derivative vectors can be labeled equal. The problem of checking for C^2 continuity arises when a piecewise cubic curve is given and one tries to convert it to B-spline format, see Section 7.7.

7.5 Finding a C^1 Parametrization

Let $\mathbf{s}$ be a C^r spline curve over the domain $u_0 < \ldots < u_L$. An affine parameter transformation $v = \alpha + \beta u$ leaves the curve invariant. We know that each curve segment $\mathbf{s}_i$ remains unchanged since Bézier curves are invariant under affine

---

[2]If two points are closer together than this tolerance, they are regarded as equal.

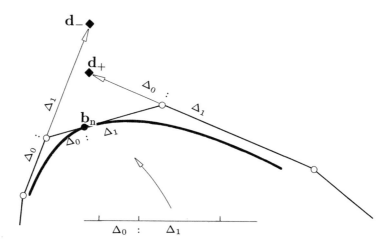

Figure 7.4: C^2 condition: the two shown segments generate different auxiliary points $\mathbf{d}_\pm$, hence they are only C^1.

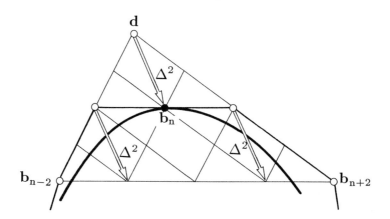

Figure 7.5: C^2 condition for uniform parameter spacing: if $\Delta^2\mathbf{b}_{n-2} = \Delta^2\mathbf{b}_n = \Delta^2$, a unique auxiliary point $\mathbf{d}$ exists. (Proof by the use of similar triangles.)

parameter transformations; see Section 3.3. The smoothness conditions (7.5) between segments are formulated in terms of the de Casteljau algorithm, which operates with constant ratios $\Delta_i : \Delta_{i+1}$ only. These ratios are left invariant under an affine parameter transformation.

This invariance property can be expressed in a different way: a C^r spline curve is determined not by *one* knot sequence, but in fact by a whole family of knot sequences that can be obtained from one another by scalings and translations (i.e., by affine maps of the real line onto itself).

We can determine these knot sequences from inspection of the piecewise Bézier polygon: if it has "corners" at the junction points, it cannot define a C^1 spline curve, and the notion of a knot sequence is meaningless. (A C^0 spline curve is C^0 over *any* knot sequence.) Suppose then that we have a piecewise Bézier polygon with $\mathbf{b}_{in-1}, \mathbf{b}_{in}, \mathbf{b}_{in+1}$ collinear for all i. We can now construct a knot sequence as follows: set $u_0 = 0$, $u_1 = 1$ and for $i = 2, \ldots, L$ determine u_i from

$$\frac{\Delta_{i-1}}{\Delta_i} = \frac{\|\Delta \mathbf{b}_{in-1}\|}{\|\Delta \mathbf{b}_{in}\|}. \tag{7.10}$$

Of course, this is only one member of the family of C^1 parametrizations of the given curve. We may for instance normalize the u_i by dividing through by u_L. This forces all u_i to be in the unit interval $[0, 1]$.

Any other choice of parameter intervals will not change the shape of the piecewise curve – that shape is uniquely determined by the Bézier polygons. However, different knot spacing *will* change the continuity class of the curve defined by its Bézier polygons; the cross plots that are shown in Figures 7.6 and 7.7 demonstrate this. We see that the continuity class of a curve is not a geometric property that is intrinsically linked to the shape of the curve – it is a result of the parametrization.

7.6 C^1 Quadratic B-spline Curves

Let us consider a C^1 piecewise quadratic spline curve $\mathbf{s}$ that is defined over L intervals $u_0 < \ldots < u_L$, as in Figure 7.8. We call the Bézier points $\mathbf{b}_{2i+1}$ *inner Bézier points*, and the $\mathbf{b}_{2i}$ *junction points*.

We can completely determine a quadratic spline curve by prescribing the knot sequence and the Bézier points

$$\mathbf{b}_0, \mathbf{b}_1, \mathbf{b}_3, \ldots, \mathbf{b}_{2i+1}, \ldots, \mathbf{b}_{2L-1}, \mathbf{b}_{2L}.$$

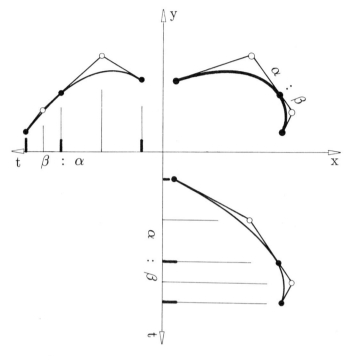

Figure 7.6: A C^1 parametrization: the piecewise quadratic Bézier curve is C^1 when the parameter intervals are chosen to be in the same ratio $\alpha : \beta$ as the Bézier points $\mathbf{b}_1, \mathbf{b}_2, \mathbf{b}_3$.

The remaining junction points are computed from the C^1 conditions

$$\mathbf{b}_{2i} = \frac{\Delta_i}{\Delta_{i-1} + \Delta_i}\mathbf{b}_{2i-1} + \frac{\Delta_{i-1}}{\Delta_{i-1} + \Delta_i}\mathbf{b}_{2i+1}. \qquad (7.11)$$

We can thus define a C^1 quadratic Bézier curve with fewer data than are necessary to define the complete piecewise Bézier polygon. The minimum amount of information that is needed is (1) the polygon $\mathbf{b}_0, \mathbf{b}_1, \mathbf{b}_3, \ldots,$ $\mathbf{b}_{2i+1} \ldots, \mathbf{b}_{2L-1}, \mathbf{b}_{2L}$, called the *B-spline polygon* or *de Boor polygon* of $\mathbf{s}$, and (2) the knot sequence $u_0, \ldots, u_L$. If the curve is described in terms of this B-spline polygon, it is sometimes called a *B-spline curve*. We also denote the quadratic B-spline polygon by $\mathbf{d}_{-1}, \mathbf{d}_0, \ldots, \mathbf{d}_{L-1}, \mathbf{d}_L$; see Figure 7.8. Each

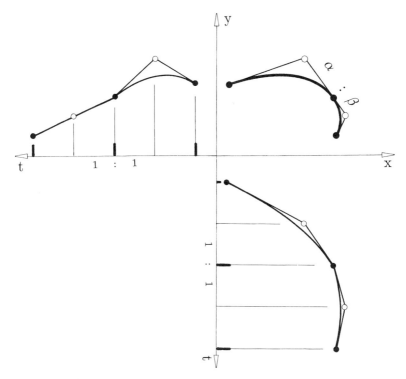

Figure 7.7: A C^0 parametrization: the piecewise Bézier curve is the same as in the previous figure. It is *not* a C^1 curve with the choice of uniform parameter intervals as indicated in the cross plot.

B-spline polygon, together with a knot sequence, determines a C^1 quadratic spline curve, and, conversely, each quadratic C^1 spline curve possesses a unique B-spline polygon.

From the definition of a quadratic B-spline polygon, we can deduce several properties, which we shall simply list since their derivation is a direct consequence of the previous definitions:

- convex hull property

- linear precision

- affine invariance

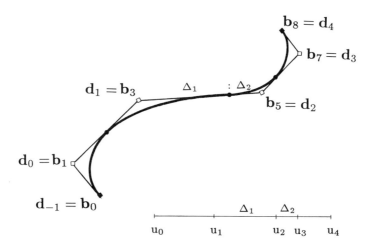

Figure 7.8: C^1 quadratic splines: the junction points $\mathbf{b}_{2i}$ are determined by the inner Bézier points and the knot sequence.

- symmetry

- endpoint interpolation

- variation diminishing property.

The last property follows since the piecewise Bézier polygon of $\mathbf{s}$ is obtained by piecewise linear interpolation of the B-spline polygon, a process that is variation diminishing, as seen in Section 2.4.

All of the above properties are shared with Bézier curves, although the convex hull property may be sharpened considerably for quadratic B-spline curves: the curve $\mathbf{s}$ lies in the union of the convex hulls of the triangles $\mathbf{b}_{2i-1}, \mathbf{b}_{2i+1}, \mathbf{b}_{2i+3}$; $i = 1 \ldots, L-2$ and the triangles $\mathbf{b}_0, \mathbf{b}_1, \mathbf{b}_3$ and $\mathbf{b}_{2L-3}, \mathbf{b}_{2L-1}, \mathbf{b}_{2L}$, as shown in Figure 7.9. A Bézier curve of degree $2L$ could only be guaranteed to lie within the convex hull of the whole polygon $\mathbf{b}_0, \ldots, \mathbf{b}_{2L}$.

By definition, quadratic B-spline curves consist of parabolic segments, i.e., planar curves. However, the B-spline control polygon may be truly three-dimensional – we thus have a method to generate C^1 space curves that are piecewise planar.

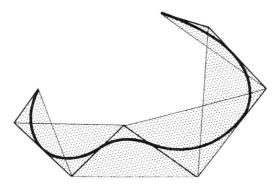

Figure 7.9: The convex hull property: a C^1 quadratic B-spline curve lies in the union of a set of triangles. The triangles are formed by triples of consecutive control vertices.

One important property that single Bézier curves do not share with B-spline curves is *local control*. If we are dealing with a single Bézier curve,[3] we know that a change of one of the control vertices affects the whole curve – it is a *global* change. Changing a control vertex of a quadratic B-spline curve, on the other hand, affects at most three curve segments. It is this local control property that made B-spline curves as popular as they are. If a part of a curve is completely designed, it is highly undesirable to jeopardize this result by changing the curve in other regions. With single Bézier curves, this is unavoidable.

As a consequence of the local control property, we may include straight line segments in a quadratic B-spline curve: if three subsequent control vertices are collinear, the quadratic segment that is determined by them must be linear. A single (higher degree) Bézier curve cannot contain linear segments unless it is itself linear; this is yet another reason why B-spline curves are much more flexible than single Bézier curves. Figure 7.10 shows a quadratic B-spline curve that includes straight line segments. Such curves occur frequently in technical design applications, as well as in font design.

From inspection of Figure 7.8, we see that the endpoints of a B-spline curve are treated in a special way. This is not the case with *closed curves*. Closed curves are defined by $\mathbf{s}(u_0) = \mathbf{s}(u_L)$. Figure 7.11 shows two closed

---

[3]In this context, we do not consider Bézier curves as parts of composite curves!

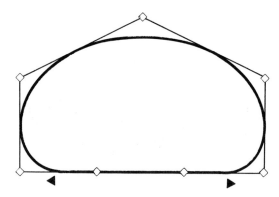

Figure 7.10: Quadratic B-spline curves: curves can be designed that include straight line segments.

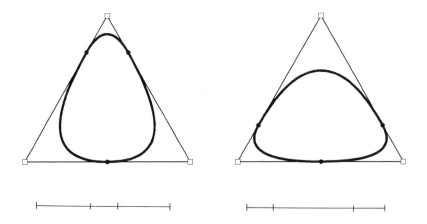

Figure 7.11: Closed curves: two closed quadratic B-spline curves are shown that have the same control polygon but different knot sequences.

quadratic B-spline curves. For such curves, C^1 continuity is defined by the additional constraint $(d/du)\mathbf{s}(u_0) = (d/du)\mathbf{s}(u_L)$.

The figure also shows that a B-spline curve depends not only on the B-spline polygon, but also on the knot sequence.

7.7 C^2 Cubic B-spline Curves

We are now interested in C^2 piecewise cubic spline curves, again defined over L intervals $u_0 < \ldots < u_L$. Consider any two adjacent curve segments $\mathbf{s}_{i-1}$ and $\mathbf{s}_i$. To be C^1 at u_i, the relevant Bézier points must be in the ratio $\Delta_{i-1} : \Delta_i$, or

$$\mathbf{b}_{3i} = \frac{\Delta_i}{\Delta_{i-1} + \Delta_i}\mathbf{b}_{3i-1} + \frac{\Delta_{i-1}}{\Delta_{i-1} + \Delta_i}\mathbf{b}_{3i+1}. \tag{7.12}$$

To be C^2 as well, an auxiliary point $\mathbf{d}_i$ must exist such that the points $\mathbf{b}_{3i-2}, \mathbf{b}_{3i-1}, \mathbf{d}_i$ and $\mathbf{d}_i, \mathbf{b}_{3i+1}, \mathbf{b}_{3i+2}$ are in the same ratio $\Delta_{i-1} : \Delta_i$, as follows from the C^2 conditions (7.9). Figure 7.12 illustrates this point.

A C^2 cubic spline curve defines the auxiliary points $\mathbf{d}_i$, which form a polygon $\mathbf{P}$. Conversely, a polygon $\mathbf{P}$ and a knot sequence $\{u_i\}$ also define a C^2 cubic spline curve. Set

$$\mathbf{b}_{3i-2} = \frac{\Delta_{i-1} + \Delta_i}{\Delta}\mathbf{d}_{i-1} + \frac{\Delta_{i-2}}{\Delta}\mathbf{d}_i, \tag{7.13}$$

$$\mathbf{b}_{3i-1} = \frac{\Delta_i}{\Delta}\mathbf{d}_{i-1} + \frac{\Delta_{i-2} + \Delta_{i-1}}{\Delta}\mathbf{d}_i, \tag{7.14}$$

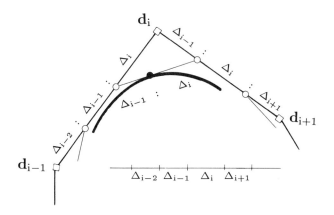

Figure 7.12: C^2 cubic B-spline curves: the auxiliary points $\mathbf{d}_i$ define the B-spline polygon of the curve.

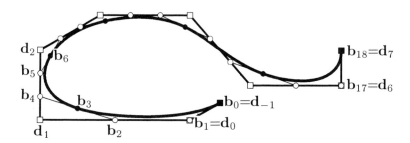

Figure 7.13: B-splines: a cubic B-spline curve with its control polygon.

where
$$\Delta = \Delta_{i-2} + \Delta_{i-1} + \Delta_i. \tag{7.15}$$

With the junction points $\mathbf{b}_{3i}$ defined in (7.12), the piecewise Bézier curve defined by the $\mathbf{d}_i$ meets the C^2 conditions at every knot u_i; $i = 2, \ldots, L - 2$.

Near the ends, things are a little more complicated. We define the cubic B-spline polygon to have vertices $\mathbf{d}_{-1}, \mathbf{d}_0, \ldots, \mathbf{d}_L, \mathbf{d}_{L+1}$ and then set

$$
\begin{aligned}
\mathbf{b}_0 &= \mathbf{d}_{-1}, \\
\mathbf{b}_1 &= \mathbf{d}_0, \\
\mathbf{b}_2 &= \frac{\Delta_1}{\Delta_0 + \Delta_1}\mathbf{d}_0 + \frac{\Delta_0}{\Delta_0 + \Delta_1}\mathbf{d}_1,
\end{aligned}
\tag{7.16}
$$

$$
\begin{aligned}
\mathbf{b}_{3L-2} &= \frac{\Delta_{L-1}}{\Delta_{L-2} + \Delta_{L-1}}\mathbf{d}_{L-1} + \frac{\Delta_{L-2}}{\Delta_{L-2} + \Delta_{L-1}}\mathbf{d}_L, \\
\mathbf{b}_{3L-1} &= \mathbf{d}_L, \\
\mathbf{b}_{3L} &= \mathbf{d}_{L+1}.
\end{aligned}
\tag{7.17}
$$

Now the spline curve is C^2 at every interior knot. This construction is due to W. Boehm [60]. An illustration is given in Figure 7.13.

If a cubic spline curve is expressed in terms of the *B-spline polygon* (the polygon consisting of the $\mathbf{d}_i$), it is usually called a C^2 cubic B-spline curve.

Cubic B-spline curves enjoy the same properties as do quadratic ones:

- convex hull property

- linear precision

- affine invariance

- symmetry

Figure 7.14: Local control: as one control vertex is moved, only the four "nearby" curve segments change.

- endpoint interpolation

- variation diminishing property

- local control.

Local control for cubic B-spline curves is not quite as local as it is for quadratic ones. If a control vertex $\mathbf{d}_i$ of a cubic B-spline curve is moved, *four* segments of the curve will be changed, as shown in Figure 7.14.

7.8 Parametrizations

"Given the de Boor polygon and the knot sequence, construct the corresponding piecewise Bézier polygon" was the topic of the last two sections. In freeform design, one creates the de Boor polygon interactively, but how does one create the knot sequence? An easy answer is to set $u_i = i$, or some other (equivalent) uniform spacing, but this method is too rigid in many cases. The jury is still out on what constitutes an "optimal" parametrization. As a rule of thumb, better curves are obtained from a given polygon if the geometry of the polygon is incorporated into the knot sequence.

For example, one may set (in the cubic case)

$$
\begin{aligned}
u_0 &= 0, \\
u_1 &= \|\mathbf{d}_1 - \mathbf{d}_{-1}\|, \\
u_i &= u_{i-1} + \|\mathbf{d}_i - \mathbf{d}_{i-1}\|; \quad i = 2, \ldots, L-1, \\
u_L &= u_{L-1} + \|\mathbf{d}_{L+1} - \mathbf{d}_{L-1}\|.
\end{aligned}
\tag{7.18}
$$

This is a *chord length parametrization* for cubic B-spline curves when the polygon is given.[4] This parametrization often produces "smoother" curves than the uniform one described above (see Sapidis [414]).

7.9   Design and Inverse Design

Quadratic B-spline curves seemed to do a pretty good job of producing complex shapes, so why increase the degree to cubic? Cubic polynomials are *true space curves*, i.e., they are not planar. For 2D shapes, piecewise quadratics might suffice, but when it comes to 3D, they can only produce piecewise 2D segments. Two examples why this is not desirable: 3D curves that are used to describe robot paths will exhibit jumps in their torsion – this is bad for the joints of the robot arm. Secondly, 3D curves that have to satisfy aesthetic requirements would simply *look* bad if described by piecewise planar shapes.[5]

Another advantage of piecewise cubics is the fact that they may have inflection points *inside* a segment. With piecewise quadratics, one would have to make sure that there is a junction point at every inflection point.

How do we design curves using cubic B-splines? The typical freeform design takes place in a 2D environment, using a mouse, light pen, or some other interactive input device. A control polygon is sketched on the terminal, the resulting curve is drawn, the control polygon is adjusted, and so on. The parametrization that is being used should be kept away from the designer, and would most likely be one of the two methods described in the previous section. To obtain a 3D curve, one would then change to another view (by rotating the curve) and continue adjusting control points. The final result might look reasonable on the terminal, but should probably undergo a final smoothing process as discussed in Chapter 23.

Sometimes designers just don't like to deal with control polygons, and prefer to manipulate the curve directly. In that case, the curve should be obtained from an interpolation process as described in Chapter 8 or 9. It may be represented internally in B-spline or piecewise Bézier form. If the designer wants to change a certain junction point $\mathbf{x}_i$ to a new location, this may easily be done *locally* using B-spline technology as follows.

---

[4]Another chord length parametrization exists if data points are given for an interpolatory spline, as described in Chapter 9.

[5]Of course, this could be overcome by using a large number of quadratic pieces. But then, why not go a step further and do everything piecewise linear, with even more pieces?

If we were to displace $\mathbf{x}(u_i)$ by a vector $\Delta\mathbf{x}_i$, then we would have to change $\mathbf{d}_i$ by a vector

$$\mathbf{e}_i = \frac{1}{c_i}\Delta\mathbf{x}_i. \tag{7.19}$$

The value of c_i may be computed to

$$c_i = \frac{1}{u_{i+1} - u_{i-1}}\left(\Delta_i\frac{u_i - u_{i-2}}{u_{i+1} - u_{i-2}} + \Delta_{i-1}\frac{u_{i+2} - u_i}{u_{i+2} - u_{i-1}}\right). \tag{7.20}$$

Thus we have solved our problem, called *inverse design*. In this mode, we would directly move the junction point $\mathbf{x}_i$ and simply hide the equivalent change in the control polygon from the designer. We need to keep in mind that this procedure (since it changes $\mathbf{d}_i$) will also change $\mathbf{x}_{i-1}$ and $\mathbf{x}_{i+1}$, although by smaller amounts.

7.10   Implementation

The following is a program for the conversion of a cubic B-spline curve to piecewise Bézier form:

```
void bspline_to_bezier(bspl,knot,l,bez)
/*    converts   cubic B-spline polygon into piecewise Bezier polygon.
Input:    bspl: B-spline control polygon
          knot: knot sequence
          l:    no. of intervals
Output:   bez:  piecewise Bezier polygon. Each junction point b_3i is
                only stored once.
Remark:   bspl starts from 0 and not -1 as in the text. All
          subscripts are therefore shifted by one. For those familiar
          with Chapter 10: don't try to use multiple knots here --
          in terms of that chapter, the end knots u_0 and u_l
          have multiplicity 3, but  all other
          knots are simple, and the curve is C2.
*/
```

This routine has to be called for each coordinate (i.e., two or three times). Speedups are therefore possible.

7.11   Problems

1. If we write the de Casteljau algorithm in the form of a triangular array as in Eq. (3.3), subdivision tells us how the three "sides" of that array are related to each other. Write explicitly how to generate the elements of one side from those of any other one.

2. Consider two Bézier curves with polygons $\mathbf{b}_0, \ldots, \mathbf{b}_n$ and $\mathbf{b}_n, \ldots, \mathbf{b}_{2n}$. Let $\mathbf{b}_{n-r} = \ldots = \mathbf{b}_n = \ldots \mathbf{b}_{n+r}$ so that both curves form one (degenerate) C^r curve. Under what conditions on $\mathbf{b}_{n-r-1}$ and $\mathbf{b}_{n+r+1}$ is that curve also C^{r+1}?

3. We are given a closed polygon. Suppose we want to make the polygon vertices the inner Bézier points $\mathbf{b}_{2i+1}$ of a piecewise quadratic and that we pick arbitrary points on the polygon legs to become the junction Bézier points $\mathbf{b}_{2i}$. Can we always find a C^1 parametrization for this (tangent continuous) piecewise quadratic curve?

4. Describe the chord length parametrization for closed B-spline curves.

5. The following amusing error occurred during the generation of the programs for this book: in the definition of **decas**, omit the auxiliary array. The routine will then destroy the input array. However, when used in **bez_to_points**, the curve was plotted correctly. Why?

Chapter 8

Piecewise Cubic Interpolation

Polynomial interpolation is a fundamental theoretical tool, but for practical purposes, better methods exist. The most popular class of methods is that of piecewise polynomial schemes. All these methods construct curves that consist of polynomial pieces of the same degree and that are of a prescribed overall smoothness. The given data are usually points and parameter values; sometimes, tangent information is added as well.

In practice, one usually encounters the use of piecewise cubic curves. They may be C^2 – the next chapter is dedicated to that case. If they are only C^1, the trade-off for the lower differentiability class is *locality*: if a data point is changed, the interpolating curve only changes in the vicinity of that data point.

This chapter can only cover the basic ideas behind piecewise cubic interpolation. A large variety of interpolation methods exists that are designed to cope with special problems. Most such methods try to preserve shape features inherent in the given data, for example, convexity or monotonicity. We mention the work by Fritsch and Carlson [197], McLaughlin [332], Foley [188], [189], McAllister and Roulier [329], and Schumaker [424].

8.1 C^1 Piecewise Cubic Hermite Interpolation

This is conceptually the simplest of all interpolants, although not the most practical one. It solves the following problem:

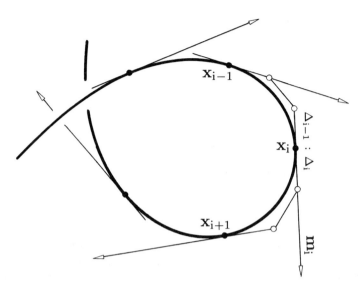

Figure 8.1: Piecewise cubic Hermite interpolation: the Bézier points are obtained directly from the data.

Given: Data points $\mathbf{x}_0, \ldots, \mathbf{x}_L$, corresponding parameter values $u_0, \ldots, u_L$, and corresponding tangent vectors $\mathbf{m}_0, \ldots, \mathbf{m}_L$.

Find: A C^1 piecewise cubic polynomial $\mathbf{s}$ that interpolates to the given data, i.e.,

$$\mathbf{s}(u_i) = \mathbf{x}_i, \qquad \frac{\mathrm{d}}{\mathrm{d}u}\mathbf{s}(u_i) = \mathbf{m}_i; \quad i = 0, \ldots, L. \tag{8.1}$$

We construct the solution as a piecewise Bézier curve, as illustrated in Figure 8.1. We find the junction Bézier points immediately: $\mathbf{b}_{3i} = \mathbf{x}_i$. To obtain the inner Bézier points, we recall the derivative formula for Bézier curves from Section 7.3:

$$\begin{aligned}
\frac{\mathrm{d}}{\mathrm{d}u}\mathbf{s}(u_i) &= \frac{3}{\Delta_{i-1}}(\mathbf{b}_{3i} - \mathbf{b}_{3i-1}) \\
&= \frac{3}{\Delta_i}(\mathbf{b}_{3i+1} - \mathbf{b}_{3i}),
\end{aligned}$$

where $\Delta_i = \Delta u_i$. Thus the inner Bézier points $\mathbf{b}_{3i+1}; i = 0, \ldots, L-1$ are given by

$$\mathbf{b}_{3i+1} = \mathbf{b}_{3i} + \frac{\Delta_i}{3}\mathbf{m}_i, \qquad (8.2)$$

and the inner Bézier points $\mathbf{b}_{3i-1}, i = 1, \ldots, L$ are

$$\mathbf{b}_{3i-1} = \mathbf{b}_{3i} - \frac{\Delta_{i-1}}{3}\mathbf{m}_i. \qquad (8.3)$$

What we have done so far is construct the piecewise Bézier form of the C^1 piecewise cubic Hermite interpolant. Of course, we can utilize the material on cubic Hermite interpolation from Section 6.5 as well. Over the interval $[u_i, u_{i+1}]$, the interpolant $\mathbf{s}$ can be expressed in terms of the cubic Hermite polynomials $\hat{H}_i^3(u)$ that were defined by (6.16). In the situation at hand, the definitions become:

$$\begin{aligned}
\hat{H}_0^3(u) &= B_0^3(t) + B_1^3(t), \\
\hat{H}_1^3(u) &= \frac{\Delta_i}{3}B_1^3(t), \\
\hat{H}_2^3(u) &= -\frac{\Delta_i}{3}B_2^3(t), \\
\hat{H}_3^3(u) &= B_2^3(t) + B_3^3(t),
\end{aligned} \qquad (8.4)$$

where $t = (u - u_i)/\Delta_i$ is the local parameter of the interval $[u_i, u_{i+1}]$. The interpolant can now be written as

$$\mathbf{s}(u) = \mathbf{x}_i\hat{H}_0^3(u) + \mathbf{m}_i\hat{H}_1^3(u) + \mathbf{m}_{i+1}\hat{H}_2^3(u) + \mathbf{x}_{i+1}\hat{H}_3^3(u). \qquad (8.5)$$

This interpolant is important for some theoretical developments; of more practical value are those developed in the following sections.

8.2 C^1 Piecewise Cubic Interpolation I

The title of this section is not very different from the one of the preceding section, and indeed the problems addressed in both sections differ only by a subtle nuance. Here, we try to solve the following problem:

Given: Data points $\mathbf{x}_0, \ldots, \mathbf{x}_L$ and tangent directions $\mathbf{l}_0, \ldots, \mathbf{l}_L$ at those data points.

Find: A C^1 piecewise cubic polynomial that passes through the given data points and is tangent to the given tangent directions there.

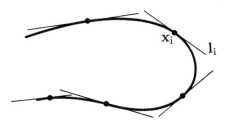

Figure 8.2: C^1 piecewise cubics: example data set.

Comparing this problem to the one in the previous section, we find that it is more vaguely formulated: the "Find" part does not contain a single formula. This reflects a typical practical situation: one is not always given parameter values u_i or tangent vectors $\mathbf{m}_i$; very often, the only available information is data points and tangent directions, as illustrated in Figure 8.2. It is important to note that we only have tangent *directions*, i.e., we have no vectors with a prescribed length. We can assume without loss of generality that the tangent directions $\mathbf{l}_i$ have been normalized to be of unit length:

$$||\mathbf{l}_i|| = 1.$$

The easiest step in finding the desired piecewise cubic is the same as before: the junction Bézier points $\mathbf{b}_{3i}$ are again given by $\mathbf{b}_{3i} = \mathbf{x}_i$, $i = 0, \ldots, L$.

For each inner Bézier point, we have a one-parameter family of solutions: we only have to ensure that each triple $\mathbf{b}_{3i-1}, \mathbf{b}_{3i}, \mathbf{b}_{3i+1}$ is collinear on the tangent at $\mathbf{b}_{3i}$ and ordered by increasing subscript in the direction of $\mathbf{l}_i$. We can then find a parametrization with respect to which the generated curve is C^1 [see Eq. (7.10)].

In general, we must determine the inner Bézier points from

$$\mathbf{b}_{3i+1} \;=\; \mathbf{b}_{3i} + \alpha_i \mathbf{l}_i, \qquad (8.6)$$

$$\mathbf{b}_{3i-1} \;=\; \mathbf{b}_{3i} - \beta_{i-1} \mathbf{l}_i, \qquad (8.7)$$

so that the problem boils down to finding reasonable values for α_i and β_i. While any nonnegative value for these numbers is a formally valid solution, values for α_i and β_i that are too small cause the curve to have a corner at $\mathbf{x}_i$, while values that are too large can create loops. There is probably no optimal choice for α_i and β_i that holds up in every conceivable application – an optimal choice must depend on the desired application.

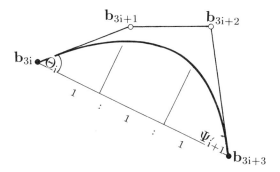

Figure 8.3: Inner Bézier points: this planar curve can be interpreted as a *function* in an oblique coordinate system with $\mathbf{b}_0, \mathbf{b}_3$ as the x-axis.

A "quick and easy" solution that has performed decently many times (but also failed sometimes) is simply to set

$$\alpha_i = \beta_i = 0.4\|\Delta\mathbf{x}_i\|. \tag{8.8}$$

(The factor 0.4 is, of course, heuristic.)

The parametrization with respect to which this interpolant is C^1 is the *chord length parametrization*. It is characterized by

$$\frac{\Delta_i}{\Delta_{i+1}} = \frac{\|\Delta\mathbf{x}_i\|}{\|\Delta\mathbf{x}_{i+1}\|}. \tag{8.9}$$

A more sophisticated solution is the following: if we consider the planar curve in Figure 8.3, we see that it can be interpreted as a function, where the parameter t varies along the straight line through $\mathbf{b}_0$ and $\mathbf{b}_3$. Then

$$\|\Delta\mathbf{b}_0\| = \frac{\|\mathbf{b}_3 - \mathbf{b}_0\|}{3\cos\Theta_i},$$

$$\|\Delta\mathbf{b}_2\| = \frac{\|\mathbf{b}_3 - \mathbf{b}_0\|}{3\cos\Psi_{i+1}}.$$

We are dealing with parametric curves, however, which are in general not planar and for which the angles Θ and Ψ could be close to 90 degrees, causing the above expressions to be undefined. But for curves with Θ_i, Ψ_{i+1} smaller than, say, 60 degrees, the above could be utilized to find reasonable values for α_i and β_i:

$$\alpha_i = \frac{1}{3\cos\Theta_i}||\Delta\mathbf{x}_i||,$$

$$\beta_i = \frac{1}{3\cos\Psi_{i+1}}||\Delta\mathbf{x}_i||.$$

Since $\cos 60° = 1/2$, we can now make a case distinction:

$$\alpha_i = \begin{cases} \frac{||\Delta\mathbf{x}_i||^2}{3 l_i \mathbf{x}_i} & \text{if } |\Theta_i| \leq 60° \\ \frac{2}{3}||\Delta\mathbf{x}_i|| & \text{otherwise} \end{cases} \tag{8.10}$$

and

$$\beta_i = \begin{cases} \frac{||\Delta\mathbf{x}_i||^2}{3 l_{i+1} \mathbf{x}_i} & \text{if } |\Psi_{i+1}| \leq 60° \\ \frac{2}{3}||\Delta\mathbf{x}_i|| & \text{otherwise.} \end{cases} \tag{8.11}$$

This method has the advantage of having *linear precision.*

Note that neither of these two methods is affinely invariant: the first method – Eq. (8.8) – does not preserve the ratios of the three points $\mathbf{b}_{3i-1}, \mathbf{b}_{3i}, \mathbf{b}_{3i+1}$ since the ratios $||\Delta\mathbf{x}_{i-1}|| : ||\Delta\mathbf{x}_i||$ are not generally invariant under affine maps.[1] The second method uses angles, which are not preserved under affine transformations. However, both methods are invariant under euclidean transformations.

8.3 C^1 Piecewise Cubic Interpolation II

Continuing with the relaxation of given constraints for the interpolatory C^1 cubic spline curve, we now address the following problem:

Given: Data points $\mathbf{x}_0, \ldots, \mathbf{x}_L$ together with corresponding parameter values $u_0, \ldots, u_L$.

Find: A C^1 piecewise cubic polynomial that passes through the given data points.

One solution to this problem is provided by C^2 (and hence also C^1) cubic splines, which are discussed in Chapter 9. We will here determine tangent directions $\mathbf{l}_i$ or tangent vectors $\mathbf{m}_i$ and then apply the methods from the previous two sections.

---

[1]Recall that only the ratio of three *collinear* points is preserved under affine maps!

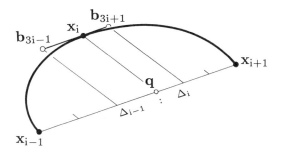

Figure 8.4: FMILL tangents: the tangent at $\mathbf{x}_i$ is parallel to the chord through $\mathbf{x}_{i-1}$ and $\mathbf{x}_{i+1}$.

Probably the simplest method for tangent estimation is known under the name FMILL. It constructs the tangent direction $\mathbf{l}_i$ at $\mathbf{x}_i$ to be parallel to the chord through $\mathbf{x}_{i-1}$ and $\mathbf{x}_{i+1}$:

$$\mathbf{l}_i = \mathbf{x}_{i+1} - \mathbf{x}_{i-1}; \quad i = 1, \ldots, L-1. \tag{8.12}$$

Once the tangent direction $\mathbf{l}_i$ has been found, the inner Bézier points are placed on it according to Figure 8.4:

$$\mathbf{b}_{3i-1} = \mathbf{b}_{3i} - \frac{\Delta_{i-1}}{3(\Delta_{i-1} + \Delta_i)}\mathbf{l}_i, \tag{8.13}$$

$$\mathbf{b}_{3i+1} = \mathbf{b}_{3i} + \frac{\Delta_i}{3(\Delta_{i-1} + \Delta_i)}\mathbf{l}_i. \tag{8.14}$$

This interpolant is also known as a Catmull-Rom spline.

This construction of the inner Bézier points does not work at $\mathbf{x}_0$ and $\mathbf{x}_L$. One way to pick the tangents there is to use Bessel tangents, as described below.

The idea behind *Bessel tangents*[2] is as follows: to find the tangent vector $\mathbf{m}_i$ at $\mathbf{x}_i$, pass the interpolating parabola $\mathbf{q}_i(u)$ through $\mathbf{x}_{i-1}, \mathbf{x}_i, \mathbf{x}_{i+1}$ with corresponding parameter values u_{i-1}, u_i, u_{i+1} and let $\mathbf{m}_i$ be the derivative of $\mathbf{q}_i$. We differentiate $\mathbf{q}_i$ at u_i:

$$\mathbf{m}_i = \frac{\mathrm{d}}{\mathrm{d}u}\mathbf{q_i}(u_i).$$

---

[2]They are sometimes attributed to Ackland [2].

Written in terms of the given data, this gives

$$\mathbf{m}_i = \frac{(1 - \alpha_i)}{\Delta_{i-1}}\Delta\mathbf{x}_{i-1} + \frac{\alpha_i}{\Delta_i}\Delta\mathbf{x}_i; \quad i = 1, \ldots, L - 1, \tag{8.15}$$

where

$$\alpha_i = \frac{\Delta_{i-1}}{\Delta_{i-1} + \Delta_i}.$$

The endpoints are treated in the same way: $\mathbf{m}_0 = \mathrm{d}/\mathrm{d}u\mathbf{q}_1(u_0), \mathbf{m}_L = \mathrm{d}/\mathrm{d}u\mathbf{q}_{L-1}(u_L)$, which gives

$$\mathbf{m}_0 = 2\frac{\Delta\mathbf{x}_0}{\Delta_0} - \mathbf{m}_1,$$

$$\mathbf{m}_L = 2\frac{\Delta\mathbf{x}_{L-1}}{\Delta_{L-1}} - \mathbf{m}_{L-1}.$$

Another interpolant that makes use of the parabolas $\mathbf{q}_i$ is known as an *Overhauser spline*, after work by A. Overhauser [354] (see also [77]). The ith segment $\mathbf{s}_i$ of such a spline is defined by

$$\mathbf{s}_i(u) = \frac{u_{i+1} - u}{\Delta_i}\mathbf{q}_i(u) + \frac{u - u_i}{\Delta_i}\mathbf{q}_{i+1}(u); \quad i = 1, \ldots, L - 2.$$

In other words, each $\mathbf{s}_i$ is a linear blend between $\mathbf{q}_i$ and $\mathbf{q}_{i+1}$. At the ends, one sets $\mathbf{s}_0(u) = \mathbf{q}_0(u)$ and $\mathbf{s}_{L-1}(u) = \mathbf{q}_{L-1}(u)$.

On closer inspection it turns out that the last two interpolants are not different at all: they both yield the same C^1 piecewise cubic interpolant (see Problems). A similar way of determining tangent vectors was developed by McConalogue [330], [331].

Finally, we mention a method created by H. Akima [4]. It sets

$$\mathbf{m}_i = (1 - c_i)\mathbf{a}_{i-1} + c_i\mathbf{a}_i,$$

where

$$\mathbf{a}_i = \frac{\Delta\mathbf{x}_i}{\Delta_i}$$

and

$$c_i = \frac{||\Delta\mathbf{a}_{i-2}||}{||\Delta\mathbf{a}_{i-2}|| + ||\Delta\mathbf{a}_i||}.$$

This interpolant appears fairly involved. It generates very good results, however, in situations where one needs curves that oscillate only minimally.

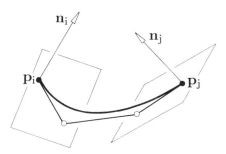

Figure 8.5: Finding cubic boundaries: while the endpoints of a boundary curve are fixed, its end tangents only have to lie in specified planes.

8.4 Point-Normal Interpolation

In a surface generation environment, one is often given a set of points $\mathbf{p}_i \in \mathbb{E}^3$ and a surface normal vector $\mathbf{n}_i$ at each data point as illustrated in Figure 8.5. Thus we only know the tangent plane of the desired surface at each data point, not the actual endpoint derivatives of the patch boundary curves.

If we know that two points $\mathbf{p}_i$ and $\mathbf{p}_j$ have to be connected, then we must construct a curve leading from $\mathbf{p}_i$ to $\mathbf{p}_j$ that is normal to $\mathbf{n}_i$ at $\mathbf{p}_i$ and to $\mathbf{n}_j$ at $\mathbf{p}_j$. A cubic will suffice to solve this generalized Hermite interpolation problem. In Bézier form, we already have $\mathbf{b}_0 = \mathbf{p}_i$ and $\mathbf{b}_3 = \mathbf{p}_j$. We still need to find $\mathbf{b}_1$ and $\mathbf{b}_2$.

There are infinitely many solutions, so we may try to pick one that is both convenient to compute and of reasonable shape in most cases. Two approaches to this problem appear in Piper [375], p. 229 and Nielson [351], p. 238.

Piper obtains $\mathbf{b}_1$ in two steps. As a first approximation to $\mathbf{b}_1$, project $\mathbf{b}_3$ into the plane defined by $\mathbf{b}_0 = \mathbf{p}_i$ and $\mathbf{n}_i$. This defines a tangent at $\mathbf{b}_0$. Place the final $\mathbf{b}_1$ anywhere on this tangent, using some of the methods described in Section 8.2. The remaining point $\mathbf{b}_2$ is then obtained analogously.

Nielson defines the tangent at $\mathbf{b}_0$ to be the intersection of the plane defined by $\mathbf{b}_0 = \mathbf{p}_i$ and $\mathbf{n}_i$ and the plane through $\mathbf{b}_0$ containing both $\mathbf{n}_i$ and $\mathbf{p}_j - \mathbf{p}_i$. Again, $\mathbf{b}_1$ would be placed on this tangent according to some measure of optimality. The same procedure is then repeated for $\mathbf{b}_2$.

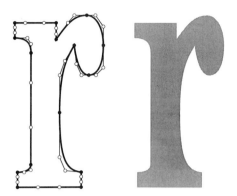

Figure 8.6: Font design: the characters in this book are stored as a sequence of cubic Bézier curves.

8.5   Font Generation

We conclude this chapter with an application of growing importance, namely *font design*. A graphics language such as PostScript has to generate characters for many different font sets – Arabic, Helvetica, boldface, just to name a few. These fonts must be scaleable, i.e., if a different font size is desired, the original fonts must be rescaled. Had the original fonts been stored as pixel maps, scaling would cause serious aliasing problems. It is common practice, therefore, to store a given character not as a pixel map, but rather to store its outline as a sequence of Bézier curves. These allow smooth arcs where desired, and also allow for sharp corners, as shown in Figure 8.6.[3] This book was printed using PostScript, so all characters have been generated as piecewise Bézier curves.

8.6   Problems

1. Show that Overhauser splines are piecewise cubics with Bessel tangents at the junction points.

2. Show that Akima's interpolant always passes a straight line segment through three subsequent points if they happen to lie on a straight line.

---

[3]This is my own rendition of the letter **r** from the front cover.

3. One can generalize the quintic Hermite interpolants from Section 6.6 to piecewise quintic Hermite interpolants. These curves need first and second derivatives as input positions. Devise ways to generate second derivative information from data points and parameter values.

4. The two methods described in Section 8.4 appear to be identical at first sight. Show that they are not.

5. Using piecewise cubic C^1 interpolation, approximate the semicircle with radius 1 to within a tolerance of $\epsilon = 0.001$. Use as few cubic segments as possible. Literature: [142], [205].

6. Program the methods from Section 8.3. Apply to the semicircle from the previous problem and compare to the special-purpose interpolant developed there.

Chapter 9

Cubic Spline Interpolation

In this chapter, we discuss what is probably *the* most popular curve scheme: C^2 cubic interpolatory splines. We have seen how polynomial Lagrange interpolation fails to produce acceptable results. On the other hand, we saw that cubic B-spline curves are a powerful modeling tool; they are able to model complex shapes easily. This "modeling" is carried out as an *approximation* process, manipulating the control polygon until a desired shape is achieved. We will see how cubic splines can also be used to fulfill the task of *interpolation*, the task of finding a spline curve passing through a given set of points. Cubic spline interpolation was introduced into the CAGD literature by J. Ferguson [183] in 1964, while the mathematical theory was studied in approximation theory (see de Boor [112] or Holladay [264]). For an outline of the history of splines, see Schumaker [423].

Because of the subject's importance, we present two entirely independent derivations of cubic interpolatory splines: the B-spline form and the Hermite form.

9.1 The B-spline Form

We are given a set of data points $\mathbf{x}_0, \ldots, \mathbf{x}_L$ and corresponding parameter values (or knots or breakpoints) $u_0, \ldots, u_L$.[1] We want a cubic B-spline curve $\mathbf{s}$, determined by the same knots and unknown control vertices $\mathbf{d}_{-1}, \ldots, \mathbf{d}_{L+1}$ such that $\mathbf{s}(u_i) = \mathbf{x}_i$; in other words, such that $\mathbf{s}$ *interpolates* to the data points.

---

[1]The knots are in general *not* given – see Section 9.4 on how to generate them.

The solution to this problem becomes obvious once one realizes the relationship between the data points $\mathbf{x}_i$ and the control vertices $\mathbf{d}_i$. Recall that we can write every B-spline curve as a piecewise Bézier curve (see Section 7.7). In that form, we have

$$\mathbf{x}_i = \mathbf{b}_{3i}; \quad i = 0, \ldots, L.$$

The inner Bézier points $\mathbf{b}_{3i\pm1}$ are related to the $\mathbf{x}_i$ by

$$\mathbf{x}_i = \frac{\Delta_i \mathbf{b}_{3i-1} + \Delta_{i-1}\mathbf{b}_{3i+1}}{\Delta_{i-1} + \Delta_i} \quad i = 1, \ldots, L-1, \tag{9.1}$$

where we have set $\Delta_i = \Delta u_i$. Finally, the $\mathbf{b}_{3i\pm1}$ are related to the control vertices $\mathbf{d}_i$ by

$$\mathbf{b}_{3i-1} = \frac{\Delta_i \mathbf{d}_{i-1} + (\Delta_{i-2} + \Delta_{i-1})\mathbf{d}_i}{\Delta_{i-2} + \Delta_{i-1} + \Delta_i}; \quad i = 2, \ldots L-1 \tag{9.2}$$

and

$$\mathbf{b}_{3i+1} = \frac{(\Delta_i + \Delta_{i+1})\mathbf{d}_i + \Delta_{i-1}\mathbf{d}_{i+1}}{\Delta_{i-1} + \Delta_i + \Delta_{i+1}}; \quad i = 1, \ldots L-2. \tag{9.3}$$

Near the endpoints of the curve, the situation is somewhat special:

$$\mathbf{b}_2 = \frac{\Delta_1 \mathbf{d}_0 + \Delta_0 \mathbf{d}_1}{\Delta_0 + \Delta_1}, \tag{9.4}$$

$$\mathbf{b}_{3L-2} = \frac{\Delta_{L-1}\mathbf{d}_{L-1} + \Delta_{L-2}\mathbf{d}_L}{\Delta_{L-2} + \Delta_{L-1}}. \tag{9.5}$$

We can now write down the relationships between the unknown $\mathbf{d}_i$ and the known $\mathbf{x}_i$, i.e., we can eliminate the $\mathbf{b}_i$:

$$(\Delta_{i-1} + \Delta_i)\mathbf{x}_i = \alpha_i \mathbf{d}_{i-1} + \beta_i \mathbf{d}_i + \gamma_i \mathbf{d}_{i+1}, \tag{9.6}$$

where we have set (with $\Delta_{-1} = \Delta_L = 0$):

$$\alpha_i = \frac{\Delta_i^2}{\Delta_{i-2} + \Delta_{i-1} + \Delta_i},$$

$$\beta_i = \frac{\Delta_i(\Delta_{i-2} + \Delta_{i-1})}{\Delta_{i-2} + \Delta_{i-1} + \Delta_i} + \frac{\Delta_{i-1}(\Delta_i + \Delta_{i+1})}{\Delta_{i-1} + \Delta_i + \Delta_{i+1}},$$

$$\gamma_i = \frac{\Delta_{i-1}^2}{\Delta_{i-1} + \Delta_i + \Delta_{i+1}}.$$

If we choose the two Bézier points $\mathbf{b}_1$ and $\mathbf{b}_{3L-1}$ arbitrarily, we obtain a linear system of the form

$$
\begin{bmatrix}
1 & & & & & \\
\alpha_1 & \beta_1 & \gamma_1 & & & \\
 & & \ddots & & & \\
 & & & \alpha_{L-1} & \beta_{L-1} & \gamma_{L-1} \\
 & & & & & 1
\end{bmatrix}
\begin{bmatrix}
\mathbf{d}_0 \\
\mathbf{d}_1 \\
\vdots \\
\mathbf{d}_{L-1} \\
\mathbf{d}_L
\end{bmatrix}
=
\begin{bmatrix}
\mathbf{r}_0 \\
\mathbf{r}_1 \\
\vdots \\
\mathbf{r}_{L-1} \\
\mathbf{r}_L
\end{bmatrix}. \qquad (9.7)
$$

Here we set

$$
\begin{aligned}
\mathbf{r}_0 &= \mathbf{b}_1, \\
\mathbf{r}_i &= (\Delta_{i-1} + \Delta_i)\mathbf{x}_i, \\
\mathbf{r}_L &= \mathbf{b}_{3L-1}.
\end{aligned}
$$

The first and last polygon vertices do not cause much of a problem:

$$
\mathbf{d}_{-1} = \mathbf{x}_0, \qquad \mathbf{d}_{L+1} = \mathbf{x}_L.
$$

This linear system can be made *symmetric*: we can multiply each equation by a common factor. In particular, we may divide the ith equation through by $\Delta_{i-1}^2 \Delta_i^2$. Also, we would have to delete the first and last rows and columns from the system, and update the right-hand side accordingly. The resulting new matrix will now be symmetric; its entries will satisfy $\alpha_{i+1} = \gamma_i$.

The coefficient matrix will be *diagonally dominant* if $\Delta_{i-2} + \Delta_{i-1} > \Delta_i$, which is easily seen from the symmetric version of the linear system. If this condition holds for all i, the system has a unique solution. The fact that it *always* has a solution, as long as the u_i are increasing, is not that easily seen. It is a consequence of the Schoenberg-Whitney theorem, for which the reader is referred to Chapter XIII of de Boor's *Guide to Splines* [114].

Note that an affine parameter transformation does not affect the linear system. Therefore it would not matter if we rescaled our parameter values u_i.

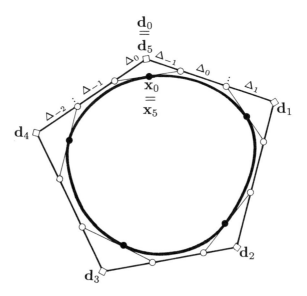

Figure 9.1: Closed curves: the interpolation problem becomes periodic.

In the special case of all Δ_i being equal, that is, for an equidistant parametrization, the system becomes even simpler:

$$
\begin{bmatrix}
1 & & & & & & \\
\frac{3}{2} & \frac{7}{2} & 1 & & & & \\
 & 1 & 4 & 1 & & & \\
 & & & \ddots & & & \\
 & & & 1 & 4 & 1 & \\
 & & & & 1 & \frac{7}{2} & \frac{3}{2} \\
 & & & & & & 1
\end{bmatrix}
\begin{bmatrix}
\mathbf{d}_0 \\
\mathbf{d}_1 \\
\mathbf{d}_2 \\
\vdots \\
\mathbf{d}_{L-2} \\
\mathbf{d}_{L-1} \\
\mathbf{d}_L
\end{bmatrix}
=
\begin{bmatrix}
\mathbf{b}_1 \\
6\mathbf{x}_1 \\
6\mathbf{x}_2 \\
\vdots \\
6\mathbf{x}_{L-2} \\
6\mathbf{x}_{L-1} \\
\mathbf{b}_{3L-1}
\end{bmatrix}.
\qquad (9.8)
$$

Frequently one must deal with *closed* curves; see Figure 9.1. The number of equations is reduced since the C^2 condition at $\mathbf{x}_0 = \mathbf{x}_L$ should not be listed twice in the linear system. It now takes the form:

$$
\begin{bmatrix}
\beta_0 & \gamma_0 & & & & & \alpha_0 \\
\alpha_1 & \beta_1 & \gamma_1 & & & & \\
& & \ddots & & & & \\
& & & \alpha_{L-2} & \beta_{L-2} & \gamma_{L-2} & \\
\gamma_{L-1} & & & & \alpha_{L-1} & \beta_{L-1}
\end{bmatrix}
\begin{bmatrix}
\mathbf{d}_0 \\
\mathbf{d}_1 \\
\vdots \\
\mathbf{d}_{L-2} \\
\mathbf{d}_{L-1}
\end{bmatrix}
=
\begin{bmatrix}
\mathbf{r}_0 \\
\mathbf{r}_1 \\
\vdots \\
\mathbf{r}_{L-2} \\
\mathbf{r}_{L-1}
\end{bmatrix}. \quad (9.9)
$$

Here, the right-hand sides are of the form

$$\mathbf{r}_i = (\Delta_{i-1} + \Delta_i)\mathbf{x}_i.$$

For these equations to make sense, we define a *periodic continuation* of the knot sequence:

$$\Delta_{-1} = \Delta_{L-1}, \Delta_{-2} = \Delta_{L-2}.$$

The matrix of this system is no longer tridiagonal; yet one does not have to resort to solving a full linear system. For details, see Ahlberg *et al.* [3], p. 15.

We conclude with a method for B-spline interpolation that occasionally appears in the literature (e.g., in Yamaguchi [475]). It is possible to solve the interpolation problem without setting up a linear system! Just do the following: construct an initial control polygon – by setting $\mathbf{d}_i = \mathbf{x}_i$, for example. This initial polygon will not define an interpolating curve. So, for i from 0 to L, correct $\mathbf{d}_i$ such that the corresponding curve passes through $\mathbf{x}_i$.[2] Repeat until the solution is found.

This method will always converge, and will not need many steps in order to do so. So why bother with linear systems? The reason is that tridiagonal systems are most effectively solved by a direct method, whereas the above iterative method amounts to solving the system via Gauss-Seidel iteration. So while geometrically appealing, the iterative method needs more computation time than the direct method.

9.2 The Hermite Form

An interpolatory C^2 piecewise cubic spline may also be written in piecewise cubic Hermite form. For $u \in (u_i, u_{i+1})$, the interpolant is of the form

$$\mathbf{x}(u) = \mathbf{x}_i H_0^3(r) + \mathbf{m}_i \Delta_i H_1^3(r) + \Delta_i \mathbf{m}_{i+1} H_2^3(r) + \mathbf{x}_{i+1} H_3^3(r), \quad (9.10)$$

---

[2]See Section 7.9 for details.

where the H_j^3 are cubic Hermite polynomials from Eq. (6.14) and $r = (u - u_i)/\Delta_i$ is the local parameter of the interval (u_i, u_{i+1}). In (9.10), the $\mathbf{x}_i$ are the known data points, while the $\mathbf{m}_i = \dot{\mathbf{x}}(u_i)$ are the unknown tangent vectors. The interpolant is supposed to be C^2; therefore,

$$\ddot{\mathbf{x}}_+(u_i) - \ddot{\mathbf{x}}_-(u_i) = \mathbf{0}. \tag{9.11}$$

We insert (9.10) into (9.11) and obtain

$$\Delta_i \mathbf{m}_{i-1} + 2(\Delta_{i-1} + \Delta_i)\mathbf{m}_i + \Delta_{i-1}\mathbf{m}_{i+1} = 3(\frac{\Delta_i \Delta \mathbf{x}_{i-1}}{\Delta_{i-1}} + \frac{\Delta_{i-1}\Delta \mathbf{x}_i}{\Delta_i}); \tag{9.12}$$
$$i = 1, \ldots, L - 1.$$

Together with two end conditions, (9.12) can be used to compute the unknown tangent vectors $\mathbf{m}_i$. Note that this formulation of the spline interpolation problem depends on the scale of the u_i; it is not invariant under affine parameter transformations. This is due to the use of the Hermite form.

The simplest end condition would be to prescribe $\mathbf{m}_0$ and $\mathbf{m}_L$, a method known as *clamped end condition*. In that case, the matrix of our linear system takes the form

$$\begin{bmatrix} 1 & & & & & \\ \alpha_1 & \beta_1 & \gamma_1 & & & \\ & & \ddots & & & \\ & & & \alpha_{L-1} & \beta_{L-1} & \gamma_{L-1} \\ & & & & & 1 \end{bmatrix} \begin{bmatrix} \mathbf{m}_0 \\ \mathbf{m}_1 \\ \vdots \\ \mathbf{m}_{L-1} \\ \mathbf{m}_L \end{bmatrix} = \begin{bmatrix} \mathbf{r}_0 \\ \mathbf{r}_1 \\ \vdots \\ \mathbf{r}_{L-1} \\ \mathbf{r}_L \end{bmatrix}, \tag{9.13}$$

where

$$\begin{aligned} \alpha_i &= \Delta_i, \\ \beta_i &= 2(\Delta_{i-1} + \Delta_i), \\ \gamma_i &= \Delta_{i-1} \end{aligned}$$

and

$$\begin{aligned} \mathbf{r}_0 &= \mathbf{m}_0, \\ \mathbf{r}_i &= 3(\frac{\Delta_i \Delta \mathbf{x}_{i-1}}{\Delta_{i-1}} + \frac{\Delta_{i-1}\Delta \mathbf{x}_i}{\Delta_i}); \quad i = 1, \ldots, L - 1, \\ \mathbf{r}_L &= \mathbf{m}_L. \end{aligned}$$

Having found the $\mathbf{m}_i$, we can easily retrieve the piecewise Bézier form of the curve according to (8.2) and (8.3).

When dealing with linear systems, it is a good idea to make sure that a solution exists and that it is unique. In our case, the coefficient matrix is *diagonally dominant*, which means that the absolute value of any diagonal element is larger than the sum of the absolute values of the remaining elements on the same row:

$$|\beta_i| > |\alpha_i| + |\gamma_i|.$$

Such matrices are always invertible; moreover, they allow Gauss elimination without pivoting (see any advanced text on numerical analysis or the fundamental spline text by Ahlberg *et al.* [3]). Thus the spline interpolation problem always possesses a unique solution (after the prescription of two consistent end conditions).

Since the coefficient matrix is *tridiagonal* (only the diagonal element and its two neighbors are nonzero), we do not have to solve a full $(L + 1) \times (L + 1)$ linear system. One forward substitution sweep and one for backward substitution is sufficient, as implemented in the programs `lu_system` and `solve_system` below. All our remarks about the linear system hold for the B-spline form as well.

9.3 End Conditions

We may have the interpolation routine select the end tangents $\mathbf{m}_0$ and $\mathbf{m}_L$ automatically instead of prescribing it ourselves. One such selection is called the *Bessel end condition*. Here, the end tangent vector $\mathbf{m}_0$ is set equal to the tangent vector at $\mathbf{x}_0$ of the interpolating parabola through the first three data points. Similarly, $\mathbf{m}_L$ is set equal to the tangent vector at $\mathbf{x}_L$ of the interpolating parabola through the last three data points. Now the right-hand side changes to

$$\mathbf{r}_0 = -\frac{2(2\Delta_0 + \Delta_1)}{\Delta_0\beta_1}\mathbf{x}_0 + \frac{\beta_1}{2\Delta_0\Delta_1}\mathbf{x}_1 - \frac{2\Delta_0}{\Delta_1\beta_1}\mathbf{x}_2 \qquad (9.14)$$

and

$$\mathbf{r}_L = \frac{2\Delta_{L-1}}{\Delta_{L-2}\beta_{L-1}}\mathbf{x}_{L-2} - \frac{\beta_{L-1}}{2\Delta_{L-2}\Delta_{L-1}}\mathbf{x}_{L-1} + \frac{2(2\Delta_{L-1} + \Delta_{L-2})}{\beta_{L-1}\Delta_{L-1}}\mathbf{x}_L. \qquad (9.15)$$

Of course, this condition may also be formulated in terms of the B-spline representation. This is done in the routine `bessel_ends` described later.

Another possibility is the *quadratic end condition*, which sets $\ddot{\mathbf{x}}(u_0) = \ddot{\mathbf{x}}(u_1)$ and $\ddot{\mathbf{x}}(u_{L-1}) = \ddot{\mathbf{x}}(u_L)$. Now the linear system changes to

$$
\begin{bmatrix}
1 & 1 & & & & \\
\alpha_1 & \beta_1 & \gamma_1 & & & \\
& & \ddots & & & \\
& & & \alpha_{L-1} & \beta_{L-1} & \gamma_{L-1} \\
& & & & 1 & 1
\end{bmatrix}
\begin{bmatrix}
\mathbf{m}_0 \\
\mathbf{m}_1 \\
\vdots \\
\mathbf{m}_{L-1} \\
\mathbf{m}_L
\end{bmatrix}
=
\begin{bmatrix}
\mathbf{r}_0 \\
\mathbf{r}_1 \\
\vdots \\
\mathbf{r}_{L-1} \\
\mathbf{r}_L
\end{bmatrix}
\tag{9.16}
$$

and

$$
\mathbf{r}_0 = \frac{2}{\Delta_0}\Delta\mathbf{x}_0, \quad \mathbf{r}_L = \frac{2}{\Delta_{L-1}}\Delta\mathbf{x}_{L-1}.
$$

A slightly more complicated end condition is provided by the *not-a-knot condition*. Using it, we force the first two and the last two polynomial segments to merge into *one* cubic piece. This means that the third derivative of $\mathbf{x}(u)$ is continuous at u_1. Writing down the conditions leads to a nontridiagonal system, which can, however, be transformed into a tridiagonal one. Its first equation is

$$
\Delta_1\beta_1\mathbf{m}_0 + \beta_1^2\mathbf{m}_1
$$
$$
= \frac{(\Delta_0)^2}{\Delta_1}\Delta\mathbf{x}_1 + \frac{\Delta_1}{\Delta_0}(3\Delta_0 + 2\Delta_1)\Delta\mathbf{x}_0;
\tag{9.17}
$$

the last one is

$$
\beta_{L-1}^2\mathbf{m}_{L-1} + \Delta_{L-2}\beta_{L-1}\mathbf{m}_L
$$
$$
= \frac{(\Delta_{L-1})^2}{\Delta_{L-2}}\Delta\mathbf{x}_{L-2} + \frac{\Delta_{L-2}}{\Delta_{L-1}}(3\Delta_{L-1} + 2\Delta_{L-2})\Delta\mathbf{x}_{L-1}.
\tag{9.18}
$$

Finally, we mention an end condition that bears the name "natural." The term stems from the fact that this condition arises "naturally" in the context of the minimum property for spline curves, as described below. The natural end condition is defined by $\ddot{\mathbf{x}}(u_0) = \ddot{\mathbf{x}}(u_L) = \mathbf{0}$. The linear system becomes

$$
\begin{bmatrix}
2 & 1 & & & & \\
\alpha_1 & \beta_1 & \gamma_1 & & & \\
& & \ddots & & & \\
& & & \alpha_{L-1} & \beta_{L-1} & \gamma_{L-1} \\
& & & & 1 & 2
\end{bmatrix}
\begin{bmatrix}
\mathbf{m}_0 \\
\mathbf{m}_1 \\
\vdots \\
\mathbf{m}_{L-1} \\
\mathbf{m}_L
\end{bmatrix}
=
\begin{bmatrix}
\mathbf{r}_0 \\
\mathbf{r}_1 \\
\vdots \\
\mathbf{r}_{L-1} \\
\mathbf{r}_L,
\end{bmatrix}
\tag{9.19}
$$

Figure 9.2: Exact clamped end condition spline.

and

$$\mathbf{r}_0 = \frac{3}{\Delta_0}\Delta\mathbf{x}_0, \quad \mathbf{r}_L = \frac{3}{\Delta_{L-1}}\Delta\mathbf{x}_{L-1}.$$

This end condition forces the curve to behave like a straight line near the endpoints; usually, this results in a poor shape of the spline curve.

The spline system becomes especially simple if the knots u_i are uniformly spaced; for example, the clamped end condition system becomes

$$
\begin{bmatrix}
1 & & & & \\
1 & 4 & 1 & & \\
& & \ddots & & \\
& & 1 & 4 & 1 \\
& & & 1 &
\end{bmatrix}
\begin{bmatrix}
\mathbf{m}_0 \\
\mathbf{m}_1 \\
\vdots \\
\mathbf{m}_{L-1} \\
\mathbf{m}_L
\end{bmatrix}
=
\begin{bmatrix}
\mathbf{r}_0 \\
\mathbf{r}_1 \\
\vdots \\
\mathbf{r}_{L-1} \\
\mathbf{r}_L
\end{bmatrix},
\qquad (9.20)
$$

where

$$
\begin{aligned}
\mathbf{r}_0 &= \mathbf{m}_0, \\
\mathbf{r}_i &= 3(\mathbf{x}_{i+1} - \mathbf{x}_{i-1}); \quad i = 1,\ldots,L-1, \\
\mathbf{r}_L &= \mathbf{m}_L.
\end{aligned}
$$

We finish this section with a few examples, using uniform parameter values in all examples.[3] Figure 9.2 shows equally spaced data points read off from a

---

[3]Due to the symmetry inherent in the data points, all parametrizations discussed later yield the same knot spacing.

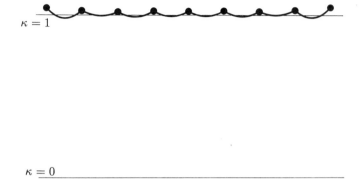

Figure 9.3: Curvature plot of exact clamped end condition spline.

Figure 9.4: Bessel end condition spline.

Figure 9.5: Curvature plot of Bessel end condition spline.

Figure 9.6: Natural end condition spline.

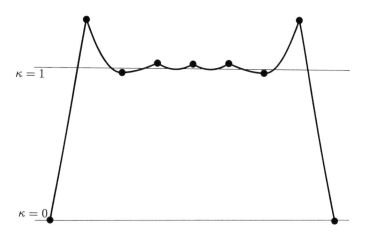

Figure 9.7: Curvature plot of natural end condition spline.

circle and the cubic spline interpolant obtained with clamped end conditions, using the exact end derivatives of the circle (the figure is scaled down in the y-direction). Figure 9.3 shows the curvature plot[4] of the spline curve. Ideally, the curvature should be constant, and the spline curvature is quite close to this ideal.

Figure 9.4 shows the same data, but now using Bessel end conditions. Near the endpoints, the curvature deviates from the ideal value, as shown in Figure 9.5.

Finally, Figure 9.6 shows the curve that is obtained using natural end conditions. The end curvatures are forced to be zero, causing considerable deviation from the ideal value, as shown in Figure 9.7.

9.4   Parametrization

The spline interpolation problem is usually stated as "given data points x_i and parameter values u_i, ..." Of course, this is the mathematician's way of describing a problem. In practice, parameter values are rarely given and therefore must be made up somehow. The easiest way to determine the u_i is simply to set $u_i = i$. This is called *uniform* or *equidistant* parametrization.

---

[4]The graph of curvature versus arc length; see also Chapter 23.

This method is too simplistic to cope with most practical situations. The reason for the overall poor[5] performance of the uniform parametrization can be blamed on the fact that it "ignores" the geometry of the data points.

The following is a heuristic explanation of this fact. We can interpret the parameter u of the curve as time. As time passes from time u_0 to time u_L, the point $\mathbf{x}(u)$ traces the curve from point $\mathbf{x}(u_0)$ to point $\mathbf{x}(u_L)$. With uniform parametrization, $\mathbf{x}(u)$ spends the same amount of time between any two adjacent data points, irrespective of their relative distances. A good analogy is a car driving along the interpolating curve. We have to spend the same amount of time between any two data points. If the distance between two data points is large, we must move with a high speed. If the next two data points are close to each other, we will overshoot since we cannot abruptly change our speed – we are moving with continuous speed and acceleration, which are the physical counterparts of a C^2 parametrization of a curve. It would clearly be more reasonable to adjust speed to the distribution of the data points.

One way of achieving this is to have the knot spacing proportional to the distances of the data points:

$$\frac{\Delta_i}{\Delta_{i+1}} = \frac{||\Delta\mathbf{x}_i||}{||\Delta\mathbf{x}_{i+1}||}. \tag{9.21}$$

A parametrization of the form of Eq. (9.21) is called *chord length parametrization*. Equation (9.21) does not uniquely define a knot sequence; rather, it defines a whole family of parametrizations that are related to each other by affine parameter transformations. In practice, the choices $u_0 = 0$ and $u_L = 1$ or $u_0 = 0$ and $u_L = L$ are reasonable options.

Chord length usually produces better results than uniform knot spacing, although not in all cases. It has been proven (Epstein [149]) that chord length parametrization (in connection with natural end conditions) cannot produce curves with corners[6] at the data points, which gives it some theoretical advantage over the uniform choice.

---

[5]There are cases in which uniform parametrization fares better than other methods. An interesting example is in Foley [189], p. 86.

[6]A corner is a point on a curve where the tangent (not necessarily the tangent vector!) changes in a discontinuous way. The special case of a change in 180 degrees is called a *cusp*; it may occur even with chord length parametrization.

Another parametrization has been named "centripetal" by E. Lee [304]. It is derived from the physical heuristics presented above. If we set

$$\frac{\Delta_i}{\Delta_{i+1}} = \left[\frac{||\Delta\mathbf{x}_i||}{||\Delta\mathbf{x}_{i+1}||}\right]^{1/2}, \tag{9.22}$$

the resulting motion of a point on the curve will "smooth out" variations in the centripetal force acting on it.

Yet another parametrization was developed by G. Nielson and T. Foley [352]. It sets

$$\Delta_i = d_i\left[1 + \frac{3}{2}\frac{\hat{\Theta}_i d_{i-1}}{d_{i-1} + d_i} + \frac{3}{2}\frac{\hat{\Theta}_{i+1} d_{i+1}}{d_i + d_{i+1}}\right], \tag{9.23}$$

where $d_i = ||\Delta x_i||$ and

$$\hat{\Theta}_i = \min(\pi - \Theta_i, \frac{\pi}{2}),$$

and Θ_i is the angle formed by $\mathbf{x}_{i-1}, \mathbf{x}_i, \mathbf{x}_{i+1}$. Thus $\hat{\Theta}_i$ is the "adjusted" exterior angle formed by the vectors $\Delta\mathbf{x}_i$ and $\Delta\mathbf{x}_{i-1}$. As the exterior angle $\hat{\Theta}_i$ increases, the interval Δ_i increases from the minimum of its chord length value up to a maximum of four times its chord length value. This method was created to cope with "wild" data sets.

We note one property that distinguishes the uniform parametrization from its competitors: it is the only one that is invariant under affine transformations of the data points. Chord length, centripetal, and the Foley methods all involve length measurements, and lengths are not preserved under affine maps. One solution to this dilemma is the introduction of a modified length measure, as described in Nielson [350].[7]

For more literature on parametrizations, see McConalogue [330], Hartley and Judd [253], [254] and Foley [189].

Figures 9.8 to 9.15[8] show the performance of the discussed parametrization methods for one sample data set. For each method, the interpolant is shown together with its curvature plot. For all methods, Bessel end conditions were chosen.

While the figures are self-explanatory, some comments are in place. Note the very uneven spacing of the data points at the marked area of the curves. Of all methods, Foley's copes best with that situation (although we add that

---

[7]The Foley parametrization was in fact first formulated in terms of that modified length measure.

[8]Kindly provided by T. Foley.

Figure 9.8: Chord length spline.

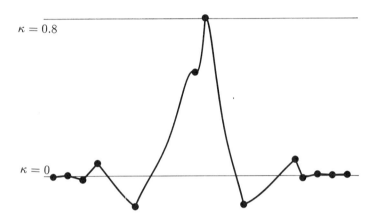

Figure 9.9: Curvature plot of chord length spline.

many examples exist where the simpler centripetal method wins out). The uniform spline curve seems to have no problems there, if one just inspects the plot of the curve itself. However, the curvature plot reveals a cusp in that region! The huge curvature at the cusp causes a scaling in the curvature plot that annihilates all other information. Also note how the chord length parametrization yields the "roundest" curve, having the smallest curvature values, but exhibiting the most marked inflection points.

There is probably no "best" parametrization, since any method can be defeated by a suitably chosen data set. The following is a (personal) recommendation. You may improve the shape of the curve, at an increase of computation time, by the following hierarchy of methods: uniform, chord length,

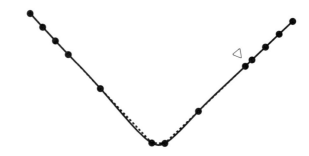

Figure 9.10: Foley spline.

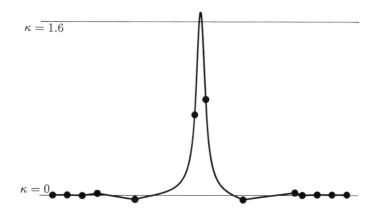

Figure 9.11: Curvature plot of Foley spline.

centripetal, Foley. The best compromise between cost and result is probably achieved by the centripetal method.

9.5   The Minimum Property

In the early days of design, say ship design in the 1800s, the problem had to be handled of how to draw (manually) a smooth curve through a given set of points. One way to obtain a solution was the following: place metal weights (called "ducks") at the data points, and then pass a thin, elastic wooden beam (called a "spline") between the ducks. The resulting curve is always

Figure 9.12: Centripetal spline.

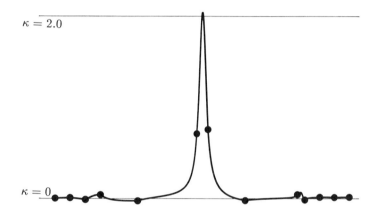

Figure 9.13: Curvature plot of centripetal spline.

Figure 9.14: Uniform spline.

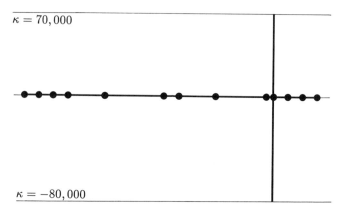

Figure 9.15: Curvature plot of uniform spline.

Figure 9.16: Spline interpolation: A plastic beam, the "spline," is forced to pass through data points, marked by metal weights, the "ducks."

very smooth and usually aesthetically pleasing. The same principle is used today when an appropriate design program is not available or for manual verification of a computer result; see Figure 9.16.

The plastic or wooden beam assumes a position that minimizes its strain energy. The mathematical model of the beam is a curve $\mathbf{s}$, and its strain energy E is given by

$$E = \int (\kappa(s))^2 \mathrm{d}s,$$

where κ denotes the curvature of the curve. The curvature of most curves involves integrals and square roots and is cumbersome to handle; therefore, one often approximates the above integral by a simpler one:

$$\hat{\mathbf{E}} = \int \left[\frac{\mathrm{d}^2}{\mathrm{d}u^2} \mathbf{s}(u) \right]^2 \mathrm{d}u. \qquad (9.24)$$

Note that $\hat{\mathbf{E}}$ is a vector; it is obtained by performing the integration on each component of $\mathbf{s}$.

The C^2 cubic interpolatory splines have the following advantage: among all C^2 functions that interpolate the given data points at the given parameter values and satisfy the same end conditions, the cubic spline yields the smallest value for each component of $\hat{\mathbf{E}}$. Let $\mathbf{s}(u)$ be the C^2 cubic spline and let $\mathbf{y}(u)$ be another C^2 interpolating curve. We can write $\mathbf{y}$ as

$$\mathbf{y}(u) = \mathbf{s}(u) + [\mathbf{y}(u) - \mathbf{s}(u)].$$

The above integrals are defined componentwise; we will show the minimum property for one component only. Let $s(u)$ and $y(u)$ be the first component of $\mathbf{s}$ and $\mathbf{y}$, respectively. The "energy integral" $\hat{E}$ of $\mathbf{y}$'s first component becomes

$$\hat{E} = \int_{u_0}^{u_L} (\ddot{s})^2 du + 2 \int_{u_0}^{u_L} \ddot{s}(\ddot{y} - \ddot{s}) du + \int_{u_0}^{u_L} (\ddot{y} - \ddot{s})^2 du.$$

We may integrate the middle term by parts:

$$\int_{u_0}^{u_L} \ddot{s}(\ddot{y} - \ddot{s}) du = \ddot{s}(\dot{y} - \dot{s})\Big|_{u_0}^{u_L} - \int_{u_0}^{u_L} \dddot{s}(\dot{y} - \dot{s}) du.$$

The first expression vanishes because of the common end conditions. In the second expression, $\dddot{s}$ is piecewise constant:

$$\int_{u_0}^{u_L} \dddot{s}(\dot{y} - \dot{s}) du = \sum_{j=0}^{L-1} \dddot{s}_j (y - s)\Big|_{u_j}^{u_{j+1}}.$$

All terms in the sum vanish because both $\mathbf{s}$ and $\mathbf{y}$ interpolate. Since

$$\int_{u_0}^{u_L} (\ddot{y} - \ddot{s})^2 du > 0$$

for continuous $\ddot{y} \neq \ddot{s}$,

$$\int_{u_0}^{u_L} (\ddot{y})^2 du \geq \int_{u_0}^{u_L} (\ddot{s})^2 du, \tag{9.25}$$

we have proved the claimed minimum property.

While the minimum property of splines is important – it has spurred substantial research activity – one should not overlook that it is artificial.

The replacement of the actual strain energy measure E by $\hat{\mathbf{E}}$ is motivated by the desire for mathematical simplicity. The curvature of a curve is given by

$$\kappa(u) = \frac{||\dot{\mathbf{x}} \wedge \ddot{\mathbf{x}}||}{||\dot{\mathbf{x}}||^3}.$$

If $||\dot{\mathbf{x}}|| \approx 1$, then $||\ddot{\mathbf{x}}||$ is a good approximation to κ. This means, however, that the curve must be parametrized according to arc length; see Eq. (11.7). This assumption is not very realistic for cubic splines; see Problems.

9.6   Implementation

The following routines produce the cubic B-spline polygon of an interpolating C^2 cubic spline curve. First, we set up the tridiagonal linear system:

```
void set_up_system(knot,l,alpha,beta,gamma)
/*     given the knot sequence, the linear system for clamped end
       condition B-spline interpolation is set up.
Input: knot:                 knot sequence (all knots are simple; but,
                             in the terminology of Chapter 10, knot[0]
                             and knot[1] are of multiplicity three.)
       points:               points to be interpolated
       l:                    number of intervals
Output:alpha, beta,gamma:    1-D arrays that constitute
                             the elements of the interpolation matrix.
Note: no data points needed so far!
*/
```

The next routine performs the LU decomposition of the matrix from the previous routine. (Note that we do not generate a full matrix but rather three linear arrays!)

```
void l_u_system(alpha,beta,gamma,l,up,low)
/*     perform LU decomposition of tridiagonal system with
       lower diagonal alpha, diagonal beta, upper diagonal gamma.

Input: alpha,beta,gamma: the coefficient matrix entries
       l:                matrix size [0,l]x[0,l]
Output:low:              L-matrix entries
       up:               U-matrix entries
*/
```

Finally, the routine that solves the system for the B-spline coefficients d_i:

```
solve_system(up,low,gamma,l,rhs,d)
/*   solve   tridiagonal linear system
     of size (l+1)(l+1) whose LU decomposition has entries up and low,
     and whose right hand side is rhs, and whose original matrix
     had gamma as  its upper diagonal. Solution is d[0],...,d[l+2].
Input: up,low,gamma:  as above.
       l:             size of system: l+1 eqs in l+1 unknowns.
       rhs:           right hand side, i.e, data points with end
                      'tangent Bezier points' in rhs[1] and rhs[l+1].
Output:d:             solution vector.
Note shift in indexing from text! Both rhs and d are from 0 to l+2.
*/
```

In case Bessel ends are desired instead of clamped ends, this is the code:

```
void bessel_ends(data,knot,l)
/*     Computes B-spline points data[1] and data[l+1]
       according to bessel end condition.

Input: data:  sequence of data coordinates data[0] to data[l+2].
              Note that data[1] and data[l+1] are expected to
              be empty, as they will be filled by this routine.
              They correspond to the Bezier points bez[1] and bez[31-1].
       knot:  knot sequence
       l:     number of intervals
Output: data: now including ''tangent Bezier points'' data[1], data[l+1].
*/
```

The centripetal parametrization is achieved by the following routine:

```
void parameters(data_x,data_y,l,knot)
/*   Finds a centripetal parametrization for a given set
     of 2D data points.
Input:      data_x, data_y:  input points, numbered from 0 to l+2.
       l:                    number of intervals.
Output:     knot:            knot sequence. Note: not (knot[1]=1.0)!
Note:       data_x[1], data_x[l+1] are not used! Same for data_y.
*/
```

A calling sequence that utilizes the above programs might look like this:

```
parameters(data_x,data_y,l,knot);

set_up_system(knot,l,alpha,beta,gamma);

l_u_system(alpha,beta,gamma,l,up,low);

bessel_ends(data_x,knot,l);
bessel_ends(data_y,knot,l);

solve_system(up,low,gamma,l,data_x,bspl_x);
solve_system(up,low,gamma,l,data_y,bspl_y);
```

Here, we solved the 2D interpolation problem with given data points in data_x, data_y, a knot sequence knot, and the resulting B-spline polygon in bspl_x, bspl_y. This calling sequence is realized in the routine c2_spline.c.

9.7 Problems

1. Show that interpolating splines reproduce straight lines – that they have *linear precision* (provided the end conditions are clamped and the tangents are read off the straight line).

2. Show that they also have quadratic and cubic precision.

3. Program the following: instead of prescribing end conditions at both ends, prescribe first and second derivatives at u_0. The interpolant can then be built segment by segment. Discuss the numerical aspects of this method.

4. Any curve may be reparametrized in terms of its arc length s. Show that a polynomial curve of degree $n > 1$ cannot be polynomial in terms of its arc length s. (See Chapter 11 for the arc length parametrization.)

5. What happens if you replace the knot sequence $u_0, u_1, \ldots, u_L$ by the reversed knot sequence $u_L, u_{L-1}, \ldots, u_0$ but leave the data points in their original ordering? Explain.

6. In the programs for spline interpolation, replace the end conditions by the requirement that a user-specified spline segment be linear. There should be no restrictions as to where this segment is located.

7. Interpolate data points from a semicircle and compare your results with those from Problems 5 and 6 in Section 8.6.

6. In the programs for spline interpolation, replace the end conditions by the requirement that a user-specified spline segment be linear. There should be no restrictions as to where this segment is located.

7. Interpolate data points from a semicircle and compare your results with those from Problems 5 and 6 in Section 8.6.

Chapter 10

B-splines

B-splines were investigated as early as the nineteenth century by N. Lobachevsky (see Renyi [392], p. 165); they were constructed as convolutions of certain probability distributions.[1] In 1946, I. J. Schoenberg [420] used B-splines for statistical data smoothing, and his paper started the modern theory of spline approximation. For the purposes of this book, the discovery of the recurrence relations for B-splines by C. de Boor [113], M. Cox [106], and L. Mansfield was one of the most important developments in this theory. The recurrence relations were first used by Gordon and Riesenfeld [229] in the context of parametric B-spline curves.

This chapter presents a theory for arbitrary degree B-spline curves. The original development of these curves makes use of divided differences and is mathematically involved (see de Boor [114]). A different approach to B-splines was taken by de Boor and Hollig [119]; they used the recurrence relations for B-splines as the starting point for the theory. In this chapter, the theory of B-splines is based on an even more fundamental concept: the Boehm knot insertion algorithm [61]. Another interesting new approach to B-splines is the *blossoming* method proposed by L. Ramshaw [386] and, in a different form, by P. de Casteljau [123], which we will discuss at the end of this chapter.

Warning: *subscripts in this chapter differ from those in Chapter 7!* For the cubic and quadratic cases special subscripts are useful, but the general theory is easier to explain with the notation used here.[2]

---

[1]However, those were only defined over a very special knot sequence.

[2]In terms of this chapter, we used end knots of multiplicity two (quadratic case) or three (cubic case) in Chapter 7. The coefficients there started with the subscript $i = -1$; here, they will start with $i = 0$.

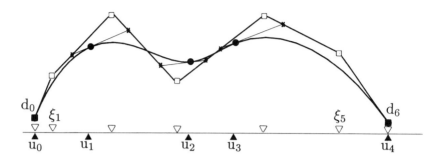

Figure 10.1: B-splines: a nonparametric C^2 cubic spline curve with its B-spline polygon. (In the context of this chapter, the end knots are of multiplicity three.)

10.1   Motivation

Figure 10.1 shows a C^2 cubic spline (nonparametric) with its B-spline polygon. The relationship between the polygon and the curve was discussed in Section 7.7. In that section, we were interested in the parametric case, whereas now we will restrict ourselves to nonparametric (functional) curves of the form $y = f(u)$. The reason is that much of the B-spline theory is explained more naturally in this setting.

In Section 5.5, we considered nonparametric Bézier curves. Recall that over the interval $[u_i, u_{i+1}]$, the abscissas of the Bézier points are $u_i + j\Delta u_i/n$; $j = 0, \dots, n$. Two cubic Bézier functions that are defined over $[u_{i-1}, u_i]$ and $[u_i, u_{i+1}]$ are C^2 at u_i if an auxiliary point $\mathbf{d}_i$ can be constructed from both curve segments as discussed in Section 7.4. Some of the points $\mathbf{d}_i$ are shown in Figure 10.1. Section 7.4 tells us how to compute the y-values of these points. Using the same reasoning for the u-coordinate $d_i^{(u)}$ (see Problems), we find the abscissa values ξ_i for $\mathbf{d}_i$ to be

$$\xi_i = d_i^{(u)} = \frac{1}{3}(u_{i-1} + u_i + u_{i+1}); \quad i = 1, 2, 3.^3 \tag{10.1}$$

We can now give an algorithm for the "design" of a cubic B-spline function:

1. Given knots u_i.

---

[3]This notation is in harmony with the cubic case of Section 7.4; we will change notation for the general case soon!

2. Find abscissas $\xi_i = \frac{1}{3}(u_{i-1} + u_i + u_{i+1})$.

3. Define real numbers d_i to obtain a polygon with vertices (ξ_i, d_i).

4. Evaluate this polygon (= piecewise linear function) at the abscissas of the inner Bézier points. This produces a refined polygon, consisting of the inner Bézier points.

5. Evaluate the refined polygon at the knots u_i, the abscissas for the junction Bézier points. We now have the junction Bézier points.

After step 5, we have generated a C^2 piecewise cubic Bézier function. In a similar manner, we could generate a C^1 piecewise quadratic Bézier function. In this chapter, we will aim for a generalization of the above definition of piecewise polynomials to include arbitrary degrees and arbitrary differentiability classes.

10.2 Knot Insertion

We will now define an algorithm to "refine" a piecewise linear function. Later, this piecewise linear function will be interpreted as a B-spline polygon, but at this point, we discuss only an algorithm that produces one piecewise linear function from another.

Suppose that we are given a number n (later the degree of the B-spline curve) and a number L (later related to the number of polynomial segments of the B-spline curve). Suppose also that we are given a nondecreasing *knot sequence*

$$u_0, \ldots, u_{L+2n-2}.$$

Not all of the u_i have to be distinct. If $u_i = \cdots = u_{i+r-1}$, i.e., if r successive knots coincide, we say that u_i has *multiplicity* r. If a knot does not coincide with any other knot, we say that it is *simple*, or that it has multiplicity one.

When we define B-spline curves later, we will use only the interval $[u_{n-1}, \ldots, u_{L+n-1}]$ as their domain. These knots are the "domain knots." We call L the potential number of curve segments – if all domain knots are simple, L denotes the number of domain intervals. For each domain knot multiplicity, the number of domain intervals drops by one. The sum of all domain knot multiplicities is related to L by

$$\sum_{i=n-1}^{L+n-1} r_i = L+1,$$

where r_i is the multiplicity of the domain knot u_i.

We shall illustrate the interplay between knot multiplicities and the number of domain intervals by means of the following examples:

Let $n = 3$, $L = 3$, and

$$\{u_0, ..., u_7\} = \{0, 1, 2, 3, 4, 5, 6, 7\}.$$

All knots are simple, so the number of domain intervals is three.

Leaving n and L unchanged, consider

$$\{u_0, ..., u_7\} = \{0, 1, 2, 3, 3, 5, 6, 7\}.$$

In this knot sequence, $u_3 = u_4$, thus $r_3 = 2$, and we only have two domain intervals.

The multiplicities of the nondomain knots do not affect the number of domain intervals. If we set

$$\{u_0, ..., u_7\} = \{0, 0, 0, 3, 4, 7, 7, 7\},$$

then u_0 and u_5 both have multiplicity three; but they are listed only once each in the domain knot sequence, which is

$$\{u_2, u_3, u_4, u_5\} = \{0, 3, 4, 7\}.$$

Thus we have three domain intervals.

We now define $n + L$ *Greville abscissas* ξ_i by

$$\xi_i = \frac{1}{n}(u_i + \ldots + u_{i+n-1}); \quad i = 0, \ldots, L+n-1. \qquad (10.2)$$

The Greville abscissas are averages of the knots. The number of Grevill abscissas equals the number of successive n-tuples of knots in the knot sequence.

We next assume that we are given ordinates d_i, also called *de Boor ordinates*, over the Greville abscissas and hence a polygon P consisting of the points (ξ_i, d_i); $i = 0, \ldots, L+n-1$. This polygon is a piecewise linear function with breakpoints at the Greville abscissas. Figure 10.2 shows some examples.

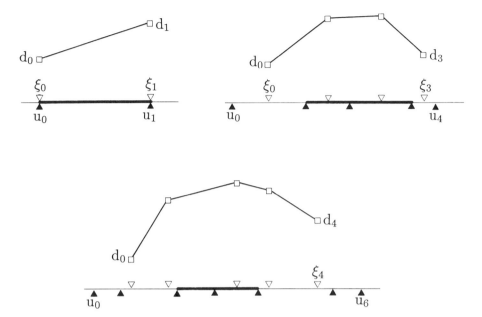

Figure 10.2: Greville abscissas: for various knot sequences and degrees, the corresponding Greville abscissas together with the polygon P are shown. Top left: $n = 1, L = 1$; top right, $n = 2, L = 2$; bottom: $n = 3, L = 2$.

Note how the polygon is not defined over the original knot sequence, but rather over the Greville abscissas.

We will now define our basic polygon manipulation technique, the knot insertion algorithm. As before, at this point we are only concerned with polygons, not with B-spline curves! Suppose a real number $u \in [u_{n-1}, \ldots, u_{L+n-1}]$ is given and we want to *insert* it into the knot sequence. We call the new knot sequence a *refined* knot sequence. It defines a new set of Greville abscissas, called ξ_i^u. Each ξ_i^u will be the abscissa for a vertex (ξ_i^u, d_i^u) of the new polygon P^u. The knot insertion algorithm is:

Knot insertion, informal: compute the Greville abscissas ξ_i^u for the refined knot sequence. Evaluate P there to obtain new ordinates $d_i^u = P(\xi_i^u)$. The *refined polygon* P^u is then formed by the points (ξ_i^u, d_i^u). The d_i^u are given by Eq. (10.4).

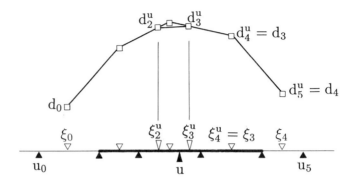

Figure 10.3: Knot insertion: the new knot is u; the new Greville abscissas are marked by larger icons. Old knot sequence: $u_0, u_1, u_2, u_3, u_4, u_5$. New sequence: $u_0, u_1, u_2, u, u_3, u_4, u_5$. In this example, $n = 2, L = 3$.

Figure 10.3 shows an example of the knot insertion procedure for the quadratic case. It is possible to insert the knot u again – it will then become a *double knot*, or a knot of multiplicity two, which simply means it is listed twice in the knot sequence. As another example of knot insertion, Figure 10.4 shows how the knot u is inserted again.

We shall formalize the knot insertion algorithm soon, but we can already deduce some properties:

- The polygon P^u is obtained from P by *piecewise linear interpolation* (see Section 2.4).

- As a consequence, knot insertion is a *variation diminishing* process: no straight line intersects P^u more often than P.

- As a further consequence, knot insertion is *convexity preserving*: if P is convex, so is P^u.

- Knot insertion is a *local* process: P differs from P^u only in the vicinity of u. (The exact definition of vicinity being a function of the degree n.)

We are now ready for an algorithmic definition of knot insertion. It is mostly intended for use in coding. The informal description above conveys the same information.

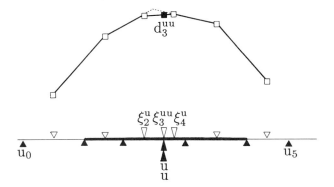

Figure 10.4: Knot insertion: the knot u is inserted again. Old knot sequence: $u_0, u_1, u_2, u, u_3, u_4, u_5$. New sequence: $u_0, u_1, u_2, u, u, u_3, u_4, u_5$.

Knot insertion algorithm:

Given: $u \in [u_{n-1}, \ldots, u_{L+n-1}]$.

Find: refined polygon P^u, defined over the refined knot sequence that includes u.

1. Find the largest I with $u_I \leq u < u_{I+1}$. If $u = u_I$ and u_I is of multiplicity n : **stop**. Else:

2. For $i = 0, \ldots, I - n + 1$, set $\xi_i^u = \xi_i$.

3. For $i = I - n + 2, \quad , I + 1$, set

$$\xi_i^u = \frac{1}{n}(u_i + \ldots + u_{i+n-2}) + \frac{1}{n}u.$$

4. For $i = I + 2, \ldots, L + n$, set $\xi_i^u = \xi_{i-1}$.

5. For $i = 0, \ldots, L + n$, set $d_i^u = P(\xi_i^u)$.

6. Renumber the knot sequence to include u as u_{I+1}.

7. Replace L by $L + 1$.

Step 5 only involves actual computation for $i = I - n + 2, \ldots, I + 1$. A formula for $P(\xi_i^u)$ is provided by (10.4). Before we can proceed further, we must verify that we have defined a *consistent* knot insertion process. If we insert two knots, the final result must be independent of the order in which the knots were inserted. More precisely, let u and v be knots to be inserted. Inserting u yields P^u; then, inserting v yields P^{uv}. Inserting v first and u second yields a polygon P^{vu}. We need to establish that $P^{uv} = P^{vu}$. The following outlines the main steps of the proof of that statement.

Let u be in the interval $[u_I, u_{I+1}]$. First, one establishes that $\xi_{i-1} \leq \xi_i^u \leq \xi_i$. Thus d_i^u is obtained by linear interpolation

$$d_i^u = \frac{\xi_i - \xi_i^u}{\xi_i - \xi_{i-1}} d_{i-1} + \frac{\xi_i^u - \xi_{i-1}}{\xi_i - \xi_{i-1}} d_i; \quad i = I - n + 2, \ldots, I + 1. \tag{10.3}$$

We invoke (10.2):

$$d_i^u = \frac{\sum_{j=i}^{i+n-1} u_j - \sum_{j=i}^{i+n-2} u_j - u}{u_{i+n-1} - u_{i-1}} d_{i-1} + \frac{\sum_{j=i}^{i+n-2} u_j + u - \sum_{j=i-1}^{i+n-2} u_j}{u_{i+n-1} - u_{i-1}} d_i$$

and simplify:

$$d_i^u = \frac{u_{i+n-1} - u}{u_{i+n-1} - u_{i-1}} d_{i-1} + \frac{u - u_{i-1}}{u_{i+n-1} - u_{i-1}} d_i; \quad i = I - n + 2, \ldots, I + 1. \tag{10.4}$$

As for the second knot v, assume that it is in the interval $[u_J, u_{J+1}]$. We obtain for the d_i^v:

$$d_i^v = \frac{u_{i+n-1} - v}{u_{i+n-1} - u_{i-1}} d_{i-1} + \frac{v - u_{i-1}}{u_{i+n-1} - u_{i-1}} d_i; \quad i = J - n + 2, \ldots, J + 1.$$

If u and v are far enough apart – more precisely if the two intervals $[u_{I-n+2}, u_{I+1}]$ and $[u_{J-n+2}, u_{J+1}]$ are disjoint – the two insertion processes do not interfere with each other and there is nothing to prove.

Otherwise, we compute

$$d_{i+1}^{uv} = \frac{u_{i+n-1} - v}{u_{i+n-1} - u_i} d_i^u + \frac{v - u_i}{u_{i+n-1} - u_i} d_{i+1}^u$$

and

$$d_{i+1}^{vu} = \frac{u_{i+n-1} - u}{u_{i+n-1} - u_i} d_i^v + \frac{u - u_i}{u_{i+n-1} - u_i} d_{i+1}^v.$$

We must show that these two expressions are equal. The proof is an exercise in algebra and is omitted here. For the more geometrically inclined, the use of Menelaos' theorem from Section 2.5 will also produce a proof.

In summary, we have shown that the knot insertion algorithm is *order independent*. This fact will be needed throughout our development of B-splines.

10.3   The de Boor Algorithm

In the previous section, we described an operation to manipulate polygons. We shall now use this operation for the definition of B-spline curves. Recall Figure 10.4, in which a knot u was reinserted so that its multiplicity was raised to two. What happens if we reinsert u again? The answer is nothing. No new Greville abscissas are generated.

In general, for degree n, repeated insertion of a knot u no longer changes the polygon after the multiplicity of u has reached n. We use this fact in the algorithmic definition of a special function, called a B-spline curve.[4] The algorithm used in this definition is called the de Boor algorithm:

de Boor algorithm, informal: To evaluate an nth degree B-spline curve (given by its de Boor ordinates and knot sequence) at a parameter value u, insert u into the knot sequence until it has multiplicity n. The corresponding polygon vertex is the desired function value.

Before we proceed further, one comment should be made. What is meant by "corresponding polygon vertex"? If a knot u_i is of multiplicity n, then one of the Greville abscissas coincides with u_i, namely, $\xi_i = \frac{1}{n}(u_i + \cdots + u_{i+n-1}) = u_i$. Consequently, the polygon has a vertex (u_i, d_i), and d_i is the function value of the B-spline curve at u_i. Figure 10.5 gives an illustration. We now realize that we have encountered an example of the de Boor algorithm earlier; see Figure 10.4 for the case $n = 2$.

Note that the de Boor algorithm needs fewer insertions if the parameter value u is already an element of the knot sequence. If it has multiplicity r, then only $n - r$ reinsertions are necessary to make u a knot of multiplicity n.

We are now ready for a formal definition. Let us denote a B-spline curve of degree n with control polygon P by $B_n P$, and its value at parameter value

---

[4]We use the term "curve" loosely to emphasize that the theory developed here carries over easily to parametric curves.

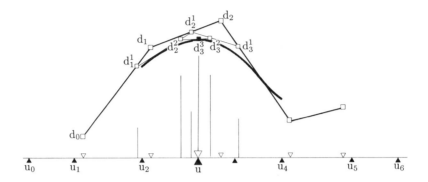

Figure 10.5: The de Boor algorithm: example with $n = 3, L = 2$.

u by $[B_n P](u)$. We will only define the curve for values of u between u_{n-1} and u_{L+n-1}.

de Boor algorithm: Let $u \in [u_I, u_{I+1}) \subset [u_{n-1}, u_{L+n-1}]$. Define

$$d_i^k(u) = \frac{u_{i+n-k} - u}{u_{i+n-k} - u_{i-1}} d_{i-1}^{k-1}(u) + \frac{u - u_{i-1}}{u_{i+n-k} - u_{i-1}} d_i^{k-1}(u) \qquad (10.5)$$

for $k = 1, \ldots, n - r$, and $i = I - n + k + 1, \ldots, I + 1$. Then

$$s(u) = [B_n P](u) = d_{I+1}^{n-r}(u) \qquad (10.6)$$

is the value of the B-spline curve at parameter value u. Here, r denotes the multiplicity of u in case it was already one of the knots. If it was not, set $r = 0$. As usual, we set $d_i^0(u) = d_i$.

C. de Boor [113] published this algorithm in 1972. It is the B-spline analog of the de Casteljau algorithm. Figure 10.6 shows schematically which d_i are involved in (10.5).

In our description of the de Boor algorithm, we did not renumber the knot sequence and the control points at each level, since our interest is only in the final result $d_{I+1}^{n-r}(u)$. Of course, at each level k, we generate a new control polygon that describes the same B-spline curve as the previous control polygon did. In particular, for $k = 1$, we obtain the knot insertion algorithm.

Figure 10.5 shows an example. We can also view that example as a case of multiple knot insertion. In that context, we have constructed several polygons that describe the same B-spline curve:

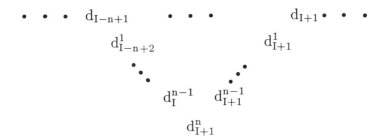

Figure 10.6: The de Boor algorithm: for $u \in [u_I, u_{I+1}]$, the scheme of generated intermediate points is shown.

k=1: the de Boor ordinates $d_0, d_1^1, d_2^1, d_3^1, d_3, d_4$ corresponding to the knot sequence $u_0, u_1, u_2, u, u_3, u_4, u_5, u_6$;

k=2: the de Boor ordinates $d_0, d_1^1, d_2^2, d_3^2, d_3^1, d_3, d_4$ corresponding to the knot sequence $u_0, u_1, u_2, u, u, u_3, u_4, u_5, u_6$;

k=3: the de Boor ordinates $d_0, d_1^1, d_2^2, d_3^3, d_3^2, d_3^1, d_3, d_4$ corresponding to the knot sequence $u_0, u_1, u_2, u, u, u, u_3, u_4, u_5, u_6$.

Let us next examine an important special case. Consider the knot sequence

$$0 = u_0 = u_1 = \cdots = u_{n-1} < u_n = u_{n+1} = \cdots = u_{2n-1} = 1.$$

Here, both u_0 and u_n have multiplicity n. We note that the Greville abscissas are given by

$$\xi_i = \frac{1}{n} \sum_{j=i}^{i+n-1} u_i = \frac{i}{n}; \quad i = 0, \ldots, n.$$

For $0 \le u \le 1$, the de Boor algorithm sets $I = n - 1$ and

$$d_i^k(u) = \frac{u_{i+n-k} - u}{u_{i+n-k} - u_{i-1}} d_{i-1}^{k-1} + \frac{u - u_{i-1}}{u_{i+n-k} - u_{i-1}} d_i^{k-1}.$$

Since $n - 1 \ge i - k \ge 0$, we have $u_{i+n-k} = 1, u_{i-1} = 0$ for all i, k; thus

$$d_i^k(u) = (1-u)d_{i-1}^{k-1} + ud_i^{k-1}; \quad k = 1, \ldots, n. \tag{10.7}$$

This is the de Casteljau algorithm![5] Schoenberg [422] first observed this in 1967, although in a different context. Riesenfeld [395] and Gordon and Riesenfeld [229] are more accessible references. We will be able to draw several important conclusions from this special case. First, we note that the restriction to the interval $[0, 1]$ is not essential: all our constructions are invariant under affine parameter transformations.

Thus, if two adjacent knots in any knot sequence both have multiplicity n, the corresponding B-spline curve is a Bézier curve between those two knots. The B-spline control polygon is the Bézier polygon; the Greville abscissas are equally spaced between the two knots.

After we inserted u until it was of multiplicity n, the initial de Boor polygon (or Bézier polygon, in this case,) was transformed into two Bézier polygons, defining the same curve as did the initial polygon. Thus we have another proof for the fact that the de Casteljau algorithm subdivides Bézier curves.

For a B-spline curve over an arbitrary knot vector, we can always reinsert the given knots into the knot sequence until each knot is of multiplicity n. The B-spline polygon corresponding to that knot sequence is the *piecewise Bézier polygon* of the curve. We have thus shown that *B-spline curves are piecewise polynomial over* $[u_{n-1}, u_{L+n-1}]$. The method of constructing the piecewise Bézier polygon from the B-spline polygon via knot insertion was developed by W. Boehm [62]. A different method was created by P. Sablonnière [406].

10.4   Smoothness of B-spline Curves

Now that we know that B-spline curves are piecewise polynomials of degree n each, we shall investigate their smoothness: how often is a B-spline curve differentiable at a point u? Obviously, we need to consider only the breakpoints – the curve is infinitely often differentiable at all other points.

To answer this question, simply reconsider the above example of Eq. (10.7). Now, let u be an existing knot of multiplicity r. Our knot sequence is:

$$\begin{aligned}
0 &= u_0 = u_1 = \cdots = u_{n-1} \\
&< u_n = u_{n+1} = \ldots = u_{n+r-1} \\
&< u_{n+r} = u_{n+r+1} = \ldots = u_{2n+r-1} = 1;
\end{aligned}$$

---

[5]The subscripts are different – this is simply a matter of notation, however.

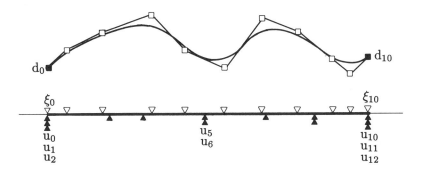

Figure 10.7: Multiple knots: the effects of multiple knots on the curve. Here, $n = 3, L = 8$.

the knot to be reinserted is $u = u_n$. The de Boor algorithm only consists of $n - r$ levels. Taking into account the multiplicities of the end knots, we have

$$d_i^k(u) = (1 - u)d_{i-1}^{k-1} + ud_i^{k-1}; \quad k = 1, \ldots, n - r. \tag{10.8}$$

These are the $n - r$ last levels in a de Casteljau algorithm. Therefore the two polynomial curve segments meeting at u are at least $n - r$ times differentiable at that point (see Sections 4.5 and 4.6).

As above, we note that the restriction to the interval $[0, 1]$ is not essential. If we want to investigate the smoothness of an arbitrary B-spline curve at a knot, we can always force its two neighbors to be of multiplicity n (without changing the curve!) and apply our arguments.

Thus *a B-spline curve is (at least) C^{n-r} at knots with multiplicity r.* In particular, the curve is $n - 1$ times continuously differentiable if all knots are simple, i.e., of multiplicity one. Figure 10.7 shows a cubic ($n = 3$) B-spline curve over a knot sequence that has several multiple entries. The triple knots at the ends force d_0 and d_{10} to be on the curve.

10.5   The B-spline Basis

Consider a knot sequence $u_0, \ldots, u_K$ and the set of piecewise polynomials of degree n defined over it, where each function in that set is $n - r_i$ times

continuously differentiable at knot u_i. All these piecewise polynomials form
a linear space, with dimension

$$\dim = (n+1) + \sum_{i=1}^{K-1} r_i. \tag{10.9}$$

For a proof, suppose we want to construct an element of our piecewise polynomial linear space. The number of independent constraints that we can impose
on an arbitrary element, or its number of *degrees of freedom*, is equal to the
dimension of the considered linear space. We may start by completely specifying the first polynomial segment, defined over $[u_0, u_1]$; we can do this in $n+1$
ways, which is the number of coefficients that we can specify for a polynomial
of degree n. The next polynomial segment, defined over $[u_1, u_2]$, must agree
with the first segment in position and $n - r_1$ derivatives at u_1, thus leaving
only r_1 coefficients to be chosen for the second segment. Continuing further,
we obtain (10.9).

We are interested in B-spline curves that are piecewise polynomials over
the special knot sequence $[u_{n-1}, u_{L+n-1}]$. The dimension of the linear space
that they form is $L + n$, which also happens to be the number of B-spline
vertices for a curve in this space. If we can define $L + n$ linearly independent
piecewise polynomials in our linear function space, we have found a basis for
this space. We proceed as follows.

Define functions $N_i^n(u)$, called *B-splines* by defining their de Boor ordinates to satisfy $d_i = 1$ and $d_j = 0$ for all $j \neq i$. The $N_i^n(u)$ are clearly elements
of the linear space formed by all piecewise polynomials over $[u_{n-1}, u_{L+n-1}]$.
They have *local support*:

$$N_i^n(u) \neq 0 \quad \text{only if} \quad u \in [u_{i-1}, u_{i+n}].$$

This follows because knot insertion, and hence the de Boor algorithm, is a
local operation; if a new knot is inserted, only those Greville abscissas that
are "close" will be affected.

B-splines also have *minimal support*: if a piecewise polynomial with the
same smoothness properties over the same knot vector has less support than
N_i^n, it must be the zero function. All piecewise polynomials defined over
$[u_{i-1}, u_{i+n}]$, the support region of N_i^n, are elements of a function space of
dimension $2n + 1$, according to (10.9). A support region that is one interval
"shorter" defines a function space of dimension $2n$. The requirement of vanishing $n - r_{i-1}$ derivatives at u_{i-1} and of vanishing $n - r_{i+n}$ derivatives at

u_{i+n} imposes $2n$ conditions on any element in the linear space of functions over $[u_{i-1}, u_{i+n-1}]$. The additional requirement of assuming a nonzero value at some point in the support region raises the number of independent constraints to $2n + 1$, too many to be satisfied by an element of the function space with dimension $2n$.

Another important property of the N_i^n is their *linear independence*. To demonstrate this independence, we must verify that

$$\sum_{j=0}^{L+n-1} c_j N_j^n(u) \equiv 0 \tag{10.10}$$

implies $c_j = 0$ for all j. It is sufficient to concentrate on one interval $[u_I, u_{I+1}]$ with $u_I < u_{I+1}$. Because of the local support property of B-splines, (10.10) reduces to

$$\sum_{j=I-n+1}^{I+1} c_j N_j^n(u) \equiv 0 \ \text{ for } u \in [u_I, u_{I+1}].$$

We have completed our proof if we can show that the linear space of piecewise polynomials defined over $[u_{I-n}, u_{I+n+1}]$ does not contain a nonzero element that vanishes over $[u_I, u_{I+1}]$. Such a piecewise polynomial cannot exist: it

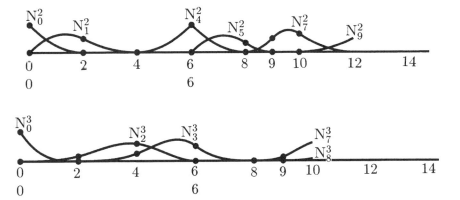

Figure 10.8: B-splines: top, some quadratic B-splines over the indicated knot sequence; bottom: some cubic ones. Note multiple knots at left end and simple knots at right end.

would have to be a nonzero local support function over $[u_{I+1}, u_{I+n+1}]$. The existence of such a function would contradict the fact that B-splines are of *minimal* local support.

Because the B-splines N_i^n are linearly independent, every piecewise polynomial s over $[u_{n-1}, u_{L+n-1}]$ may be written uniquely in the form

$$s(u) = \sum_{j=0}^{L+n-1} d_i N_i^n(u). \tag{10.11}$$

The B-splines thus form a *basis* for this space. This reveals the origin of their name, which is short for *B*asis splines.

If we set all $d_i = 1$ in (10.11), the function $s(u)$ will be identically equal to one, thus asserting that B-splines form a *partition of unity*.

Figure 10.8 gives examples of quadratic and cubic basis functions.

10.6   Two Recursion Formulas

We have defined B-spline basis functions in a constructive way: the B-spline N_i^n is defined by the knot sequence and the Greville abscissa ξ_i. The function N_i^n is given by its B-spline control polygon with de Boor ordinates $d_j = \delta_{i,j}$; $j = 0, \ldots, L + n - 1$. From it, we can construct the piecewise Bézier polygon by inserting every knot until it is of multiplicity n. We can then compute values of $N_i^n(u)$ by applying the de Casteljau algorithm to the Bézier polygon corresponding to the interval that contains u. There is a more direct way, which we now discuss.

To further explore B-splines, let us investigate how they "react" to knot insertion. Let $\hat{u}$ be a new knot inserted into a given knot sequence. Denote the B-splines over the "old" knot sequence by N_i^n, those over the "new" knot sequence by $\hat{N}_i^n$. Note that there is one more element in the set of $\hat{N}_i^n$ than in that of the N_i^n. In fact, the linear space of all piecewise polynomials over the old knot sequence is a subspace of the linear space of all piecewise polynomials over the new sequence. Let N_l^n be an "old" basis function that has $\hat{u}$ in its support. Its B-spline polygon is defined by $d_j = \delta_{j,l}$, where j ranges from 0 to $L + n - 1$ and δ denotes the Kronecker delta. Its B-spline polygon with respect to the new knot sequence is obtained by the knot insertion process.

Only two of the new de Boor ordinates will be different from zero. Equation (10.4) yields

$$\hat{d}_l = \frac{u_{l+n-1} - \hat{u}}{u_{l+n-1} - u_{l-1}} \cdot 0 + \frac{\hat{u} - u_{l-1}}{u_{l+n-1} - u_{l-1}} \cdot 1,$$

$$\hat{d}_{l+1} = \frac{u_{l+n} - \hat{u}}{u_{l+n} - u_l} \cdot 1 + \frac{\hat{u} - u_l}{u_{l+n} - u_l} \cdot 0.$$

(Recall that $d_l = 1$, whereas all other $d_j = 0$.) Hence

$$\hat{d}_l = \frac{\hat{u} - u_{l-1}}{u_{l+n-1} - u_{l-1}},$$

$$\hat{d}_{l+1} = \frac{u_{l+n} - \hat{u}}{u_{l+n} - u_l}.$$

Thus we can write N_l^n in terms of $\hat{N}_l^n$ and $\hat{N}_{l+1}^n$:

$$N_l^n(u) = \frac{\hat{u} - u_{l-1}}{u_{l+n-1} - u_{l-1}} \hat{N}_l^n(u) + \frac{u_{l+n} - \hat{u}}{u_{l+n} - u_l} \hat{N}_{l+1}^n(u). \qquad (10.12)$$

This result is due to W. Boehm [61]. It allows us to write B-splines as linear combinations of B-splines over a refined knot sequence.

For the second important recursion formula, we must define an additional B-spline function,[6] N_i^0:

$$N_i^0(u) = \begin{cases} 1 & \text{if } u_{i-1} \le u < u_i \\ 0 & \text{else} \end{cases}. \qquad (10.13)$$

The announced recursion formula relates B-splines of degree n to B-splines of degree $n - 1$:

$$N_l^n(u) = \frac{u - u_{l-1}}{u_{l+n-1} - u_{l-1}} N_l^{n-1}(u) + \frac{u_{l+n} - u}{u_{l+n} - u_l} N_{l+1}^{n-1}(u). \qquad (10.14)$$

To prove (10.14), we shall prove the following more general statement:

$$s(u) = \sum_{j=i+r-n+1}^{i+1} d_j^r N_j^{n-r}(u) \qquad (10.15)$$

---

[6]See Problem 4 at the end of the chapter!

Figure 10.9: The B-spline recursion: top, two linear B-splines yield a quadratic one; bottom, two quadratic B-splines yield a cubic one.

for all $r \in [0, n]$. For its proof, we first check that it is true for $r = n$; this follows from (10.13). By the de Boor algorithm, (10.15) is equivalent to

$$s(u) = \sum_{j=i-n+1+r}^{i+1} (1 - \alpha_j^r) d_{j-1}^{r-1} N_j^{n-r}(u) + \sum_{j=i-n+1+r}^{i+1} \alpha_j^r d_j^{r-1} N_j^{n-r}(u),$$

where

$$\alpha_j^r = \frac{u - u_{i-1}}{u_{i+n-r} - u_{i-1}}.$$

An index transformation yields

$$s(u) = \sum_{j=i-n+r}^{i} (1 - \alpha_{j+1}^r) d_j^{r-1} N_{j+1}^{n-r}(u) + \sum_{j=i-n+1+r}^{i+1} \alpha_j^r d_j^{r-1} N_j^{n-r}(u).$$

Because of the local support of the N_j^{n-r}, this may be changed to

$$s(u) = \sum_{j=i-n+r}^{i+1} (1 - \alpha_{j+1}^r) d_j^{r-1} N_{j+1}^{n-r}(u) + \sum_{j=i-n+r}^{i+1} \alpha_j^r d_j^{r-1} N_j^{n-r}(u).$$

Hence, by the inductive hypothesis,

$$s(u) = \sum_{j=i-n+r}^{i+1} [\alpha_j^r N_j^{n-r}(u) + (1 - \alpha_{j+1}^r) N_{j+1}^{n-r}(u)] d_j^{r-1}.$$

This step completes the proof of (10.15), since we have now shown that (10.15) holds for $r-1$ provided that it holds for r. The recurrence (10.14) now follows from comparing (10.15) and (10.6). The development of equation (10.14) is due to L. Mansfield, C. de Boor, and M. Cox; see de Boor [113] and Cox [106]. For an illustration of (10.14), see Figure 10.9.

The recursion formula (10.14) shows that a B-spline of degree n is a strictly convex combination of two lower degree ones; it is therefore a very stable formula from a numerical viewpoint. If B-spline curves must be evaluated repeatedly at the same parameter values u_k, it is a good idea to compute the values for $N_i^n(u_k)$ using (10.14) and then to store them.

10.7 Repeated Knot Insertion

We may insert more and more knots into the knot sequence; let us now investigate the effect of such a process. A B-spline curve of degree n is defined over $u_{n-1}, \ldots, u_{L+n-1}$. Let us set $a = u_{n-1}$, $b = u_{L+n-1}$. Now insert more knots u_i^r; here r counts the overall number of insertions and i denotes the number of u_i^r in the new knot sequence. After r knot insertions, we have a new polygon P^r that describes the same curve as did the original polygon P. As we insert more and more knots, so as to become dense in $[a, b]$, the sequence of polygons P^r converges to the curve that they all define:

$$\lim_{r \to \infty} P^r = [B_n P]. \tag{10.16}$$

To begin, we recall that a B-spline curve depends only on $d_k, \ldots, d_{k+n}$ over the interval $[u_k, u_{k+1}]$. Then for $u \in [u_k, u_{k+1}]$,

$$\min(d_k, \ldots, d_{k+n}) \leq [B_n P](u) \leq \max(d_k, \ldots, d_{k+n})$$

by the strong convex hull property.

We need to show that for any ϵ, we can find an r such that

$$|P^r(u) - [B_n P](u)| \leq \epsilon \quad \text{for all } u.$$

We know that for any ϵ, we can find an r such that

$$|\xi_{i+1}^r - \xi_i^r| \leq \delta$$

and

$$|P^r(\xi_{i+1}^r) - P^r(\xi_i^r)| \leq \epsilon,$$

since each P^r is continuous. Thus

$$\max[P^r(\xi_i^r), \ldots, P^r(\xi_{i+n}^r)] - \min[P^r(\xi_i^r), \ldots, P^r(\xi_{i+n}^r)] \leq n\epsilon$$

for those i that are relevant to the interval $[u_k, u_{k+1}]$. But we also know that

$$\min(d_i^r) \leq [B_n P](u) \leq \max(d_i^r).$$

Thus,

$$|[B_n P](u) - P_j^r| \leq n\epsilon; \quad j \in [i, \ldots, i+n],$$

which finally yields

$$|[B_n P](u) - P^r(u)| \leq n\epsilon,$$

proving our convergence claim.

The use of repeated knot insertion lies in the *rendering* of B-spline curves. If sufficiently many knots have been inserted into the knot sequence, the resulting control polygon will be arbitrarily close to the curve. Then, instead of plotting the curve directly, one simply plots the refined polygon. To obtain an *adaptive* rendering method, one would control the knot insertion process by inserting more knots where the curve is of high curvature and fewer knots where it is flat.

Of course, B-spline curves may also be *parametric*. All we have to do is use functional B-spline curves (all over the same knot vector) for each component of the parametric curve:

$$\mathbf{x}(u) = \sum_{i=0}^{L+n-1} \mathbf{d}_i N_i^n(u) = \sum_{i=0}^{L+n-1} \begin{bmatrix} d_i^x \\ d_i^y \\ d_i^z \end{bmatrix} N_i^n(u).$$

The curves for $n = 2$ and $n = 3$ were already described in Chapter 7, although with a different notation that especially suited those cases. General degree B-spline curves enjoy all the properties of the lower degree ones, such as affine invariance and the convex hull property.

An interesting application of repeated knot insertion is due to G. Chaikin [86]. Although this scheme may be described in the context of functional curves, we prefer the more intuitive parametric version. Consider a quadratic B-spline curve over a uniform knot sequence. Insert a new knot at the midpoint of every interval of the knot sequence. If the "old" curve had control vertices $\mathbf{d}_i$ and those of the new one are $\mathbf{d}_i^{(1)}$, it is easy to show that

$$\mathbf{d}_{2i-1}^{(1)} = \frac{3}{4}\mathbf{d}_i + \frac{1}{4}\mathbf{d}_{i-1} \quad \text{and} \quad \mathbf{d}_{2i+1}^{(1)} = \frac{3}{4}\mathbf{d}_i + \frac{1}{4}\mathbf{d}_{i+1}.$$

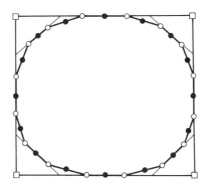

Figure 10.10: Chaikin's algorithm: starting with the (closed) control polygon of a quadratic B-spline curve, two levels of the Chaikin iteration are shown.

If this procedure is repeated infinitely often, the resulting sequence of polygons will converge to the original curve, as follows from our previous considerations. Figure 10.10 shows the example of a closed quadratic B-spline curve; two levels of the iteration are shown.

Chaikin's algorithm may be described as *corner-cutting*: At each step, we chisel away the corners of the initial polygon. This process is, on a high level, similar to that of degree elevation for Bézier curves, which is also a convergent process (see Section 5.2). One may ask if corner-cutting processes will always converge to a smooth curve. The answer is "yes," with some mild stipulations on the corner-cutting process, and was first proved by de Boor [116]. One may thus use a corner-cutting procedure to *define* a curve – and only very few of the curves thus generated are piecewise polynomial! Recent work has been carried out by Prautzsch and Micchelli [384] and [334], based on earlier results by de Rham [125], [126].

R. Riesenfeld [396] realized that Chaikin's algorithm actually generates uniform quadratic B-spline curves. A general algorithm for the simultaneous insertion of several knots into a B-spline curve has been developed by Cohen, Lyche, and Riesenfeld [99]. This so-called "Oslo algorithm" needs a theory of discrete B-splines for its development (see Bartels *et al.* [42]). The knot insertion algorithm as developed in this chapter is more intuitive and equally powerful.

10.8   Additional Material

After the more theoretical developments of the previous two sections, let us examine some of the properties that we can now derive for B-spline curves.

Linear precision: If $l(u)$ is a straight line of the form $l = au + b$, and if we read off values at the Grevill abscissas, the resulting B-spline curve reproduces the straight line:

$$\sum l(\xi_i) N_i^n(u) = l(u).$$

This property is a direct consequence of the de Boor algorithm. It was originally obtained by T. Greville [241], [240] in a different context. The original Greville result is the motivation for the term "Greville abscissas."

Strong convex hull property: Each point on the curve lies in the convex hull of no more than $n + 1$ nearby control points.

Variation diminishing property: The curve is not intersected by any straight line more often than is the polygon. This result has a very simple proof, presented by Lane and Riesenfeld [299]: we may insert every knot until it is of full multiplicity. This is a variation diminishing process, since it is piecewise linear interpolation. Once all knots are of full multiplicity, the B-spline control polygon is the piecewise Bézier polygon, for which we showed the variation diminishing property in Section 5.3.

The parametric case: In the parametric case, it is desirable to have u_0 and u_{L+n-1} both of full multiplicity n. This condition forces the first and last control points $\mathbf{d}_0$ and $\mathbf{d}_{L+n-1}$ to lie on the endpoints of the curve. In this way, one has better control of the behavior of the curve at the ends. The spline curves that we discussed in Chapter 7 are all described in this form, although we did not formally make use of knot multiplicities there. If the end knots are allowed to be of lower (even simple) multiplicity, the first and last control vertices do not lie on the curve, and are called "phantom vertices" by Barsky and Thomas [40]. Figure 10.11 gives several examples of B-spline curves over various knot sequences.

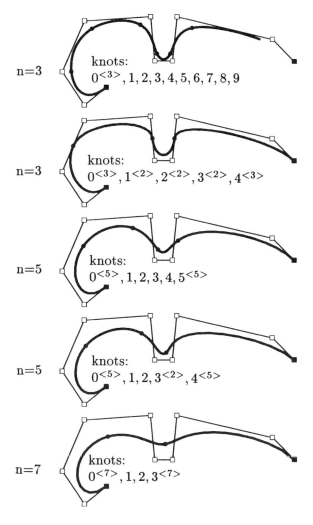

Figure 10.11: Parametric B-spline curves: several examples, all with the same control polygon but with different degrees and knot sequences.

The de Boor algorithm allows a nice geometric description in parametric form. Formally, we perform the algorithm for all components of the control polygon. Geometrically, we may "engrave" parts of the knot

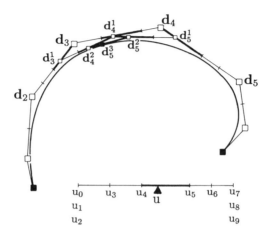

Figure 10.12: The de Boor algorithm in parametric form: all groups of $n+1-r$ intervals that contain $[u_4, u_5]$ are mapped onto polygon legs.

sequence on each polygon leg: map the first $n + 1$ subsequent knots (starting at u_0) onto $\mathbf{d}_0\mathbf{d}_1$, the next subsequent $n + 1$ knots onto $\mathbf{d}_1\mathbf{d}_2$, and so on, until the last subsequent $n + 1$ knots are mapped to the last polygon leg. For example, in Figure 10.12, with $n = 3$ and $L = 5$, the knots $[u_2, u_3, u_4, u_5]$ are mapped to $\mathbf{d}_2\mathbf{d}_3$, whereas $[u_0, u_1, u_2, u_3]$ are mapped to $\mathbf{d}_0\mathbf{d}_1$. The interval $[u_I, u_{I+1}]$, which contains the evaluation parameter u, is thus mapped to n polygon legs by n affine maps, which are equivalent to linear interpolation as outlined in Section 2.3. These affine maps take u to the $\mathbf{d}_j^1(u)$. The same procedure is then repeated: consider all sets of subsequent n knots that contain u_I, u_{I+1}. Map them by affine maps to the polygon formed by the $\mathbf{d}_j^1$ and denote the images of u by $\mathbf{d}_j^2$, etc. The final step is to map the interval u_I, u_{I+1} onto $\mathbf{d}_I^{n-1}, \mathbf{d}_{I+1}^{n-1}$ to obtain the point $\mathbf{d}_{I+1}^n$ on the curve.

Finally, a note on how to *store* B-spline curves. It is not convenient to store the knot sequence $\{u_i\}$ and simply list multiple knots as often as indicated by their multiplicity, following the approach taken by Schumaker [423]. Roundoff may produce knots that are a small distance apart, yet meant to be identical. It is wiser to store only distinct knots and to note their multiplicities in a second array. From these two arrays, one may compute the original knot sequence when required, e.g., by the de Boor algorithm.

10.9 B-spline Blossoms

In Section 3.4, we generalized the de Casteljau algorithm by allowing evaluations at n different arguments $t_1, \ldots, t_n$, thus arriving at the blossom $\mathbf{b}[t_0, \ldots, t_n]$ of a polynomial curve $\mathbf{b}(t)$. The same priciple may be applied to B-spline curves. For the sake of concreteness, let us consider the case of a cubic B-spline curve, and also restrict ourselves to the parameter interval $[u_4, u_5]$.

We will now modify the standard de Boor algorithm: at each level $k; k = 1, 2, 3$, we will evaluate at a different argument $v_k \in [u_4, u_5]$. The relevant control points for our interval are $\mathbf{d}_2, \ldots, \mathbf{d}_5$, and we obtain the following scheme:

$$
\begin{array}{llll}
\mathbf{d}_2 & & & \\
\mathbf{d}_3 & \mathbf{d}_3^1[v_1] & & \\
\mathbf{d}_4 & \mathbf{d}_4^1[v_1] & \mathbf{d}_4^2[v_1, v_2] & \\
\mathbf{d}_5 & \mathbf{d}_5^1[v_1] & \mathbf{d}_5^2[v_1, v_2] & \mathbf{d}_5^3[v_1, v_2, v_3].
\end{array}
$$

Again, it is easy to see that it does not matter in which order we "feed" the the v_i into this scheme. Also, if all v_i agree, we recover the standard de Boor algorithm. We shall use the notation $\mathbf{d}_4[v_1, v_2, v_3]$ for the point $\mathbf{d}_5^3[v_1, v_2, v_3]$, indicating that we are dealing with the interval $[u_4, u_5]$.

In general, for a parameter interval $[u_I, u_{I+1}]$, we shall define as the *B-spline blossom* – or just blossom – the function $\mathbf{d}_I[v_1, \ldots, v_n]$, obtained by applying the de Boor algorithm for the interval $[u_I, u_{I+1}]$ to the control points $\mathbf{d}_{I-n+1}, \ldots, \mathbf{d}_{I+1}$, and using a (possibly) different argument v_k at level k of the algorithm. Thus the blossom corresponding to the interval $[u_I, u_{I+1}]$ is a function of n variables $v_1, \ldots, v_n$; in fact, it is a multivariate *polynomial* function. The whole B-spline curve possesses as many blossoms as it has domain intervals.

Note that we have not restricted the v_i to be in the interval $[u_I, u_{I+1}]$! In fact, a very interesting result arises if we evaluate for v_i outside that interval.

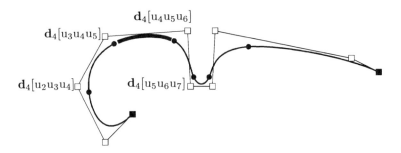

Figure 10.13: B-spline blossoms (cubic): The knot sequence is $u_0 = u_1 = u_2 < u_3 < u_4 < u_5 < u_6 < u_7 < u_8 = u_9 = u_{10}$. The control points corresponding to $[u_4, u_5]$ are shown as blossom values.

Returning to our cubic example above, let us set $[v_1, v_2, v_3] = [u_2, u_3, u_4]$. The scheme becomes:

$$
\begin{array}{llll}
\mathbf{d}_2 & & & \\
\mathbf{d}_3 & \mathbf{d}_2 & & \\
\mathbf{d}_4 & \bullet & \mathbf{d}_2 & \\
\mathbf{d}_5 & \bullet & \bullet & \mathbf{d}_2 = \mathbf{d}_4[u_2, u_3, u_4].
\end{array}
$$

The •-entries in the scheme were not computed, because their values do not contribute to the final result. We have, similar to the Bézier case, recovered one of the control points! The algorithm no longer uses only convex combinations; instead extrapolation is used several times.

To illustrate the principle of control point recovery, we try one more set of arguments, namely $[v_1, v_2, v_3] = [u_3, u_4, u_5]$. The scheme becomes:

$$
\begin{array}{llll}
\mathbf{d}_2 & & & \\
\mathbf{d}_3 & \bullet & & \\
\mathbf{d}_4 & \mathbf{d}_3 & \bullet & \\
\mathbf{d}_5 & \bullet & \mathbf{d}_3 & \mathbf{d}_3 = \mathbf{d}_4[u_3, u_4, u_5].
\end{array}
$$

Again, we have recovered a control point. Continuing in this manner, we find that $\mathbf{d}_4 = \mathbf{d}_4[u_4, u_5, u_6]$ and $\mathbf{d}_5 = \mathbf{d}_4[u_5, u_6, u_7]$. Figure 10.13 illustrates these examples.

More generally, we have the following: the curve segment defined over to $[u_I, u_{I+1}]$ needs $n+1$ control points for its definition, namely $\mathbf{d}_{I-n+1}, \ldots, \mathbf{d}_{I+1}$. They can be obtained as blossom values:

$$\mathbf{d}_{I-n+1+k} = \mathbf{d}_I[u_{I-n+1+k}, \ldots, u_{I+k}]; \quad k = 0, \ldots, n. \tag{10.17}$$

The arguments of $\mathbf{d}_I$ on the right-hand side are all n-tuples of subsequent knots that contain either u_I or u_{I+1}.

We will now use the blossoming principle to discuss *degree elevation*. Formally, we can write any n^{th} degree piecewise polynomial curve over a given knot sequence as one of degree $n+1$. It will not be over the same knot sequence: instead, we will have to increase the multiplicity of each knot by one. We denote these new knots by $\hat{u}_j$. The task is then to find the B-spline control points of the degree elevated curve, similar to the process of degree elevation for Bézier curves as described in Section 5.1.

The degree elevated curve $\hat{\mathbf{d}}$ has $\hat{\mathbf{d}}_K[v_1, \ldots, v_{n+1}]$ as the blossom of the interval $[\hat{u}_K, \hat{u}_{K+1}] = [u_I, u_{I+1}]$. How can we write it as a combination of blossoms of n variables? We can try the following:

$$\hat{\mathbf{d}}_K[v_1, \ldots, v_{n+1}] = \frac{1}{n+1} \sum_{j=1}^{n+1} \mathbf{d}_I[v_1, \ldots, v_{n+1}|v_j], \tag{10.18}$$

where $[v_1, \ldots, v_n|v_j]$ is the argument sequence $[v_1, \ldots, v_{n+1}]$ with v_j removed from it. This simple attempt already yields the solution: $\hat{\mathbf{d}}_K$ is a blossom, being a barycentric combination of blossoms. Also, it is symmetric and multiaffine, and it yields a point on the given curve for the case of all v_j being equal.

To be more specific: we know that the control points $\hat{\mathbf{d}}_j$ are the blossom values

$$\hat{\mathbf{d}}_{K-n+r} = \hat{\mathbf{d}}_K[\hat{u}_{K\ n|r}, \ldots, \hat{u}_{K|r}]; \quad r = 0, \ldots, n+1$$

by application of (10.17). Using (10.18), we now have the desired result:

$$\hat{\mathbf{d}}_{K-n+r} = \frac{1}{n+1} \sum_{j=1}^{n+1} \mathbf{d}_I[\hat{u}_{K-n+r}, \ldots, \hat{u}_{K+r}|\hat{u}_{K-n+r+j-1}]; \quad r = 0, \ldots, n+1. \tag{10.19}$$

Thus the new B-spline vertices can be found by evaluating blossoms at n arguments of the new knot sequence and then taking their average. The corresponding formula for the basis functions is given in Section 10.10. A specific case is discussed in Example 10.1.

Literature on B-spline blossoms: [442], [439], [437], [440], [387].

Let a cubic B-spline curve be defined over $\{u_0 = u_1 = u_2, u_3, \ldots\}$. Then the interval $[u_4, u_5]$ corresponds to $[\hat{u}_7, \hat{u}_8]$. The new control point $\hat{\mathbf{d}}_4$ is computed as follows:

$$
\begin{aligned}
\hat{\mathbf{d}}_4 &= \hat{\mathbf{d}}_7[\hat{u}_4, \hat{u}_5, \hat{u}_6, \hat{u}_7] \\
&= \frac{1}{4}(\mathbf{d}_4[\hat{u}_4, \hat{u}_5, \hat{u}_6] + \mathbf{d}_4[\hat{u}_4, \hat{u}_5, \hat{u}_7] + \mathbf{d}_4[\hat{u}_4, \hat{u}_6, \hat{u}_7] + \mathbf{d}_4[\hat{u}_5, \hat{u}_6, \hat{u}_7]) \\
&= \frac{1}{2}(\mathbf{d}_4[u_3, u_3, u_4] + \mathbf{d}_4[u_3, u_4, u_4]).
\end{aligned}
$$

For the last step, we have utilized $\hat{u}_4 = \hat{u}_5 = u_3$ and $\hat{u}_6 = \hat{u}_7 = u_4$.

Example 10.1: B-spline degree elevation and blossoms.

10.10   B-spline Basics

Here, we present a collection of the most important formulas and definitions of this chapter. As before, n is the (maximal) degree of each polynomial segment, L is the number of curve segments if all knots in the domain are simple, and, more generally, $L+1$ is the sum of all domain knot multiplicities.

Knot sequence: $\{u_0, \ldots, u_{L+2n-2}\}$.

Domain: Curve is only defined over $[u_{n-1}, \ldots, u_{L+n-1}]$.

Greville abscissas: $\xi_i = \frac{1}{n}(u_i + \cdots + u_{i+n-1})$.

Support: N_i^n is nonnegative over $[u_{i-1}, u_{i+n}]$.

Control polygon P: (ξ_i, d_i); $i = 0, \ldots, L + n - 1$.

Knot insertion: To insert $u_I \leq u < u_{I+1}$: (1) Find new Greville abscissas $\hat{\xi}_i$. (2) Set new $d_i = P(\hat{\xi}_i)$.

de Boor algorithm: Given $u_I \leq u < u_{I+1}$, set

$$
d_i^k(u) = \frac{u_{i+n-k} - u}{u_{i+n-k} - u_{i-1}} d_{i-1}^{k-1}(u) + \frac{u - u_{i-1}}{u_{i+n-k} - u_{i-1}} d_i^{k-1}(u)
$$

for $k = 1, \ldots, n - r$, and $i = I - n + k + 1, \ldots, I + 1$. Here, r denotes the multiplicity of u. (Normally, u is not already in the knot sequence; then, $r = 0$.)

Boehm recursion: Let $\hat{u}$ be a new knot; then,

$$N_l^n(u) = \frac{\hat{u} - u_{l-1}}{u_{l+n-1} - u_{l-1}} \hat{N}_l^n(u) + \frac{u_{l+n} - \hat{u}}{u_{l+n} - u_l} \hat{N}_{l+1}^n(u).$$

Mansfield, de Boor, Cox recursion:

$$N_l^n(u) = \frac{u - u_{l-1}}{u_{l+n-1} - u_{l-1}} N_l^{n-1}(u) + \frac{u_{l+n} - u}{u_{l+n} - u_l} N_{l+1}^{n-1}(u).$$

Derivative:

$$\frac{\mathrm{d}}{\mathrm{d}u} N_l^n(u) = \frac{n}{u_{n+l-1} - u_{l-1}} N_l^{n-1}(u) - \frac{n}{u_{l+n} - u_l} N_{l+1}^{n-1}.$$

Derivative of B-spline curve:

$$\frac{\mathrm{d}}{\mathrm{d}u} s(u) = n \sum_{i=1}^{L+n-1} \frac{\Delta d_{i-1}}{u_{n+i-1} - u_{i-1}} N_i^{n-1}(u).$$

Degree elevation:

$$N_i^n(u) = \frac{1}{n} \sum_{j=i-1}^{n+i} N_i^{n+1}(u|u_j),$$

where $N_i^{n+1}(u|u_j)$ is defined over the original knot sequence except that the knot u_j has its multiplicity increased by one. This identity was discovered by H. Prautzsch in 1984 [383]. Another reference is Barry and Goldman [31].

10.11 Implementation

Here is the header for the de Boor algorithm code:

```
float deboor(degree,coeff,knot,u,i)
/*   uses de Boor algorithm to compute one
     coordinate on B-spline curve for param. value u in interval i.
Input: degree:      polynomial degree of each piece of curve
       coeff:       B-spline control points
       knot:        knot sequence
       u:           evaluation abscissa
       i:           u's interval: u[i]<= u < u[i+1]
Output:             coordinate value.
*/
```

This program does not need to know about L. The next program generates a set of points on a whole B-spline curve – for one coordinate, to be honest – so it has to be called twice for a 2D curve and three times for a 3D curve.

```
bspl_to_points(degree,l,coeff,knot,dense,points,point_num)
/*      generates points on B-spline curve. (one coordinate)
Input: degree:        polynomial degree of each piece of curve
       l:             number of active  intervals
       coeff:         B-spline control points
       knot:          knot sequence: knot[0]...knot[l+2*degree-2]
       dense:         how many points per segment
Output:points:        output array with function values.
       point_num:     how many points are generated. That number is
                      easier computed here than in the calling program:
                      no points are generated between multiple knots.
*/
```

The main program `deboormain.c` generates a postscript plot of a B-spline curve. A sample input file is in `bspl.dat`; it creates the outline of the letter **r** from Figure 8.6.

As a second example, the input data for the y-values of the curve in Figure 10.1 are:

degree = 3; l = 4; coeff = 0.8, 2.8, 5.7, 2.6, 5.7, 4.0, 0.6; knot = 0.0, 0.0, 0.0, 2.6, 7.7, 9.9, 17.8, 17.8, 17.8; dense = 10.

10.12   Problems

1. Prove (10.1). Hint: use similar triangles.

2. Find the Bézier points of the closed B-spline curves of degree four whose control polygons consist of the edges of a square and have (a) uniform knot spacing and simple knots, (b) uniform knot spacing, and knots all with multiplicity two.

3. For the case of a planar parametric B-spline curve, does symmetry of the polygon with respect to the y-axis imply that same symmetry for the curve?

4. If one uses the recursive relation (10.14) for the evaluation of the B-spline curve, it appears that one needs two additional knots u_{-1} and u_{L+2n-1}. Why does this not really pose a problem?

5. Use the procedure **deboor** to write a knot insertion routine. Then write a program to convert an arbitrary B-spline curve to its piecewise Bézier form.

6. Take the routine **deboor** and generalize it to a blossom routine. Experiment with "interesting" arguments.

7. After solving Problem 6, use your blossoming routine to program degree elevation for B-spline curves.

Chapter 11

W. Boehm: Differential Geometry I

Differential geometry is based largely on the pioneering work of L. Euler (1707–1783), C. Monge (1746–1818), and C. F. Gauss (1777–1855). One of their concerns was the description of local curve and surface properties such as curvature. These concepts are also of interest in modern computer-aided geometric design. The main tool for the development of general results is the use of local coordinate systems, in terms of which geometric properties are easily described and studied. This introduction discusses local properties of curves independent of a possible imbedding into a surface.

11.1 Parametric Curves and Arc Length

A curve in $I\!R^3$ is given by the parametric representation

$$\mathbf{x} = \mathbf{x}(t) = \begin{bmatrix} x(t) \\ y(t) \\ z(t) \end{bmatrix}, \quad t \in \lfloor a, b \rfloor \subset I\!R, \qquad (11.1)$$

where its cartesian coordinates x, y, z are differentiable functions of t. (We have encountered a variety of such curves already, among them Bézier and B-spline curves.) To avoid potential problems concerning the parametrization of the curve, we shall assume that

$$\dot{\mathbf{x}}(t) = \begin{bmatrix} \dot{x}(t) \\ \dot{y}(t) \\ \dot{z}(t) \end{bmatrix} \neq \mathbf{0}, \quad t \in [a, b], \qquad (11.2)$$

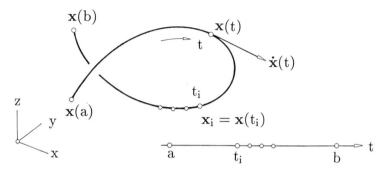

Figure 11.1: Parametric curve in space.

where dots denote derivatives with respect to t. Such a parametrization is called *regular*.

A change $\tau = \tau(t)$ of the parameter, where τ is a differentiable function of t, will not change the shape of the curve. This *reparametrization* will be regular if $\dot{\tau} \neq 0$ for all $t \in [a, b]$, i.e., we can find the inverse $t = t(\tau)$. Let

$$s = s(t) = \int_a^t ||\dot{\mathbf{x}}|| dt \qquad (11.3)$$

be such a parametrization. Because

$$\dot{\mathbf{x}} dt = \frac{d\mathbf{x}}{d\tau} \frac{d\tau}{dt} dt = \frac{d\mathbf{x}}{d\tau} d\tau,$$

s is independent of any regular reparametrization. It is an invariant parameter and is called *arc length* parametrization of the curve. One also calls $ds = ||\dot{\mathbf{x}}|| dt$ the *arc element* of the curve.

Remark 1:   Arc length may be introduced more intuitively as follows: let $t_i = a + i\Delta t$ and let $\Delta t > 0$ be an equidistant partition of the t-axis. Let $\mathbf{x}_i = \mathbf{x}(t_i)$ be the corresponding sequence of points on the curve. *Chord length* is then defined by

$$S = \sum_i ||\Delta \mathbf{x}_i|| = \sum_i ||\frac{\Delta \mathbf{x}_i}{\Delta t}|| \Delta t, \qquad (11.4)$$

where $\Delta \mathbf{x}_i = \mathbf{x}_{i+1} - \mathbf{x}_i$. It is easy to check that for $\Delta t \to 0$, chord length S converges to arc length s, while $\Delta \mathbf{x}_i / \Delta t$ converges to the tangent vector $\dot{\mathbf{x}}_i$ at $\mathbf{x}_i$.

Remark 2: Although arc length is an important concept, it is primarily used for theoretical considerations and for the development of curve algorithms. If, for some application, computation of the arc length is unavoidable, it may be approximated by the chord length (11.4).

11.2 The Frenet Frame

We will now introduce a special local coordinate system, linked to a point $\mathbf{x}(t)$ on the curve, that will significantly facilitate the description of local curve properties at that point. Let us assume that all derivatives needed below do exist. The first terms of the Taylor expansion of $\mathbf{x}(t + \Delta t)$ at t are given by

$$\mathbf{x}(t + \Delta t) = \mathbf{x} + \dot{\mathbf{x}}\Delta t + \ddot{\mathbf{x}}\frac{1}{2}\Delta t^2 + \dddot{\mathbf{x}}\frac{1}{6}\Delta t^3 + \ldots.^1$$

Let us assume that the first three derivatives are linearly independent. Then $\dot{\mathbf{x}}, \ddot{\mathbf{x}}, \dddot{\mathbf{x}}$ form a local affine coordinate system with origin $\mathbf{x}$. In this system, $\mathbf{x}(t)$ is represented by its *canonical coordinates*

$$\begin{bmatrix} \Delta t + \ldots \\ \frac{1}{2}\Delta t^2 + \ldots \\ \frac{1}{6}\Delta t^3 + \ldots \end{bmatrix},$$

where "..." denotes terms of degree four and higher in Δt.

From this local affine coordinate system, one easily obtains a local cartesian (orthonormal) system with origin $\mathbf{x}$ and axes $\mathbf{t}, \mathbf{m}, \mathbf{b}$ by the Gram-Schmidt process of orthonormalization, as shown in Figure 11.2:

$$\mathbf{t} = \frac{\dot{\mathbf{x}}}{||\dot{\mathbf{x}}||}, \quad \mathbf{m} = \mathbf{b} \wedge \mathbf{t}, \quad \mathbf{b} = \frac{\dot{\mathbf{x}} \wedge \ddot{\mathbf{x}}}{||\dot{\mathbf{x}} \wedge \ddot{\mathbf{x}}||}, \tag{11.5}$$

where "$\wedge$" denotes the cross product.

The vector $\mathbf{t}$ is called *tangent vector* (see Remark 1), $\mathbf{m}$ is called *main normal vector*,[2] and $\mathbf{b}$ is called *binormal vector*. The frame (or trihedron) $\mathbf{t}$, $\mathbf{m}$, $\mathbf{b}$ is called the *Frenet frame*; it varies its orientation as t traces out the curve.

---

[1] We use the abbreviation $\Delta t^2 = (\Delta t)^2$.

[2] Warning: one often sees the notation $\mathbf{n}$ for this vector. We use $\mathbf{m}$ to avoid confusion with surface normals, which are discussed later.

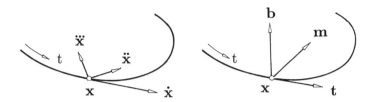

Figure 11.2: Local affine system (left) and Frenet frame (right).

The plane spanned by the point $\mathbf{x}$ and the two vectors $\mathbf{t}$, $\mathbf{m}$ is called the *osculating plane* $\mathbf{O}$. Its equation is

$$\det \begin{bmatrix} \mathbf{y} & \mathbf{x} & \dot{\mathbf{x}} & \ddot{\mathbf{x}} \\ 1 & 1 & 0 & 0 \end{bmatrix} = \det[\mathbf{y} - \mathbf{x}, \dot{\mathbf{x}}, \ddot{\mathbf{x}}] = 0,$$

where $\mathbf{y}$ denotes any point on $\mathbf{O}$. Its parametric form is

$$\mathbf{O}(u, v) = \mathbf{x} + u\dot{\mathbf{x}} + v\ddot{\mathbf{x}}.$$

Remark 3:　The process of orthonormalization yields

$$\mathbf{m} = \frac{\dot{\mathbf{x}}\ddot{\mathbf{x}} \cdot \ddot{\mathbf{x}} - \ddot{\mathbf{x}}\ddot{\mathbf{x}} \cdot \dot{\mathbf{x}}}{\|\dot{\mathbf{x}}\ddot{\mathbf{x}} \cdot \ddot{\mathbf{x}} - \ddot{\mathbf{x}}\ddot{\mathbf{x}} \cdot \dot{\mathbf{x}}\|}.$$

This equation may also be used for planar curves, where the binormal vector $\mathbf{b} = \mathbf{t} \wedge \mathbf{m}$ agrees with the normal vector of the plane.

11.3　Moving the Frame

Letting the Frenet frame vary with t provides a good idea of the curve's behavior in space. It is a fundamental idea in differential geometry to express the local change of the frame in terms of the frame itself. The resulting formulas are particularly simple if one uses arc length parametrization. We denote differentiation with respect to arc length by a prime. Since $\mathbf{x}' = \mathbf{t}$ is a unit vector, one finds the following two identities:

$$\mathbf{x}' \cdot \mathbf{x}' = 1 \quad \text{and} \quad \mathbf{x}' \cdot \mathbf{x}'' = 0.$$

The first identity states that the curve is traversed with *unit speed*; the second one states that the tangent vector is perpendicular to the second derivative vector, provided the curve is parametrized with respect to arc length.

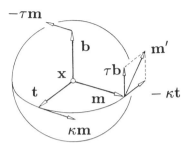

Figure 11.3: The geometric meaning of the Frenet-Serret formulas.

Some simple calculations yield the so-called *Frenet-Serret* formulas:

$$
\begin{aligned}
\mathbf{t}' &= & +\kappa\mathbf{m} & \\
\mathbf{m}' &= & -\kappa\mathbf{t} & & +\tau\mathbf{b}, \\
\mathbf{b}' &= & & -\tau\mathbf{m}
\end{aligned}
\tag{11.6}
$$

where the terms κ and τ, called *curvature* and *torsion*, may be defined both in terms of arc length s and in terms of the actual parameter t. We give both definitions:

$$
\kappa = \kappa(s) = ||\mathbf{x}''||,
$$

$$
\kappa = \kappa(t) = \frac{||\dot{\mathbf{x}} \wedge \ddot{\mathbf{x}}||}{||\dot{\mathbf{x}}||^3},
\tag{11.7}
$$

$$
\tau = \tau(s) = \frac{1}{\kappa^2}\det[\mathbf{x}', \mathbf{x}'', \mathbf{x}'''],
$$

$$
\tau = \tau(t) = \frac{\det[\dot{\mathbf{x}}, \ddot{\mathbf{x}}, \dddot{\mathbf{x}}]}{||\dot{\mathbf{x}} \wedge \ddot{\mathbf{x}}||^2}.
\tag{11.8}
$$

Figure 11.3 illustrates the formulas of Eqs. (11.6).

Curvature and torsion have an intuitive geometric meaning: consider a point $\mathbf{x}(s)$ on the curve and a "consecutive" point $\mathbf{x}(s + \Delta s)$. Let $\Delta\alpha$ denote the angle between the two tangent vectors $\mathbf{t}$ and $\mathbf{t}(s + \Delta s)$ and let $\Delta\beta$ denote the angle between the two binormal vectors $\mathbf{b}$ and $\mathbf{b}(s + \Delta s)$, both angles measured in radians. It is easy to verify that $\Delta\alpha = \kappa\Delta s + \ldots$ and $\Delta\beta = -\tau\Delta s + \ldots$, where "$\ldots$" denotes terms of higher degree in Δs. Thus, when $\Delta s \to ds$, one finds that

$$\kappa = \frac{d\alpha}{ds}, \quad \tau = -\frac{d\beta}{ds}.$$

In other words, κ and $-\tau$ are the angular velocities of $\mathbf{t}$ and $\mathbf{b}$, respectively, because the frame is moved according to the parameter s.

   Remark 4:   Note that κ and τ are independent of the current parametrization of the curve. They are euclidean invariants of the curve, i.e., they are not changed by a rigid body motion of the curve. Moreover, any two continuous functions $\kappa = \kappa(s) > 0$ and $\tau = \tau(s)$ define uniquely (except for rigid body motions) a curve that has curvature κ and torsion τ.

   Remark 5:   The curve may be written in canonical form in terms of the Frenet frame. Then it has the form

$$\mathbf{x}(s + \Delta s) = \begin{bmatrix} \Delta s & -\frac{1}{6}\kappa^2\Delta s^3 + \ldots \\ \frac{1}{2}\kappa\Delta s^2 & +\frac{1}{6}\kappa'\Delta s^3 + \ldots \\ & \frac{1}{6}\kappa\tau\Delta s^3 + \ldots \end{bmatrix},$$

where "$\ldots$" again denotes terms of higher degree in Δs.

11.4   The Osculating Circle

The circle that has second-order contact with the curve at $\mathbf{x}$ is called the *osculating circle* (Figure 11.4). Its center is $\mathbf{c} = \mathbf{x} + \rho\mathbf{m}$, and its radius $\rho = \frac{1}{\kappa}$ is called the *radius of curvature*. We shall provide a brief development of these facts. Using the Frenet-Serret formulas of Eqs. (11.6), the Taylor expansion of $\mathbf{x}(s + \Delta s)$ can be written as

$$\mathbf{x}(s + \Delta s) = \mathbf{x}(s) + \mathbf{t}\Delta s + \frac{1}{2}\kappa\mathbf{m}\Delta s^2 + \ldots.$$

Let ρ^* be the radius of the circle that is tangent to $\mathbf{t}$ at $\mathbf{x}$ and passes through the point $\mathbf{y} = \mathbf{x} + \Delta\mathbf{x}$, where $\Delta\mathbf{x} = \mathbf{t}\Delta s + \frac{1}{2}\kappa\mathbf{m}\Delta s^2$ (see Figure 11.5). Note that $\mathbf{y}$ lies in the osculating plane $\mathbf{O}$. Inspection of the figure reveals that $(\frac{1}{2}\Delta\mathbf{x} - \rho^*\mathbf{m})\Delta\mathbf{x} = 0$, i.e., one obtains

$$\rho^* = \frac{1}{2}\frac{(\Delta\mathbf{x})^2}{\mathbf{m}\Delta\mathbf{x}}.$$

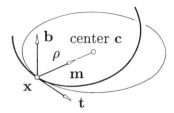

Figure 11.4: The osculating circle.

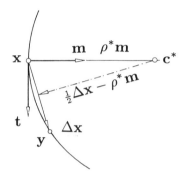

Figure 11.5: Construction of the osculating circle.

From the definition of $\Delta\mathbf{x}$ one obtains $(\Delta\mathbf{x})^2 = \Delta s^2 + \ldots$ and $\mathbf{m}\Delta\mathbf{x} = \frac{1}{2}\kappa(\Delta s)^2$. Thus $\rho^* = \frac{1}{\kappa} + \ldots$. In particular, $\rho = \frac{1}{\kappa}$ as $\Delta s \to 0$. Obviously, this circle lies in the osculating plane.

Remark 6:   Let $\mathbf{x}$ be a rational Bézier curve of degree n as defined in Chapter 15. Its curvature and torsion at $\mathbf{b}_0$ are given by

$$\kappa = \frac{n-1}{n}\frac{w_0w_2}{w_1^2}\frac{b}{a^2}, \quad \tau = \frac{n-2}{n}\frac{w_0w_3}{w_1w_2}\frac{c}{ab}, \tag{11.9}$$

where a is the distance between $\mathbf{b}_0$ and $\mathbf{b}_1$, b is the distance of $\mathbf{b}_2$ to the tangent spanned by $\mathbf{b}_0$ and $\mathbf{b}_1$, and c is the distance of $\mathbf{b}_3$ from the osculating plane spanned by $\mathbf{b}_0, \mathbf{b}_1,$ and $\mathbf{b}_2$ (Figure 11.6). Note that these formulas can be used to calculate curvature and torsion at arbitrary points $\mathbf{x}(t)$ of a Bézier curve after subdividing it there (see Section 15.2).

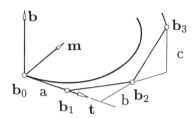

Figure 11.6: Frenet frame and geometric meaning of a, b, c.

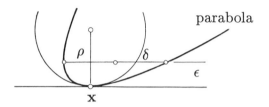

Figure 11.7: Curvature of a parabola.

Remark 7:   An immediate application of (11.9) is the following: let $\mathbf{x}$ be a point on an integral quadratic Bézier curve, i.e., a parabola. Let 2δ denote the length of a chord parallel to the tangent at $\mathbf{x}$, and let ϵ be the distance between the chord and the tangent. The radius of curvature at $\mathbf{x}$ is then $\rho = \frac{\delta^2}{2\epsilon}$ (see Figure 11.7).

Remark 8: An equivalent way to formulate (11.9) is given by

$$\kappa = 2\frac{n-1}{n}\frac{w_0 w_2}{w_1^2}\frac{\text{area}[\mathbf{b}_0, \mathbf{b}_1, \mathbf{b}_2]}{\text{dist}^3[\mathbf{b}_0, \mathbf{b}_1]} \qquad (11.10)$$

and

$$\tau = \frac{3}{2}\frac{n-2}{n}\frac{w_0 w_3}{w_1 w_2}\frac{\text{volume}[\mathbf{b}_0, \mathbf{b}_1, \mathbf{b}_2, \mathbf{b}_3]}{\text{area}^2[\mathbf{b}_0, \mathbf{b}_1, \mathbf{b}_2]}. \qquad (11.11)$$

The advantage of this formulation is that it can be generalized to "higher order curvatures" of curves that span $I\!\!R^d$, $3 < d \le n$ (see Remark 12). An application of this possible generalization is addressed in Remark 13.

11.5 Nonparametric Curves

Let $y = y(t)$; $t \in [a, b]$ be a function. The planar curve $\begin{bmatrix} t \\ y(t) \end{bmatrix}$ is called the graph of $y(t)$ or a *nonparametric curve*. From the above, one derives the following:

the arc element:
$$ds = \sqrt{1 + \dot{y}^2}dt,$$

the tangent vector:
$$\mathbf{t} = \frac{1}{\sqrt{1 + \dot{y}^2}} \begin{bmatrix} 1 \\ \dot{y} \end{bmatrix},$$

the curvature:
$$\kappa = \frac{\ddot{y}}{[1 + \dot{y}^2]^{\frac{3}{2}}},$$

and the center of curvature:
$$\mathbf{c} = \mathbf{x} + \frac{1 + \dot{y}^2}{\ddot{y}} \begin{bmatrix} -\dot{y} \\ 1 \end{bmatrix}.$$

Remark 9: Note that κ has a sign here. Any planar parametric curve can be given a *signed curvature*, for instance, by using the sign of $\det(\dot{\mathbf{x}}, \ddot{\mathbf{x}})$ [see also Eq. (23.1)].

Remark 10: For a nonparametric Bézier curve (see Section 5.5),
$$y(u) = b_0 B_0^n(t) + \cdots + b_n B_n^n(t).$$

Where $u = u_0 + t\Delta u$ is a global parameter, we obtain
$$a = \frac{1}{n}\sqrt{\Delta u^2 + n^2(\Delta b_0)^2}, \quad b = -\frac{\Delta u}{n}\frac{\Delta^2 b_0}{a},$$

as illustrated in Figure 11.8.

11.6 Composite Curves

A curve can be composed of several segments; we have seen spline curves as an example. Let $\mathbf{x}_-$ denote the right endpoint of a segment and $\mathbf{x}_+$ the left

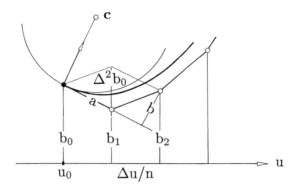

Figure 11.8: Curvature of nonparametric Bézier curve.

endpoint of the adjacent segment. (We will consider only continuous curves, so that $\mathbf{x}_- = \mathbf{x}_+$ always.) Let t be a global parameter of the composite curve and let dots denote derivatives with respect to t. Obviously, the curve is tangent continuous if

$$\dot{\mathbf{x}}_+ = \alpha\dot{\mathbf{x}}_-. \qquad (11.12)$$

Moreover, it follows from (11.9) that it is curvature and osculating plane continuous if in addition

$$\ddot{\mathbf{x}}_+ = \alpha^2\ddot{\mathbf{x}}_- + \alpha_{21}\dot{\mathbf{x}}_-, \qquad (11.13)$$

and it is torsion continuous if in addition

$$\dddot{\mathbf{x}}_+ = \alpha^3\dddot{\mathbf{x}}_- + \alpha_{32}\ddot{\mathbf{x}}_- + \alpha_{31}\dot{\mathbf{x}}_- \qquad (11.14)$$

and vice versa. Since we require the parametrization to be regular, it follows that $\alpha > 0$, while the α_{ij} are arbitrary parameters.

It is interesting to note that curvature and torsion continuous curves exist that are not κ' continuous[3] (see Remark 4). Conversely,

$$\mathbf{x}''' = \mathbf{t}'' = \kappa'\mathbf{m} + \kappa(-\kappa\mathbf{t} + \tau\mathbf{b})$$

implies that $\mathbf{x}'''$ is continuous if κ' is and vice versa. To ensure $\mathbf{x}'''_- = \mathbf{x}'''_+$, the coefficients α and α_{ij} must be the result of the application of the chain rule;

---

[3]Recall that $\kappa' = \mathrm{d}\kappa(s)/\mathrm{d}s$, where the prime denotes differentiation with respect to arc length s of the (composite) curve. A formula for κ' is provided by Eq. (23.2).

i.e., with $\alpha_{21} = \beta$ and $\alpha_{31} = \gamma$, one finds that $\alpha_{32} = 3\alpha\beta$. Now, as above, the curve is tangent continuous if

$$\dot{\mathbf{x}}_+ = \alpha\dot{\mathbf{x}}_-, \quad \alpha > 0,$$

it is curvature and osculating plane continuous if in addition

$$\ddot{\mathbf{x}}_+ = \alpha^2\ddot{\mathbf{x}}_- + \beta\dot{\mathbf{x}}_-,$$

but it is κ' continuous if in addition

$$\dddot{\mathbf{x}}_+ = \alpha^3\dddot{\mathbf{x}}_- + 3\alpha\beta\ddot{\mathbf{x}}_- + \gamma\dot{\mathbf{x}}_-$$

and vice versa.

Remark 11: For planar curves, torsion continuity is a vacuous condition, but κ' continuity is meaningful.

Remark 12: The above results may be used for the definition of higher order *geometric continuity*. A curve is said to be G^r, or r^{th} order geometrically continuous, if a regular reparametrization exists after which it is C^r. This definition is obviously equivalent to the requirement of C^{r-2} continuity of κ and C^{r-3} continuity of τ. As a consequence, geometric continuity may be defined by using the chain rule, as in the example for $r = 3$ above.

Remark 13: The geometric invariants curvature and torsion may be generalized for higher dimensional curves. Continuing the process mentioned in Remark 8, one finds that a d-dimensional curve has $d - 1$ geometric invariants. Continuity of these invariants only makes sense in $I\!\!E^d$, as was demonstrated for $d = 2$ in Remark 11.

Remark 14: Note that although curvature and torsion are euclidean invariants, curvature and torsion continuity (as well as the generalizations discussed in Remarks 12 and 13) are affinely invariant properties of a curve. Both are also projectively invariant properties; see Boehm [68] and Goldman and Micchelli [212].

Remark 15: If two curve segments meet with a continuous tangent and have (possibly different) curvatures κ_- and κ_+ at the common point, then the ratio κ_-/κ_+ is also a projectively invariant quantity. This is known as Memke's theorem, see Bol [76].

Chapter 12

Geometric Continuity I

12.1 Motivation

Before we explain in detail the concept of geometric continuity, we will give an example of a curve that is *curvature continuous* yet *not twice differentiable*. Such curves (and, later, surfaces) are the objects that we will label *geometrically continuous*.

Consider Figure 12.1. It shows two symmetric parabolas that are combined to form a composite curve c. This curve is C^1 over the knot sequence $\{0, 1, 2\}$ (or any affine map thereof). However, it does not form a C^2 piecewise polynomial curve over those two intervals: for two parabolas to form a C^2 piecewise quadratic curve, they must both be part of *one* global parabola (see Section 4.6). That cannot be the case here, since c has parallel tangents at b_0 and b_4, an impossibility for parabolas.

Now let us check for c's curvature at b_2. The two parabolic arcs are symmetric with respect to the perpendicular bisector of b_1 and b_3. They are also tangent continuous at b_2; hence, their centers of curvature at b_2 agree. It follows that c is curvature continuous at b_2. In fact, it is more than that: it possesses a continuously varying normal vector. Another example of a curve that is not C^2 yet has continuously varying curvature and normal vector is the rational representation of the circle as shown later in Figure 14.11.

Differential geometry teaches us that c can be *reparametrized* so that the new parameter is arc length. With that new parametrization, the curve will actually be C^2. Details are explained in Chapter 11.

We shall adopt the term G^2 curves (second-order geometrically continuous) for curves that are *twice differentiable with respect to arc length but*

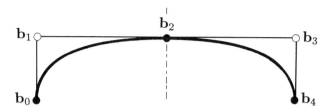

Figure 12.1: Geometric continuity: a curve **c** is shown that consists of two parabolic arcs, both shown with their Bézier polygons. This curve has continuous curvature but is not twice differentiable at $\mathbf{b}_2$.

not necessarily twice differentiable with respect to their current parametrization. Note that curves with a zero tangent vector cannot be G^2 under this definition.

12.2   A Characterization of G^2 Curves

We shall give a general characterization of G^2 curves. We make no assumptions about the actual form of **x** (piecewise polynomial, piecewise trigonometric, ...); we assume only that **x** has a global parameter u and that **x** is a C^1 curve with respect to that parametrization. We also make the assumption that $\dot{\mathbf{x}}(u) \neq \mathbf{0}$.

Let $+$ and $-$ as subscripts denote right and left limits, respectively. Differentiation with respect to arc length s will be denoted by a prime; differentation with respect to the given parameter u will be denoted by a dot.

For **x** to be G^2 at a parameter value u, we must have

$$\mathbf{x}''_+(s) - \mathbf{x}''_-(s) = \mathbf{0}.$$

Since the parameter u can be viewed as a function of arc length s, a change of variable and application of the chain rule yields

$$(u')^2\ddot{\mathbf{x}}_+(u) + u''_+\dot{\mathbf{x}}(u) - (u')^2\ddot{\mathbf{x}}_-(u) - u''_-\dot{\mathbf{x}}(u) = \mathbf{0}. \qquad (12.1)$$

Let us introduce a function $\nu(u)$ by

$$\nu(u) = \frac{u''_- - u''_+}{(u')^2}. \qquad (12.2)$$

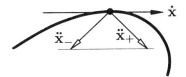

Figure 12.2: G^2 continuity: a curve is not C^2 at a point if its right and left second derivatives do not agree there. It is G^2 if their difference is parallel to the tangent vector.

Note that the denominator in the definition of $\nu(u)$ does not vanish, since we excluded curves with zero tangent vectors from our considerations.

We can now formulate our characterization: a C^1 curve $\mathbf{x}(u)$ is G^2 if and only if there exists a function $\nu(u)$ such that

$$\ddot{\mathbf{x}}_+(u) - \ddot{\mathbf{x}}_-(u) = \nu(u)\dot{\mathbf{x}}(u). \tag{12.3}$$

The function $\nu(u)$ is uniquely determined by the parametrization $s(u)$. The essence of (12.3) is presented in Figure 12.2.

Every C^2 curve (with nonvanishing tangent vectors) is also G^2, and for C^2 curves the function $\nu(u)$ is seen to be identically zero. Of more interest in this context are G^2 piecewise polynomial curves, which are piecewise analytic, and hence certainly piecewise C^2. For these curves, real numbers ν_i must exist such that

$$\ddot{\mathbf{x}}_+(u_i) - \ddot{\mathbf{x}}_-(u_i) = \nu_i\dot{\mathbf{x}}(u_i). \tag{12.4}$$

To see how this ties in with (12.3), we simply define the function $\nu(u)$ as

$$\nu(u) = \begin{cases} \nu_i & u = u_i \\ 0 & \text{else} \end{cases}. \tag{12.5}$$

A warning concerning (12.4): the ν_i in this equation are not invariant under affine parameter transformations. If we reparametrize a curve by $v = au + b$, then $\nu(v) = a\nu(u)$, as can be seen from (12.3).

Equation (12.4) was derived by Nielson [346] in 1974 as a property of ν-splines, C^1 piecewise cubic interpolatory splines that are G^2 but generally (for $\nu_i \neq 0$) not C^2. However, (12.4) actually holds for any piecewise polynomial G^2 curve; that is, we can find the ν_i from any particular representation of the curve.

Equation (12.1) provides one handy characterization for G^2 curves. It is worthwhile to point out that the use of *arc length* is not at all important for this purpose; it is used for its proximity to differential geometry. In fact, we can say that a curve is G^2 if it is C^2 with respect to *some* parametrization, a definition used by several authors (see Barsky [32], Farin [155], and Manning [323]).

12.3   Nu-splines

Nielson [346] derived the equations that determine a G^2 interpolating spline from a particular variational formulation of the interpolation problem (see the end of this section). The characterization of Eq. (12.4) allows a more straightforward derivation of the defining equations.

Let $L+1$ data points $\mathbf{x}_i$; $i = 0, \ldots, L$ be given together with $L+1$ distinct parameter values u_i. We could pass the unique (except for end conditions) interpolating C^2 spline through the data points, as described in Chapter 9. Knowing that G^2 curves are more general than C^2 ones, we could exploit the added generality to define a wider class of interpolating curves. With some luck, this class will contain interpolants that are more attractive in some sense than are C^2 splines.

An interpolatory G^2 piecewise cubic spline may be written in piecewise cubic Hermite form: for $u \in (u_i, u_{i+1})$, the interpolant can be written

$$\mathbf{x}(u) = \mathbf{x}_i H_0^3(r) + \mathbf{m}_i \Delta_i H_1^3(r) + \Delta_i \mathbf{m}_{i+1} H_2^3(r) + \mathbf{x}_{i+1} H_3^3(r), \qquad (12.6)$$

where the H_j^3 are cubic Hermite polynomials from Eq. (6.14) and $r = (u - u_i)/\Delta_i$ is the local parameter of the interval (u_i, u_{i+1}). In (12.6), the $\mathbf{x}_i$ are the known data points, while the $\mathbf{m}_i$ are as yet unknown tangent vectors. The interpolant is supposed to be G^2; it is therefore characterized by (12.4), more specifically,

$$\ddot{\mathbf{x}}_+(u_i) - \ddot{\mathbf{x}}_-(u_i) = \nu_i \mathbf{m}_i \qquad (12.7)$$

for some constants ν_i, where $\mathbf{m}_i = \dot{\mathbf{x}}(u_i)$. The ν_i are constants that can be used to manipulate the shape of the interpolant; they will be discussed soon.

We insert (12.6) into (12.7) and obtain

$$3\left(\frac{\Delta_i \Delta \mathbf{x}_{i-1}}{\Delta_{i-1}} + \frac{\Delta_{i-1} \Delta \mathbf{x}_i}{\Delta_i}\right) = \Delta_i \mathbf{m}_{i-1} + (2\Delta_{i-1} + 2\Delta_i + \tfrac{1}{2}\Delta_{i-1}\Delta_i\nu_i)\mathbf{m}_i$$
$$+\Delta_{i-1}\mathbf{m}_{i+1}; i = 1, \ldots, L - 1. \qquad (12.8)$$

Together with two end conditions, (12.8) can be used to compute the unknown tangent vectors $\mathbf{m}_i$. The simplest end condition is prescribing $\mathbf{m}_0$ and $\mathbf{m}_L$, but any other end condition from Chapter 9 may be used as well. Note that this formulation of the ν-spline interpolation problem depends on the scale of the u_i; it is not invariant under affine parameter transformations. This results from the use of the Hermite form.

It is now time to investigate the advantages of ν-spline interpolation over standard C^2 spline interpolation. We have created an interpolating piecewise polynomial curve that is more general than the C^2 interpolating spline; setting all ν_i equal to zero reduces the ν-spline system (12.8) to the tridiagonal system for C^2 splines (see Chapter 9). Letting some of the ν_i differ from zero will produce a different curve.

The effect of the ν_i on the shape of the interpolant is best studied by considering a limiting case. Suppose we let the kth of the ν_i tend to $+\infty$. Then the kth equation in (12.8) reduces to $\mathbf{m}_k = \mathbf{0}$. The effect of increasing the value of ν_k is therefore a reduction in length of the tangent vector $\mathbf{m}_k$. If we let all ν_i tend to infinity, all tangent vectors will converge to zero length, and the interpolant will look like a piecewise linear curve through the data points. This limiting case is not G^2, since we did not admit curves with zero tangent vectors into the class of G^2 curves. Figure 12.3 shows that although the limiting ν-spline interpolant *looks* like a piecewise linear curve, it is actually C^1.

The effect of the ν_i on the shape of the interpolating ν-spline curve leads to their description as *tension parameters*. The higher the values of the tension parameters, the "tighter" the curve. Figure 12.4 gives an example. Nu-spline curves are best used in an interactive computer graphics environment: as a first pass, the interpolating C^2 spline to a given data set is computed. Next, one might increase tension values in regions where the spline curve exhibits unwanted undulations. This process is interactive and must be repeated until a satisfactory shape is obtained.

While the original development of ν-splines considered only nonnegative tension values, it seems that negative ν_i can also influence the shape of an interpolating ν-spline curve in a beneficial way (Farin [158]). Finding "good" tension values is not easy, however, and an automatic method for doing so is not yet known.

We have neglected the question of whether the linear system (12.8) always possesses a solution. For the case of all ν_i being zero, we know that

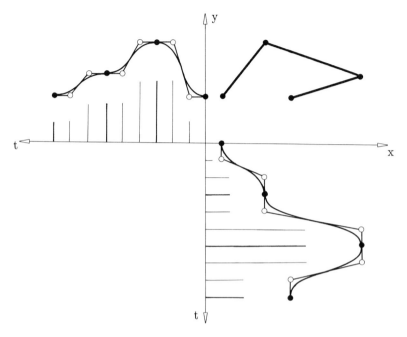

Figure 12.3: Nu-splines: in the limiting case $\nu_i \to \infty$, the interpolant approaches the polygon through the data points. The cross plot shows that it is still C^1.

the coefficient matrix is positive definite, since this is simply the case of C^2 cubic splines. For that case, we always have a unique solution. If all ν_i are greater than zero, the diagonal elements of the coefficient matrix become larger, and positive definiteness – and thus the existence of a unique solution – is maintained. If some of the ν_i become negative, however, we cannot make a statement about the existence of a solution (or of a unique solution).

A theoretical analysis of the solvability of (12.8) involves the eigenvalues of the coefficient matrix: if they are nonzero, then it is nonsingular. A preliminary analysis using Gershgorin's circle theorem is presented by Barsky [33]. At the time this book was written, no general *a priori* check was known that determines if a given set of tension parameters will cause the matrix to be singular.

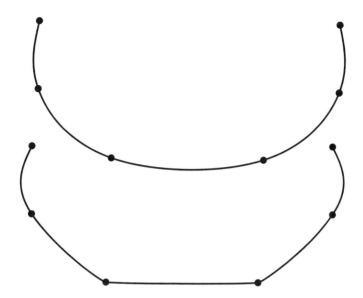

Figure 12.4: Nu-splines: the top curve is a C^2 cubic spline interpolant to the data points. The bottom curve results from selecting high values for ν_2 and ν_3.

We finish our discussion of ν-spline interpolation with a minimum property that was the starting point for this theory. Standard C^2 splines minimize the functional

$$\int_{u_0}^{u_L} [\ddot{\mathbf{x}}(t)]^2 \mathrm{d}t,$$

as explained in Section 9.5. The ν-spline interpolant minimizes the more general functional

$$\int_{u_0}^{u_L} [\ddot{\mathbf{x}}(t)]^2 \mathrm{d}t + \sum \nu_i [\mathbf{m}_i]^2. \tag{12.9}$$

If several ν_i are chosen to be very large, this functional forces the corresponding tangent vectors $\mathbf{m}_i$ to become small.

12.4 G^2 Piecewise Bézier Curves

Equation (12.4) provides the condition for a general piecewise polynomial to be G^2. Let us now investigate what happens if we represent such a curve in piecewise Bézier form.

It will be useful to rewrite (12.4). We cross-product both sides with $\dot{\mathbf{x}}$ and obtain

$$\ddot{\mathbf{x}}_+ \wedge \dot{\mathbf{x}} = \ddot{\mathbf{x}}_- \wedge \dot{\mathbf{x}} \qquad (12.10)$$

as another characterization of G^2 continuity. In terms of differential geometry, this equation states that $\mathbf{x}$ possesses both continuously varying curvature and a binormal vector.

Now let $\mathbf{b}_0, \ldots, \mathbf{b}_n$ and $\mathbf{c}_0, \ldots, \mathbf{c}_n$ be the Bézier polygons of two n^{th} degree polynomials. We assume[1] that both curve segments form a C^1 curve over a knot partition u_0, u_1, u_2, implying

$$\text{ratio}(\mathbf{b}_{n-1}, \mathbf{b}_n, \mathbf{c}_1) = \frac{\Delta_0}{\Delta_1} \qquad (12.11)$$

where $\Delta_0 = u_1 - u_0$, $\Delta_1 = u_2 - u_1$. Recalling the formulas for first and second derivatives of Bézier curves, we can express (12.10) in terms of the Bézier points that are involved:

$$\frac{1}{\Delta_0^3}\Delta^2\mathbf{b}_{n-2} \wedge \Delta\mathbf{b}_{n-1} = \frac{1}{\Delta_1^3}\Delta^2\mathbf{c}_0 \wedge \Delta\mathbf{c}_0. \qquad (12.12)$$

Invoking the geometric interpretation of the second difference vector Δ^2 (see Figure 4.3), we can rewrite (12.12) and obtain a *characterization of G^2 continuity for Bézier curves* (following Farin [155]):

$$\frac{1}{\Delta_0^3}\Delta\mathbf{b}_{n-2} \wedge \Delta\mathbf{b}_{n-1} = \frac{1}{\Delta_1^3}\Delta\mathbf{c}_0 \wedge \Delta\mathbf{c}_1, \qquad (12.13)$$

now only using first differences. A useful modification, utilizing (12.11), is the following:

$$\frac{1}{\Delta_0^2}\Delta\mathbf{b}_{n-2} \wedge (\mathbf{c}_1 - \mathbf{b}_{n-1}) = \frac{1}{\Delta_1^2}(\mathbf{c}_1 - \mathbf{b}_{n-1}) \wedge \Delta\mathbf{c}_1. \qquad (12.14)$$

The absolute value of the cross product of two vectors equals the area of the parallelogram spanned by the two vectors. Thus the two preceding formulas yield:

$$\frac{\text{area}(\mathbf{b}_{n-2}, \mathbf{b}_{n-1}, \mathbf{b}_n)}{\text{area}(\mathbf{c}_0, \mathbf{c}_1, \mathbf{c}_2)} = \frac{\Delta_0^3}{\Delta_1^3} \qquad (12.15)$$

and

$$\frac{\text{area}(\mathbf{b}_{n-2}, \mathbf{b}_{n-1}, \mathbf{c}_1)}{\text{area}(\mathbf{b}_{n-1}, \mathbf{c}_1, \mathbf{c}_2)} = \frac{\Delta_0^2}{\Delta_1^2}. \qquad (12.16)$$

---

[1]Remember that one can always find a knot partition such that a tangent continuous piecewise polynomial becomes a C^1 curve!

In both cases, we denote by area$(\mathbf{a}, \mathbf{b}, \mathbf{c})$ the area of the triangle with vertices $\mathbf{a}, \mathbf{b}, \mathbf{c}$. These two formulas are a characterization for *curvature continuity* of two Bézier curves. This is a weaker requirement than G^2 continuity, since it does not guarantee continuity of the osculating plane.

Note that, since $\mathbf{b}_{n-2}, \mathbf{b}_{n-1}, \mathbf{b}_n = \mathbf{c}_0, \mathbf{c}_1, \mathbf{c}_2$ have to be coplanar for (12.12) to hold, it makes sense to assign a *sign* to the above areas. A consequence of G^2 continuity is then that $\mathbf{b}_{n-2}$ and $\mathbf{c}_2$ must be on the same side of the tangent through $\mathbf{b}_{n-1}$ and $\mathbf{c}_1$. Otherwise, (12.16) would involve square roots of negative numbers.

Another consequence of the coplanarity of $\mathbf{b}_{n-2}, \mathbf{b}_{n-1}, \mathbf{b}_n = \mathbf{c}_0, \mathbf{c}_1, \mathbf{c}_2$ is *affine invariance* of the G^2 condition of Eq. (12.16). The area ratio of two coplanar triangles is not affected by affine maps, and so the left-hand side of (12.16) is not affected by an affine map: *affine maps take G^2 curves to G^2 curves.*

Proof that G^2 continuity is a genuine generalization of the concept of second-order differentiability involves consideration of two examples.

Figure 12.5 shows two adjacent Bézier curves. They form a C^1 curve over the knot partition determined by the ratio of the three collinear points

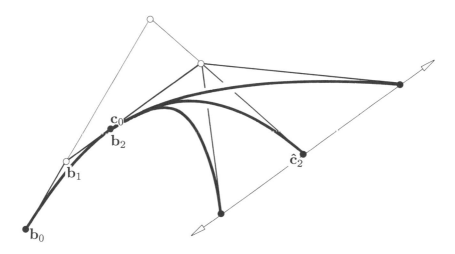

Figure 12.5: G^2 continuity: a family of curvature continuous piecewise quadratic Bézier curves. The point $\hat{\mathbf{c}}_2$ may vary parallel to the tangent at $\mathbf{c}_0$.

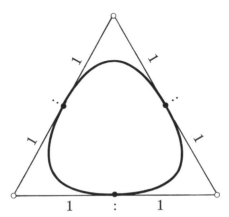

Figure 12.6: G^2 curves: a closed quadratic G^2 spline.

$\mathbf{b}_2, \mathbf{b}_3 = \mathbf{c}_0, \mathbf{c}_1$. The location of $\mathbf{b}_1$ would uniquely determine $\hat{\mathbf{c}}_2$ if we wanted our two quadratics to form one C^2 curve (see the C^r condition in Section 7.2). If we relax that requirement and are content with G^2 continuity, we may place $\mathbf{c}_2$ anywhere on the straight line through $\hat{\mathbf{c}}_2$ and parallel to the tangent through $\mathbf{b}_3$.

The next example: let $\mathbf{a}$, $\mathbf{b}$, $\mathbf{c}$ be the vertices of a triangle and let $\mathbf{d}$, $\mathbf{e}$, $\mathbf{f}$ be the edge midpoints as in Figure 12.6. We can interpret these points as Bézier points of a piecewise quadratic curve (Figure 12.6). This closed curve is G^2. It is a C^1 curve over any uniform knot sequence, for example, $0, 1, 2$. However, one easily verifies that it is not a twice differentiable piecewise quadratic with respect to any knot sequence.

12.5   Direct G^2 Cubic Splines

We have seen in Section 7.7 how to construct the piecewise Bézier polygon of a C^2 cubic spline from a given knot sequence $u_0, \ldots, u_L$ and a given polygon $\mathbf{d}_{-1}, \ldots, \mathbf{d}_{L+1}$. The procedure was to find the inner Bézier points $\mathbf{b}_{3i\pm1}$ on the control polygon using the C^2 conditions and then find the junction points $\mathbf{b}_{3i}$ from the C^1 conditions. The analogous construction for a G^2 spline is less restricted. In particular, one does not have to prescribe a knot sequence in addition to the control polygon. It will be selected automatically by the following algorithm (see [155]).

We are given a control polygon $\mathbf{d}_{-1}, \ldots, \mathbf{d}_{L+1}$, called the G^2 control polygon. We want to construct a G^2 piecewise cubic Bézier curve with L segments and so must determine $\mathbf{b}_0, \mathbf{b}_1, \ldots \mathbf{b}_{3L}$. We proceed as follows:

- In the case of an open polygon, set
  $\mathbf{b}_0 = \mathbf{d}_{-1}, \mathbf{b}_1 = \mathbf{d}_0, \mathbf{b}_{3L-1} = \mathbf{d}_L, \mathbf{b}_{3L} = \mathbf{d}_{L+1}$.
- Next, select $\mathbf{b}_2$ anywhere on the polygon leg $\mathbf{d}_0, \mathbf{d}_1$ and
  select $\mathbf{b}_{3L-2}$ anywhere on the polygon leg $\mathbf{d}_{L-1}, \mathbf{d}_L$.
- Choose the inner points $\mathbf{b}_{3i-2}, \mathbf{b}_{3i-1}$
  anywhere on the polygon leg $\mathbf{d}_{i-1}, \mathbf{d}_i$.
- Find the junction points $\mathbf{b}_{3i}$ from (12.16).

The above "anywheres" should be interpreted in the sense that a program needs to be supplied extra data, either in the form of interactive input by mouse or lightpen, or by input through a data file.

For the last step, we must compute a ratio Δ_0/Δ_1 for each junction point. It is computable from the inner Bézier points. We may set $\Delta_0 = 1$ without loss of generality, and obtain

$$\Delta_1 = \sqrt{\frac{\text{area}(\mathbf{b}_{3i-2}, \mathbf{b}_{3i-1}, \mathbf{b}_{3i+1})}{\text{area}(\mathbf{b}_{3i-1}, \mathbf{b}_{3i+1}, \mathbf{b}_{3i+2})}},$$

and finally the desired junction point $\mathbf{b}_{3i}$:

$$\mathbf{b}_{3i} = \frac{\Delta_1 \mathbf{b}_{3i-1} + \mathbf{b}_{3i+1}}{1 + \Delta_1}. \tag{12.17}$$

This construction yields the piecewise Bézier polygon of a G^2 cubic spline curve. We note that the osculating plane (see Chapter 11) at $\mathbf{b}_{3i}$ is spanned by $\mathbf{d}_{i-1}, \mathbf{d}_i, \mathbf{d}_{i+1}$. For closed polygons, the first two steps in the above construction are omitted. Numbering would then be mod L.

In interactive design, one would utilize G^2 cubic splines in a two-step procedure. The design of the G^2 control polygon may be viewed as a rough sketch. The program would estimate the inner Bézier points automatically, and the designer could fine-tune the curve shape by readjusting them where necessary. At no time does he or she have to worry about the knot sequence – it is computed automatically from the G^2 conditions.

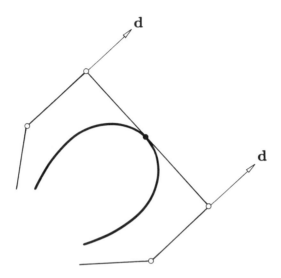

Figure 12.7: G^2 splines: these two cubics are a G^2 spline but do not possess a G^2 control polygon.

There is one interesting difference between the above construction for a G^2 spline and the corresponding construction for a C^2 spline: every cubic C^2 possesses a B-spline control polygon – but not every G^2 piecewise cubic curve possesses a G^2 control polygon. The two cubics in Figure 12.7 are curvature continuous, yet they cannot be obtained with the above construction: the control point $\mathbf{d}_1$ would have to be at infinity.

12.6   Implementation

We include a program that utilizes Eq. (12.17). It assumes that the piecewise Bézier polygon has been determined except for the junction points $\mathbf{b}_{3i}$, which will be computed:

```
void direct_gspline(l,bez_x,bez_y)
/* From given interior Bezier points,
   the junction Bezier points b3i are found from the G2 conditions.
Input: l:           no of cubic pieces.
       bez_x,bez_y: interior Bezier points b_{3i+1}, b_{3i-1}.
Output:bez_x,bez_y: completed piecewise Bezier polygon.
```

```
Note:                  b_0 and b_{31+3} should be provided, too!
*/
```

The routine **height**, which is used by **direct_gspline**, has the header

```
float height(px,py,ax,ay,bx,by)
/* find the height of point (px,py) over
   straight line thru (ax,ay) and (bx,by)
*/
```

12.7 Problems

1. Figure 12.6 shows a triangle and an inscribed piecewise quadratic curve. Find the ratio of the areas enclosed by the curve and the triangle.

2. G^2 piecewise cubics, when constructed as direct G^2 splines, may contain straight line segments that are written as cubics. Discuss osculating plane continuity for such curves.

3. The G^2 piecewise cubic from Figure 12.7 cannot be represented as a direct G^2 spline. Can it be obtained from a ν-spline interpolation problem?

4. Find cases where the solution to a ν-spline interpolation problem is not a G^2 curve, even for finite values of ν_i.

Chapter 13

Geometric Continuity II

The preceding chapter contained the basic algorithms for G^2 curves: an interpolation algorithm (ν-splines) and a design algorithm (direct G^2 splines). In this chapter, we shall add a theoretical framework to these algorithms. See also the remarks in Chapter 11 concerning geometric continuity.

13.1 Gamma-splines

The direct G^2 cubic splines from Section 12.5 may be a handy tool in interactive design, but they are not amenable to mathematical analysis. A more formal description reveals the relationship between G^2 splines and classical B-splines, as described in Chapter 7. This formalization, developed by W. Boehm [64], is concerned with the same G^2 curves as above, but because of the different treatment, G^2 splines in this context are called γ-splines.

Consider Figure 13.1. The two points $\mathbf{d}_-$ and $\mathbf{d}_+$ are the auxiliary points for the C^2 condition of standard cubic B-spline curves (see Section 7.4). Since they do not agree, the curve is not twice differentiable at $\mathbf{b}_{3i}$. Some notation:

$$A_- = \text{area}(\mathbf{b}_{3i-2}, \mathbf{b}_{3i-1}, \mathbf{b}_{3i+1}),$$

$$A_+ = \text{area}(\mathbf{b}_{3i-1}, \mathbf{b}_{3i+1}, \mathbf{b}_{3i+2}),$$

$$A = \text{area}(\mathbf{b}_{3i-1}, \mathbf{d}_-, \mathbf{b}_{3i+1}) = \text{area}(\mathbf{b}_{3i-1}, \mathbf{d}_+, \mathbf{b}_{3i+1}).$$

The last statement needs justification; it is provided through

$$A_- \frac{\Delta_i}{\Delta_{i-1}} = A_+ \frac{\Delta_{i-1}}{\Delta_i} = A,$$

which follows directly from (12.16).

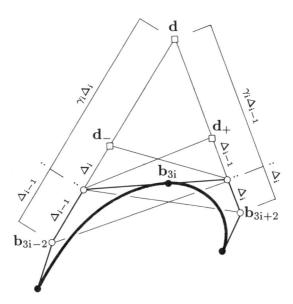

Figure 13.1: Gamma-splines: the G^2 conditions can be formulated in terms of triangle areas.

Let $\mathbf{d}_i$ be the intersection of the two straight lines through $\mathbf{b}_{3i-2}, \mathbf{d}_-$ and $\mathbf{d}_+, \mathbf{b}_{3i+2}$. A number γ_i exists then such that

$$\text{area}(\mathbf{b}_{3i-1}, \mathbf{d}_i, \mathbf{b}_{3i+1}) = \gamma_i A.$$

It follows that

$$\text{ratio}(\mathbf{b}_{3i-2}, \mathbf{b}_{3i-1}, \mathbf{d}_i) = \frac{\Delta_{i-1}}{\gamma_i \Delta_i}, \tag{13.1}$$

$$\text{ratio}(\mathbf{d}_i, \mathbf{b}_{3i+1}, \mathbf{b}_{3i+2}) = \frac{\gamma_i \Delta_{i-1}}{\Delta_i}. \tag{13.2}$$

We can now formulate an algorithm for γ-spline curves: given a control polygon $\mathbf{d}_{-1}, \ldots, \mathbf{d}_{L+1}$, a knot sequence $u_0, \ldots, u_L$, and a set of *shape parameters* $\gamma_1, \ldots, \gamma_{L-1}$, find the piecewise Bézier polygon of the corresponding γ-spline curve. We proceed as indicated in Figure 13.2. We first determine the inner Bézier points $\mathbf{b}_{3i\pm1}$ and then find the junction points $\mathbf{b}_{3i}$.

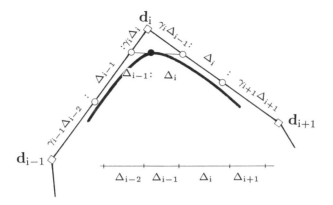

Figure 13.2: Gamma-splines: the Bézier points are connected to the G^2 control polygon by the shown ratios.

For the inner Bézier points we get:

$$\mathbf{b}_{3i-2} = \frac{\Delta_{i-1} + \gamma_i \Delta_i}{\Delta} \mathbf{d}_{i-1} + \frac{\gamma_{i-1} \Delta_{i-2}}{\Delta} \mathbf{d}_i \qquad (13.3)$$

$$\mathbf{b}_{3i-1} = \frac{\gamma_i \Delta_i}{\Delta} \mathbf{d}_{i-1} + \frac{\gamma_{i-1} \Delta_{i-2} + \Delta_{i-1}}{\Delta} \mathbf{d}_i, \qquad (13.4)$$

where

$$\Delta = \gamma_{i-1} \Delta_{i-2} + \Delta_{i-1} + \gamma_i \Delta_i. \qquad (13.5)$$

For the junction points we find

$$\mathbf{b}_{3i} = \frac{\Delta_i}{\Delta_{i-1} + \Delta_i} \mathbf{b}_{3i-1} + \frac{\Delta_{i-1}}{\Delta_{i-1} + \Delta_i} \mathbf{b}_{3i+1}. \qquad (13.6)$$

This equation is identical to the corresponding one for C^2 cubic B-spline curves, Eq. (7.12).

For $\gamma_i = 1$, we recapture the familiar construction for C^2 cubic splines. The value for γ_i may also be negative, giving rise to curves that may have loops. In these cases, the curve does not necessarily stay within the convex hull of the G^2 control polygon (see Farin, Hansford, and Worsey [167] for more details). Typically, one would use values for the γ_i that are between 0 and 1. As $\gamma_i \to 0$, the curve comes closer to $\mathbf{d}_i$, giving it a sharper bend there. If we let all γ_i approach zero, the resulting curves will converge to the control polygon, a situation comparable to letting all ν_i tend to infinity in ν-spline interpolation. Figure 13.3 gives an example.

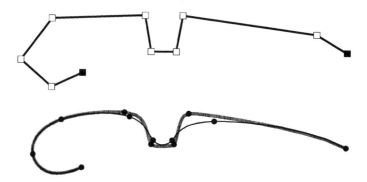

Figure 13.3: Gamma-splines: two curves are shown that are defined by the same control polygon (top). The thin black curve is the C^2 uniform B-spline curve; the heavy gray curve corresponds to small values for $\gamma_3, \gamma_4, \gamma_5, \gamma_6$.

13.2 Local Basis Functions for G^2 Splines

In the previous sections, we developed two methods for the generation of G^2 splines: the direct construction from Section 12.5 and the γ-spline development in the preceding section. Clearly, the direct approach is superior for design in that it is based only on geometric entities, whereas for γ-splines, one needs to specify a knot sequence and a sequence of γ_i, both being numerical input. The strength of the γ-spline approach is not in the design field but in the analysis of G^2 splines.

The following property is easy to prove for γ-spline curves, but it is not obvious how to even formulate it for the direct G^2 approach. Consider two γ-spline curves $\mathbf{g}$ and $\hat{\mathbf{g}}$ over the same knot sequence and with the same γ_i. Denote the G^2 control vertices of $\mathbf{g}$ by $\mathbf{d}_i$, those of $\hat{\mathbf{g}}$ by $\hat{\mathbf{d}}_i$. We observe that the barycentric combination

$$\mathbf{h}(u) = (1 - \alpha)\mathbf{g}(u) + \alpha\hat{\mathbf{g}}(u)$$

is again a γ-spline curve. Moreover, the G^2 control polygon for $\mathbf{h}$ consists of the points $(1 - \alpha)\mathbf{d}_i + \alpha\hat{\mathbf{d}}_i$. A glance at Figure 13.4 reveals the truth of this statement: the points $\mathbf{d}_{i-1}, \mathbf{d}_i, \hat{\mathbf{d}}_{i-1}, \hat{\mathbf{d}}_i$ form a bilinear surface. Thus the Bézier points and the G^2 control vertices of $\mathbf{h}$ are related to each other in the same ratios as those of $\mathbf{g}$ and $\hat{\mathbf{g}}$, assuring that $\mathbf{h}$ is again a γ-spline curve.

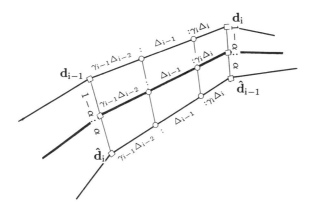

Figure 13.4: Gamma-splines: a barycentric combination of two γ-splines is obtained by forming the barycentric combination of their G^2 control polygons.

A consequence of this linearity property is that all γ-splines over the same knot sequence and with the same γ_i form a linear space whose dimension, $L + 3$, equals the number of control vertices of each γ-spline in that space. Each element of that space then has a basis representation

$$\mathbf{x}(u) = \sum_{i=-1}^{L+1} \mathbf{d}_i M_i(u). \qquad (13.7)$$

We are slightly negligent here: actually, the M_i do not only depend on u but also on the u_i and the γ_i.

We shall now develop several properties of the M_i until we are finally able to give an explicit form for them. As the geometry of the γ-spline construction reveals, they have the following properties:

Partition of unity: This follows since the affine invariance of the γ-spline construction implies that (13.7) is a barycentric combination:

$$\sum_{i=-1}^{L+1} M_i(u) \equiv 1. \qquad (13.8)$$

Positivity: For $\gamma_i \geq 0$, the γ-spline curve lies in the convex hull of the control polygon. Thus (13.7) is a convex combination:

$$M_i(u) \geq 0. \qquad (13.9)$$

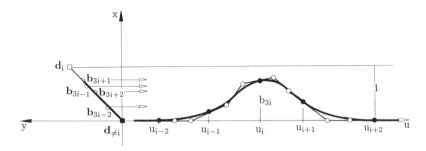

Figure 13.5:   Local basis for G^2 splines: a basis function M_i is obtained through the cross plot technique. Only the plot for $x(u)$ is shown, the one for $y(u)$ being identical.

Local support: If we change one $\mathbf{d}_i$, the curve is only changed over the four intervals $(u_{i-2}, \ldots, u_{i+2})$. This is illustrated in Figure 7.14 in the context of C^2 B-spline curves. Thus the corresponding basis function $M_i(u)$ must vanish outside this region:

$$M_i(u) = 0 \quad \text{for} \quad u \notin [u_{i-2}, u_{i+2}]. \tag{13.10}$$

Equation (13.10) is a consequence of the fact that a change in $\mathbf{d}_i$ does not affect $\mathbf{b}_j$ with $j \leq 3i - 6$ or with $j \geq 3i + 6$. That change does not affect $\mathbf{b}_{3i\pm5}$ and $\mathbf{b}_{3i\pm4}$ either – therefore, the first and second derivatives of the curve at u_{i-2} and u_{i+2} remain unchanged. As a consequence,

$$\frac{\mathrm{d}}{\mathrm{d}u} M_i(u_{i\pm2}) = \frac{\mathrm{d}^2}{\mathrm{d}u^2} M_i(u_{i\pm2}) = 0. \tag{13.11}$$

With these properties at hand, we can now construct M_i. Consider the control polygon that is obtained by setting $\mathbf{d}_i = \begin{bmatrix} 1 \\ 1 \end{bmatrix}$ while setting all other vertices $\mathbf{d}_j = \mathbf{0}$. The graph of this polygon is quite degenerate – only one control point is nonzero. Its usefulness stems from the fact that the cross plot of the corresponding γ-spline curve consists of $\begin{bmatrix} M_i(u) \\ M_i(u) \end{bmatrix}$; in other words, it singles out exactly one basis function. We can therefore construct the Bézier points of M_i by the use of a cross plot (see Figure 13.5); if necessary, consult

Sections 5.5 and 5.6. The Bézier ordinates of M_i are now a simple consequence of (13.3), (13.4), and (13.6):

$$b_{3i-2} = \frac{\gamma_{i-1}\Delta_{i-2}}{\Gamma_1}, \tag{13.12}$$

$$b_{3i-1} = \frac{\gamma_{i-1}\Delta_{i-2} + \Delta_{i-1}}{\Gamma_1}, \tag{13.13}$$

$$b_{3i+1} = \frac{\Delta_i + \gamma_{i+1}\Delta_{i+1}}{\Gamma_2}, \tag{13.14}$$

$$b_{3i+2} = \frac{\gamma_{i+1}\Delta_{i+1}}{\Gamma_2}, \tag{13.15}$$

where

$$\Gamma_1 = \gamma_{i-1}\Delta_{i-2} + \Delta_{i-1} + \gamma_i\Delta_i$$

and

$$\Gamma_2 = \gamma_i\Delta_{i-1} + \Delta_i + \gamma_{i+1}\Delta_{i+1}.$$

For the junction ordinate b_{3i} we find

$$b_{3i} = \frac{\Delta_i}{\Delta_{i-1} + \Delta_i}b_{3i-1} + \frac{\Delta_{i-1}}{\Delta_{i-1} + \Delta_i}b_{3i+1}. \tag{13.16}$$

All remaining Bézier ordinates of M_i are zero.

The basis functions M_i are C^1 by construction. They are not twice differentiable, not even G^2. It seems a bit of a miracle, then, that a linear combination of *curvature discontinuous* functions can generate a parametric curve that is *curvature continuous*.[1] Let us investigate this phenomenon more closely. A γ-spline, being a G^2 curve, must satisfy the G^2 condition of Eq. (12.4). This condition must hold for both the x- and the y-component of the curve (and for the z-component if it happens to be a space curve). For the special example in Figure 13.5, both components are given by $M_i(u)$. Thus

$$\frac{\mathrm{d}^2}{\mathrm{d}u^2}M_i(u_j+) - \frac{\mathrm{d}^2}{\mathrm{d}u^2}M_i(u_j-) = \nu_j\frac{\mathrm{d}}{\mathrm{d}u}M_i(u_j), \quad \text{all } j \tag{13.17}$$

for some ν_j, which are nonzero in the general case.

When we talk about G^2 continuity of M_i, we really mean G^2 continuity of its graph, which is a 2D curve $\begin{bmatrix} u \\ M_i(u) \end{bmatrix}$. To be G^2, both components of

---

[1] As long as $\dot{\mathbf{x}}(u) \neq \mathbf{0}$ for any u.

this curve must satisfy (12.4). The second one does because of (13.17), but the first one, u, does not if the ν_j are allowed to be nonzero, as in this case. The graph of M_i is therefore not curvature continuous.

Equation (13.17) may be used to derive a connection between the γ_j and the ν_j. One obtains (Boehm [64]):

$$\nu_j = 2\left(\frac{1}{\Delta_{j-1}} + \frac{1}{\Delta_j}\right)\left(\frac{1}{\gamma_j} - 1\right). \tag{13.18}$$

Let us summarize the historical development of these splines. The first local basis for G^2 splines was developed by G. Nielson and J. Lewis [310] in 1975. In 1981, B. Barsky [32] developed a local basis for so-called β-splines, which are, in the context of this chapter, γ-splines with constant $\gamma_i = \gamma$ and a distorted knot sequence with $\Delta_i = \beta\Delta_{i-1}$. Later, local bases were developed for β-spline curves that are equivalent to γ-splines (Bartels and Beatty [41]).

13.3   Beta-splines

Our derivation of γ-splines was partly based on the fact that every tangent continuous piecewise cubic may be endowed with a knot sequence such that it becomes a C^1 piecewise cubic curve (see Section 7.5). In particular, the γ-splines $M_i(u)$ are C^1 piecewise cubics.

A slight modification of the γ-splines M_i gives rise to so-called β-splines $\hat{M}_i$. The modification is not to the function values of M_i but rather to the knot sequence. The idea is to rescale each interval so that it is of unit length. As an example, consider Figures 7.6 and 7.7. The coordinate function $x(u)$ of a C^1 piecewise cubic $\mathbf{x}(u)$ is itself a C^1 piecewise cubic over the u-axis, as shown in Figure 7.6. Rescaling each interval so that it is of unit length causes the graph of $x(u)$ to become tangent discontinuous. Given the tangent discontinuous graph $x(u)$ of Figure 7.7, we can only recover the parametric curve in the cross plot if we know the rescaling factors that made each interval of unit length. These rescaling factors are called $\beta_{1,i}$ and are defined by

$$\beta_{1,i} = \frac{\Delta_{i+1}}{\Delta_i}.$$

To obtain β-splines $\hat{M}_i$ from γ-splines M_i, we thus have to rescale the u-axis so that each interval Δ_i is scaled to be of unit length. The resulting β-spline may be tangent discontinuous; see Figure 13.6.

Plate I.
An automobile.
(Courtesy of Mercedes-Benz, FRG.)

Plate II.
Color rendering of the
hood. *(Courtesy of
Mercedes-Benz, FRG.)*

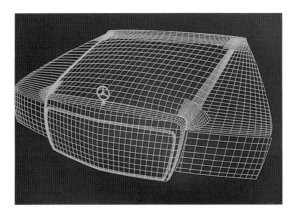

Plate III.
Wire frame rendering of the
hood *(Courtesy of
Mercedes-Benz, FRG.)*

Plate IV. In a database, the hood is stored as an assembly of bicubic spline surfaces. The B-spline net of one of the surfaces is shown. *(Courtesy of Mercedes-Benz, FRG.)*

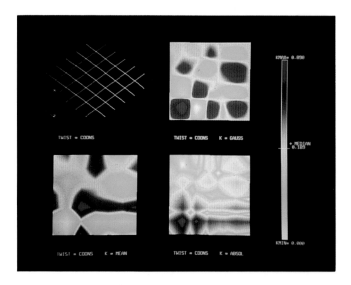

Plate V. A wire frame rendering of a surface (top left) and its Gaussian (top right), mean (bottom left), and absolute (bottom right) curvatures.

Plate VI.
The same curvatures
for a different surface.

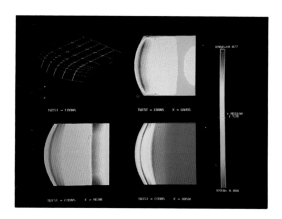

Plate VII.
The surface from Plate VI,
but now with several twist
vectors. (Also, a different color
map is used.)

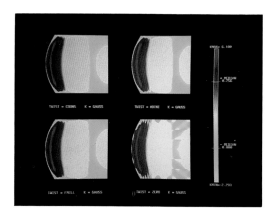

Plate VIII.
The surface from Plate VI,
after perturbations have
been applied.

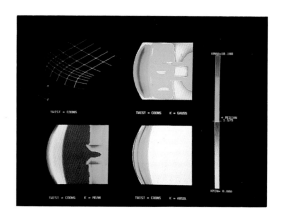

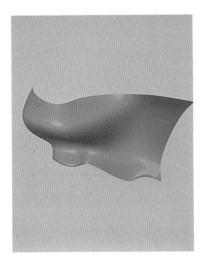

Plate IX.
A bicubic B-spline surface.
(Courtesy of Silicon Graphics.)

Plate X.
The surface from Plate IX,
now with its B-spline net.
(Courtesy of Silicon Graphics.)

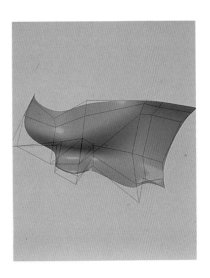

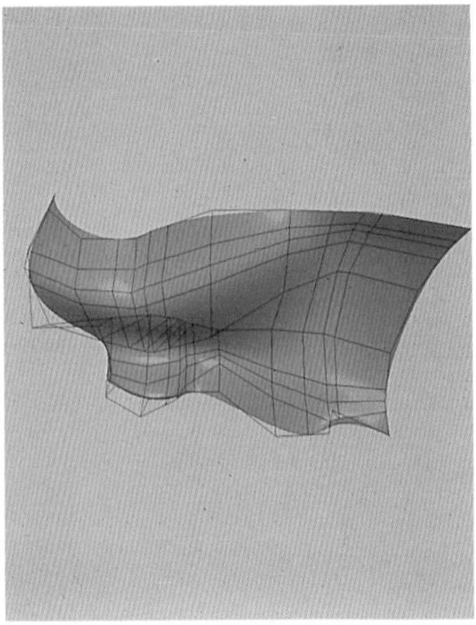

Plate XI.
The same surface, now
with the piecewise Bézier net.
(Courtesy of Silicon Graphics.)

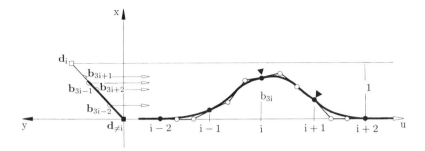

Figure 13.6: β-splines: a basis function is shown. It is defined over a uniform knot partition and is potentially tangent discontinuous. Here, tangent discontinuities occur at the marked points.

In the context of β-splines, the shape parameters γ_i are replaced by the ν_i from the G^2 condition (12.4) and called $\beta_{2,i}$:

$$\beta_{2,i} = \nu_i.$$

Since the γ_i and the ν_i are related by (13.18), we can now write a piecewise cubic G^2 curve in either γ-spline or β-spline form. More material about this relationship is presented in DeRose and Barsky [136]. Additional information on β-splines can be found in Goodman and Unsworth [224], [223].

It should be noted that several definitions of β-spline curves exist. The previous one is the only one that is capable of modeling the full variety of G^2 piecewise cubics. Another kind of β-spline is obtained by setting all $\beta_{1,i} = \beta_1$ and all $\beta_{2,i} = \beta_2$, thus reducing the number of shape parameters available for the whole curve to two *global* shape parameters. This approach to G^2 piecewise cubics, as described in Barsky [32], [34], has gained popularity in the graphics community. For the purpose of designing real objects, however, these curves are not flexible enough: instead of being a true generalization of B-spline curves, they abandon the most useful property of B-spline curves, namely *local control*.

13.4   Higher Order Geometric Continuity

Just as we can define higher order parametric continuity C^r, we may also define higher order geometric continuity. We say that a curve is r^{th} order

Figure 13.7: G^r continuity: a segment of a C^r curve may be reparametrized. The resulting curve is not C^r any more, but still G^r.

geometrically continuous, or G^r, at a given point, if it can be reparametrized such that it will become C^r (see Remark 12 in Section 11.6). In particular, the new parameter might be arc length.

To derive conditions for G^r continuity, we start with a composite C^r curve $\mathbf{x}(u)$ with a global parameter u. At a given parameter value u, derivatives from the left and from the right agree:

$$\frac{d^i}{du^i}\mathbf{x}_- = \frac{d^i}{du^i}\mathbf{x}_+; \quad i = 0, \ldots, r. \tag{13.19}$$

Now let us reparametrize the right segment by introducing a new parameter $t = t(u)$; see Figure 13.7. By our above definition, the resulting composite curve will be G^r, while it is clearly not C^r any more. We will now study the conditions for G^r continuity using this composite G^r curve.

Modifying Eq. (13.19) so as to incorporate the new parametrization yields:

$$\frac{d^i}{du^i}\mathbf{x}_- = \frac{d^i}{du^i}\mathbf{x}(t)_+; \quad i = 0, \ldots, r. \tag{13.20}$$

The terms on the right-hand side of this equation may be expanded using the chain rule. For $i = 1$, we obtain

$$\mathbf{x}'_- = \dot{\mathbf{x}}_+ \frac{dt}{du}, \tag{13.21}$$

where a prime denotes differentiation with respect to u, and a dot denotes differentiation with respect to t. For $i = 2$, we have to apply both the chain and the product rule to the right-hand side of (13.21):

$$\mathbf{x}_-'' = \ddot{\mathbf{x}}_+\left(\frac{dt}{du}\right)^2 + \dot{\mathbf{x}}_+\frac{d^2t}{du^2}. \tag{13.22}$$

For the case $i = 3$:

$$\mathbf{x}_-{}''' = \dddot{\mathbf{x}}_+ \left(\frac{\mathrm{d}t}{\mathrm{d}u}\right)^3 + 3\ddot{\mathbf{x}}_+ \frac{\mathrm{d}t}{\mathrm{d}u}\frac{\mathrm{d}^2 t}{\mathrm{d}u^2} + \dot{\mathbf{x}}_+ \frac{\mathrm{d}^3 t}{\mathrm{d}u^3}. \tag{13.23}$$

Let us define $\alpha_i = \mathrm{d}^i t / \mathrm{d}u^i$. Then the above equations may be written in matrix form:

$$\begin{bmatrix} \mathbf{x}'_- \\ \mathbf{x}_-{}'' \\ \mathbf{x}_-{}''' \end{bmatrix} = \begin{bmatrix} \alpha_1 & 0 & 0 \\ \alpha_2 & \alpha_1^2 & 0 \\ \alpha_3 & 3\alpha_1\alpha_2 & \alpha_1^3 \end{bmatrix} \begin{bmatrix} \dot{\mathbf{x}}_+ \\ \ddot{\mathbf{x}}_+ \\ \dddot{\mathbf{x}}_+ \end{bmatrix}. \tag{13.24}$$

The lower triangular matrix in (13.24) is called a *connection matrix*; it connects the derivatives of one segment to that of the other. For r^{th} order geometric continuity, the connection matrix is a lower triangular $r \times r$ matrix; for more details see Gregory [235] or Goodman [214]. See also the related discussion in Section 11.6. The connection matrix is a powerful theoretical tool, and has been used to derive variation diminishing properties of geometrically continuous curves (Dyn and Micchelli [148]), to show the projective invariance of torsion continuity (Boehm [68]), and for other theoretical pursuits (Goldman and Micchelli [212]).

The above definition of geometric continuity has been used by Manning [323], Barsky [32], Barsky and DeRose [37], Degen [127], Pottman [379], [380], and Farin [155]. In terms of classical differential geometry, the concept of "G^2" is called "order two of contact"; see do Carmo [140]. It was used in a constructive context by G. Geise [201] as early as 1962.

An interesting phenomenon arises if we consider geometric continuity of order higher than two. Consider a G^3 space curve. It is easy to verify that it possesses continuous curvature and torsion. But the converse is not true: there are space curves with continuous curvature and torsion that are not G^3 (Farin [155]). This more general class of curves, called *Frenet frame continuous*, has been studied by Boehm [66]; see also Section 11.6 and Hagen [243], [244]. They are characterized by a more general connection matrix than that for G^3 continuity; it is given by

$$\begin{bmatrix} \alpha_1 & 0 & 0 \\ \alpha_2 & \alpha_1^2 & 0 \\ \alpha_3 & \beta & \alpha_1^3 \end{bmatrix},$$

where β is an arbitrary constant. For higher order Frenet frame continuity, one has to resort to higher dimensional spaces; this has been carried out by

Dyn and Micchelli [148], Goodman [214], Goldman and Micchelli [212], and
Pottmann [378]; see also the survey by Gregory [235]. An even more general
concept than that of Frenet frame continuity has been discussed recently by
H. Pottmann [379].

A condition for torsion continuity of two adjacent Bézier curves with polygons $\mathbf{b}_0, \ldots, \mathbf{b}_n$ and $\mathbf{c}_0, \ldots, \mathbf{c}_n$ is given by

$$\frac{\text{volume}[\mathbf{b}_{n-3}, \ldots, \mathbf{b}_n]}{||\Delta \mathbf{b}_{n-1}||^6} = \frac{\text{volume}[\mathbf{c}_0, \ldots, \mathbf{c}_3]}{||\Delta \mathbf{c}_0||^6}. \tag{13.25}$$

See Boehm [65], Farin [155], or Hagen [243]. Note the similarity to (12.15)!

A nice geometric interpretation of the fact that torsion continuity is more
general than G^3 continuity is due to W. Boehm [65]. If $\mathbf{b}_{n-3}, \ldots, \mathbf{b}_n$ and
$\mathbf{c}_0, \ldots, \mathbf{c}_3$ are given such that the two curves are G^3, can we vary $\mathbf{c}_3$ and still
maintain G^3 continuity? The answer is yes, and $\mathbf{c}_3$ may be displaced by any
vector parallel to the tangent spanned by $\mathbf{b}_{n-1}$ and $\mathbf{c}_1$ – very similar, in other
words, to the situation depicted in Figure 12.5. But we may displace $\mathbf{c}_3$ by
any vector parallel to the osculating plane spanned by $\mathbf{b}_{n-2}, \mathbf{b}_n, \mathbf{c}_2$ and still
maintain torsion continuity!

13.5 Implementation

The following program incorporates Eqs. (13.3) and (13.4) (using uniform
Δ_i), and has to be called once for each coordinate. It does *not* compute the
$\mathbf{b}_{3i}$ from (13.6). The idea is that a user might still want to adjust the $\mathbf{b}_{3i\pm1}$
further before computing the junction points. They can be computed using
the procedure `direct_gspline` from Section 12.6.

```
void  gamma_spline(bspl,l,gamma,bez)
/*   from B-spline polygon and gammas, find interior
     Bezier points b3i+-1.
Input: bspl:    control polygon (one coordinate only).
       l:       number of cubic pieces.
       gamma:   gamma_i's.
Output:bez:     piecewise Bezier polygon, but without the
                junction points b3i.
*/
```

13.6 Problems

1. Prove (13.18).

2. Suppose two cubic segments form a G^2 curve, but are not C^2 cubics with respect to their parametrization. We know that we can reparametrize both so that they are C^2 afterward. In particular, that reparametrization may be piecewise polynomial. What is the minimum degree?

3. We derived B-spline interpolation from the C^2 conditions in Section 9.1. Derive interpolatory γ-splines from Eqs. (13.3), (13.4), and (13.6).

4. Show that the average of two G^2 piecewise cubics over the same knot sequence is in general not a G^2 curve, i.e., if $\mathbf{x}_1(u)$ and $\mathbf{x}_2(u)$ are two G^2 cubics over the same knot sequence (but possibly with different shape parameters), then $(1 - \alpha)\mathbf{x}_1(u) + \alpha\mathbf{x}_2(u)$ is in general not G^2.

5. Find an example of a torsion continuous curve that is not G^3.

6. Let a G^3 curve consist of two cubic Bézier curves. The derivatives of the two curves at the junction point are related by a connection matrix. Work out the corresponding connection matrix for the Bézier points.

Chapter 14

Conic Sections

Conic sections (short: conics) have received the most attention throughout the centuries of any known curve type. Today, they are an important design tool in the aircraft industry; they are also used in areas such as font design. A great many algorithms for the use of conics in design were developed in the 1940s; Liming [313] and [314] are two books with detailed descriptions of those methods.

The first person to consider conics in a CAD environment was S. Coons [101]. Later, R. Forrest [190] further investigated conics and also rational cubics. We shall treat conics in the rational Bézier form; a good reference for this approach is Lee [303]. We present conics partly as a subject in its own right, but also as a first instance of rational Bézier and B-spline curves, to be discussed later.

14.1 Projective Maps of the Real Line

Polynomial curves, as studied before, bear a close relationship to affine geometry. Consequently, the de Casteljau algorithm makes use of ratios, which are the fundamental invariant of affine maps. Thus the class of polynomial curves is invariant under affine transformations: an affine map maps a polynomial curve onto another polynomial curve.

Conic sections, and later rational polynomials, are invariant under a more general type of map: the so-called *projective maps*. These maps are studied in *projective geometry*. This is not the place to outline the ideas of that kind of geometry; the interested reader is referred to the text by Penna and Patterson [361]. All we need here is the concept of a projective map.

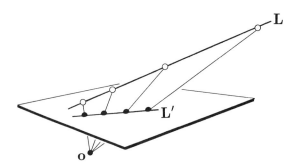

Figure 14.1: Projections: a straight line **L** is mapped onto another straight line **L'** by a projection. Note how ratios of corresponding triples of points are distorted.

We start with a map that is familiar to everybody with a background in computer graphics: the *projection*. Consider a plane (called image plane) **P** and a point **o** (called center or origin of projection) in $I\!E^3$. A point **p** is projected onto **P** through **o** by finding the intersection $\hat{\mathbf{p}}$ between the straight line through **o** and **p** with **P**. For a projection to be well-defined it is necessary that **o** is not in **P**. Any object in $I\!E^3$ can be projected into **P** in this manner.

In particular, we can project a straight line, **L**, say, onto **P**, as shown in Figure 14.1. We clearly see that our projection is not an affine map: the ratios of corresponding points on **L** and **L'** are not the same. But a projection leaves another geometric property unchanged: the *cross ratio* of four collinear points.

The cross ratio, cr, of four collinear points is defined as a ratio of ratios [ratios are defined by Eq. (2.7)]:

$$\mathrm{cr}(\mathbf{a}, \mathbf{b}, \mathbf{c}, \mathbf{d}) = \frac{\mathrm{ratio}(\mathbf{a}, \mathbf{b}, \mathbf{d})}{\mathrm{ratio}(\mathbf{a}, \mathbf{c}, \mathbf{d})}. \qquad (14.1)$$

This particular definition is only one of several equivalent ones; any permutation of the four points gives rise to a new (equivalent) definition. Our convention (14.1) has the advantage of being symmetric: $\mathrm{cr}(\mathbf{a}, \mathbf{b}, \mathbf{c}, \mathbf{d}) = \mathrm{cr}(\mathbf{d}, \mathbf{c}, \mathbf{b}, \mathbf{a})$. Cross ratios were first studied by C. Brianchon and F. Moebius, who proved their invariance under projective maps in 1827; see [337].

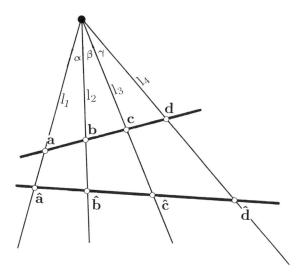

Figure 14.2: Cross ratios: the cross ratios of $\mathbf{a}, \mathbf{b}, \mathbf{c}, \mathbf{d}$ and $\hat{\mathbf{a}}, \hat{\mathbf{b}}, \hat{\mathbf{c}}, \hat{\mathbf{d}}$ only depend on the angles shown and are thus equal.

Let us now prove this invariance claim. We have to show, with the notation from Figure 14.2, that

$$\mathrm{cr}(\mathbf{a}, \mathbf{b}, \mathbf{c}, \mathbf{d}) = \mathrm{cr}(\hat{\mathbf{a}}, \hat{\mathbf{b}}, \hat{\mathbf{c}}, \hat{\mathbf{d}}). \tag{14.2}$$

This fact is called the *cross ratio theorem*.

For a proof, consider Figure 14.2. Denote the area of a triangle with vertices $\mathbf{p}$, $\mathbf{q}$, $\mathbf{r}$ by $\Delta(\mathbf{p}, \mathbf{q}, \mathbf{r})$. We note that for instance

$$\mathrm{ratio}(\mathbf{a}, \mathbf{b}, \mathbf{c}) = \Delta(\mathbf{a}, \mathbf{b}, \mathbf{o})/\Delta(\mathbf{b}, \mathbf{c}, \mathbf{o}).$$

This gives

$$
\begin{aligned}
\mathrm{cr}(\mathbf{a}, \mathbf{b}, \mathbf{c}, \mathbf{d}) &= \frac{\Delta(\mathbf{a}, \mathbf{b}, \mathbf{o})/\Delta(\mathbf{b}, \mathbf{d}, \mathbf{o})}{\Delta(\mathbf{a}, \mathbf{c}, \mathbf{o})/\Delta(\mathbf{c}, \mathbf{d}, \mathbf{o})} \\
&= \frac{l_1 l_2 \sin\alpha / l_2 l_4 \sin(\beta + \gamma)}{l_1 l_3 \sin(\alpha + \beta) / l_3 l_4 \sin\gamma} \\
&= \frac{\sin\alpha / \sin(\beta + \gamma)}{\sin(\alpha + \beta) / \sin\gamma}.
\end{aligned}
$$

Thus the cross ratio of the four points **a**, **b**, **c**, **d** only depends on the angles at **o**. The four rays emanating from **o** may therefore be intersected by any straight line; the four points of intersection will have the same cross ratio, regardless of the choice of the straight line. All such straight lines are related by projections, and we can therefore say that projections leave the cross ratio of four collinear points invariant. Since the cross ratio is the same for any straight line intersecting the given four straight lines, one also calls it the cross ratio of the four given lines.

A concept that is slightly more abstract than that of projections is that of *projective maps*. Going back to Figure 14.1, we can interpret both **L** and **L′** as copies of the real line. Then the projection of **L** onto **L′** can be viewed as a map of the real line onto itself. With this interpretation a projection defines a projective map of the real line onto itself. On the real line, a point is given by a real number, so we can assume a correspondence between the point **a** and a real number a.

An important observation about projective maps of the real line to itself is that they are defined by three preimage and three image points. To observe this, we inspect Figure 14.2. The claim is that **a**, **b**, **d** and their images $\hat{\mathbf{a}}, \hat{\mathbf{b}}, \hat{\mathbf{d}}$ determine a projective map. It is true since if we pick an arbitrary fourth point **c** on **L**, its image $\hat{\mathbf{c}}$ on **L′** is determined by the cross ratio theorem.

A projective map of the real line onto itself is thus determined by three preimage numbers a, b, c and three image numbers $\hat{a}, \hat{b}, \hat{c}$. The projective image $\hat{t}$ of a point t can then be computed from

$$\operatorname{cr}(a, b, t, c) = \operatorname{cr}(\hat{a}, \hat{b}, \hat{t}, \hat{c}).$$

Setting $\rho = (b - a)/(c - b)$ and $\hat{\rho} = (\hat{b} - \hat{a})/(\hat{c} - \hat{b})$, this is equivalent to

$$\frac{\rho}{(t - a)/(c - t)} = \frac{\hat{\rho}}{(\hat{t} - \hat{a})/(\hat{c} - \hat{t})}.$$

Solving for $\hat{t}$:

$$\hat{t} = \frac{(t - a)\hat{\rho}\hat{c} + (c - t)\hat{a}\rho}{\rho(c - t) + \hat{\rho}(t - a)}. \tag{14.3}$$

A convenient choice for the image and preimage points is $a = \hat{a} = 0, c = \hat{c} = 1$. Equation (14.3) then takes on the simpler form

$$\hat{t} = \frac{t\hat{\rho}}{\rho(1 - t) + \hat{\rho}t}. \tag{14.4}$$

Thus a projective map of the real line onto itself corresponds to a *rational linear transformation.* It is left for the reader to verify that the projective map becomes an affine map in the special case that $\rho = \hat{\rho}$.

14.2 Conics as Rational Quadratics

Many equivalent ways exist to define a conic section; for our purposes the following one is very useful: *A conic section in $I\!E^2$ is the projection of a parabola in $I\!E^3$ into a plane.*

When it comes to the formulation of conics as rational curves, it is customary to abandon the principle of being independent of a fixed coordinate system. One typically chooses the center of the projection to be the origin **0** of a 3D cartesian coordinate system. The plane into which one projects is taken to be the plane $z = 1$. Since we will study planar curves in this section, we may think of this plane as a copy of $I\!E^2$, thus identifying points $[\ x \quad y\ ]^T$ with $[\ x \quad y \quad 1\ ]^T$. Our special projection is characterized by

$$\begin{bmatrix} x \\ y \\ z \end{bmatrix} \rightarrow \begin{bmatrix} x/z \\ y/z \\ 1 \end{bmatrix}.$$

Note that a point $[\ x \quad y\ ]^T$ is the projection of a whole family of points: every point on the straight line $[\ wx \quad wy \quad w\ ]^T$ projects to $[\ x \quad y\ ]^T$. In the following, we will use the shorthand notation $[\ w\mathbf{x} \quad w\ ]^T$ with $\mathbf{x} \in I\!E^2$ for $[\ wx \quad wy \quad w\ ]^T$.[1] An illustration of this special projection is given in Figure 14.3.

Figure 14.4 gives an example of how to obtain a conic using the projection method.

Back to conics: let $\mathbf{c}(t) \in I\!E^2$ be a point on a conic. Then numbers $w_0, w_1, w_2 \in I\!R$ and points $\mathbf{b}_0, \mathbf{b}_1, \mathbf{b}_2 \in I\!E^2$ exist such that

$$\mathbf{c}(t) = \frac{w_0 \mathbf{b}_0 B_0^2(t) + w_1 \mathbf{b}_1 B_1^2(t) + w_2 \mathbf{b}_2 B_2^2(t)}{w_0 B_0^2(t) + w_1 B_1^2(t) + w_2 B_2^2(t)}. \tag{14.5}$$

Let us prove (14.5). We may identify $\mathbf{c}(t) \in I\!E^2$ with $[\ \mathbf{c}(t) \quad 1\ ]^T \in I\!E^3$. This point is the projection of a point $[\ w(t)\mathbf{c}(t) \quad w(t)\ ]^T$, which lies on a

---

[1]Sometimes the set of all points $[\ wx \quad wy \quad w\ ]^T$ is called the *homogeneous form* or *homogeneous coordinates* of $[\ x \quad y\ ]^T$.

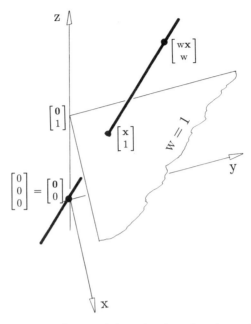

Figure 14.3: Projections: the special projection that is used to write objects in the plane $z = 1$ as projections of objects in $I\!\!E^3$.

3D parabola. The third component $w(t)$ of this 3D point must be a quadratic function in t, and may be expressed in Bernstein form:

$$w(t) = w_0 B_0^2(t) + w_1 B_1^2(t) + w_2 B_2^2(t).$$

Having determined $w(t)$, we may now write

$$w(t) \begin{bmatrix} \mathbf{c}(t) \\ 1 \end{bmatrix} = \begin{bmatrix} \mathbf{c}(t) \sum w_i B_i^2(t) \\ \sum w_i B_i^2(t) \end{bmatrix}.$$

Since the left-hand side of this equation denotes a parabola, we may write

$$\sum_{i=0}^{2} \begin{bmatrix} \mathbf{p}_i \\ w_i \end{bmatrix} B_i^2(t) = \begin{bmatrix} \mathbf{c}(t) \sum w_i B_i^2(t) \\ \sum w_i B_i^2(t) \end{bmatrix}$$

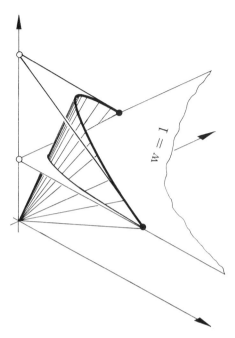

$w = 1$

Figure 14.4: Conic sections: a parabolic arc in three-space is projected into the plane $z = 1$; the result, in this example, is part of a hyperbola.

with some points $\mathbf{p}_i \in I\!\!E^2$. Thus

$$\sum_{i=0}^{2} \mathbf{p}_i B_i^2(t) = \mathbf{c}(t) \sum_{i=0}^{2} w_i B_i^2(t), \qquad (14.6)$$

and hence

$$\mathbf{c}(t) = \frac{\mathbf{p}_0 B_0^2(t) + \mathbf{p}_1 B_1^2(t) + \mathbf{p}_2 B_2^2(t)}{w_0 B_0^2(t) + w_1 B_1^2(t) + w_2 B_2^2(t)}.$$

Setting $\mathbf{p}_i = w_i \mathbf{b}_i$ now proves (14.5).

We call the points $\mathbf{b}_i$ the *control polygon* of the conic $\mathbf{c}$; the numbers w_i are called *weights* of the corresponding control polygon vertices. Thus the conic control polygon is the projection of the control polygon with vertices $[\ w_i \mathbf{b}_i \quad w_i\ ]^{\mathrm{T}}$, which is the control polygon of the 3D parabola that we projected onto $\mathbf{c}$.

The form (14.5) is called the *rational quadratic form* of a conic section. If all weights are equal, we recover nonrational quadratics, i.e., parabolas. The influence of the weights on the shape of the conic is illustrated in Figure 14.5. In that figure, we have chosen

$$\mathbf{b}_0 = \begin{bmatrix} 0 \\ 1 \end{bmatrix}, \mathbf{b}_1 = \begin{bmatrix} 0 \\ 0 \end{bmatrix}, \mathbf{b}_2 = \begin{bmatrix} 1 \\ 0 \end{bmatrix}.$$

Note that a common nonzero factor in the w_i does not affect the conic at all. If $w_0 \neq 0$, one may therefore always achieve $w_0 = 1$ by a simple scaling of all w_i. There are other changes of the weights that leave the curve shape unchanged: these correspond to *rational linear parameter transformations*. Let us set

$$t = \frac{\hat{t}}{\hat{\rho}(1 - \hat{t}) + \hat{t}}, \quad (1 - t) = \frac{\hat{\rho}(1 - \hat{t})}{\hat{\rho}(1 - \hat{t}) + \hat{t}}$$

[corresponding to the choice $\rho = 1$ in (14.4)]. We may insert this into (14.5) and obtain:

$$\mathbf{c}(\hat{t}) = \frac{\hat{\rho}^2 w_0 \mathbf{b}_0 B_0^2(\hat{t}) + \hat{\rho} w_1 \mathbf{b}_1 B_1^2(\hat{t}) + w_2 \mathbf{b}_2 B_2^2(\hat{t})}{\hat{\rho}^2 w_0 B_0^2(\hat{t}) + \hat{\rho} w_1 B_1^2(\hat{t}) + w_2 B_2^2(\hat{t})}. \tag{14.7}$$

Thus the curve shape is not changed if each weight w_i is replaced by $\hat{w}_i = \hat{\rho}^{2-i} w_i$ (for an early reference, see Forrest [190]). If, for a given set of weights w_i, we select

$$\hat{\rho} = \sqrt{\frac{w_2}{w_0}},$$

we obtain $\hat{w}_0 = w_2$, and, after dividing all three weights through by w_2, we even have $\hat{w}_0 = \hat{w}_2 = 1$. A conic that satisfies this condition is said to be in *standard form*. All conics with $w_0, w_2 \neq 0$ may be rewritten in standard form with the above choice of $\hat{\rho}$, provided, of course, that $w_2/w_0 \geq 0$.

If in standard form, i.e., $w_0 = w_2 = 1$, the point $\mathbf{s} = \mathbf{c}(\frac{1}{2})$ is called the *shoulder point*. The shoulder point tangent is parallel to $\mathbf{b}_0 \mathbf{b}_2$. If we set $\mathbf{m} = (\mathbf{b}_0 + \mathbf{b}_2)/2$, then the ratio of the three collinear points $\mathbf{m}, \mathbf{s}, \mathbf{b}_1$ is given by

$$\text{ratio}(\mathbf{m}, \mathbf{s}, \mathbf{b}_1) = w_1.$$

We finish this section with a theorem that will be useful in the later development of rational curves: *Any four tangents to a conic intersect each other in the same cross ratio.* The theorem is illustrated in Figure 14.6. The

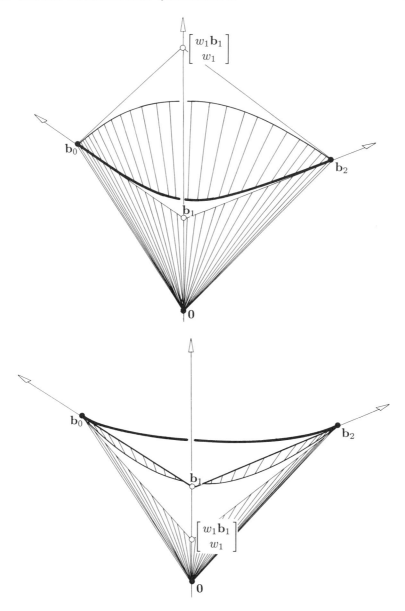

Figure 14.5: Conic sections: in the two examples shown, $w_0 = w_2 = 1$. As w_1 becomes larger, i.e., as $[w_1\mathbf{b}_1, w_1]$ moves "up" on the z-axis, the conic is "pulled" toward $\mathbf{b}_1$.

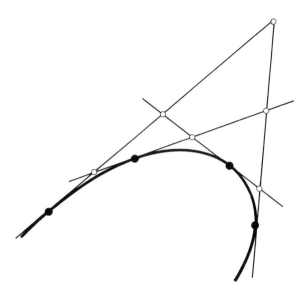

Figure 14.6: The four tangent theorem: four points are marked on each of the four tangents to the shown conic. The four cross ratios generated by them are all equal.

proof of this four tangent theorem is simple: one shows that it is true for parabolas (see Problems). It then follows for all conics by their definition as a projection of a parabola and by the fact that cross ratios are invariant under projections. This theorem is due to J. Steiner. It is a projective version of the three tangent theorem from Section 3.1.

14.3   A de Casteljau Algorithm

We may evaluate (14.5) by evaluating the numerator and the denominator separately and then dividing through. A more geometric algorithm is obtained by projecting each intermediate de Casteljau point $[\ w_i^r \mathbf{b}_i^r \quad w_i^r\ ]^{\mathrm{T}}$ into $I\!\!E^2$:

$$\mathbf{b}_i^r(t) = (1-t)\frac{w_i^{r-1}}{w_i^r}\mathbf{b}_i^{r-1} + t\frac{w_{i+1}^{r-1}}{w_i^r}\mathbf{b}_{i+1}^{r-1}, \qquad (14.8)$$

where

$$w_i^r(t) = (1-t)w_i^{r-1}(t) + tw_{i+1}^{r-1}(t). \qquad (14.9)$$

This is the desired implicit form, since the barycentric coordinates u, v, w of $\mathbf{c}(t)$ are given by

$$
\tau_0 = \frac{\begin{vmatrix} c^x & b_1^x & b_2^x \\ c^y & b_1^y & b_2^y \\ 1 & 1 & 1 \end{vmatrix}}{\begin{vmatrix} b_0^x & b_1^x & b_2^x \\ b_0^y & b_1^y & b_2^y \\ 1 & 1 & 1 \end{vmatrix}}, \quad \tau_1 = \frac{\begin{vmatrix} b_0^x & c^x & b_2^x \\ b_0^y & c^y & b_2^y \\ 1 & 1 & 1 \end{vmatrix}}{\begin{vmatrix} b_0^x & b_1^x & b_2^x \\ b_0^y & b_1^y & b_2^y \\ 1 & 1 & 1 \end{vmatrix}}, \quad \tau_2 = \frac{\begin{vmatrix} b_0^x & b_1^x & c^x \\ b_0^y & b_1^y & c^y \\ 1 & 1 & 1 \end{vmatrix}}{\begin{vmatrix} b_0^x & b_1^x & b_2^x \\ b_0^y & b_1^y & b_2^y \\ 1 & 1 & 1 \end{vmatrix}}.
$$

The implicit form has an important application: suppose we are given a conic section $\mathbf{c}$ and an arbitrary point $\mathbf{x} \in I\!\!E^2$. Does $\mathbf{x}$ lie on $\mathbf{c}$? This question is hard to answer if $\mathbf{c}$ is given in the parametric form (14.5). Using the implicit form, this question is answered easily. First, compute the barycentric coordinates τ_0, τ_1, τ_2 of $\mathbf{x}$ with respect to $\mathbf{b}_0, \mathbf{b}_1, \mathbf{b}_2$. Then insert τ_0, τ_1, τ_2 into (14.20). If (14.20) is satisfied, $\mathbf{x}$ lies on the conic (but see Problems).

Any conic section is uniquely determined by five distinct points in the plane. If the points have coordinates $(x_1, y_1), \ldots, (x_5, y_5)$, the implicit form of the interpolating conic is given by

$$
f(x, y) = \begin{vmatrix} x^2 & xy & y^2 & x & y & 1 \\ x_1^2 & x_1 y_1 & y_1^2 & x_1 & y_1 & 1 \\ x_2^2 & x_2 y_2 & y_2^2 & x_2 & y_2 & 1 \\ x_3^2 & x_3 y_3 & y_3^2 & x_3 & y_3 & 1 \\ x_4^2 & x_4 y_4 & y_4^2 & x_4 & y_4 & 1 \\ x_5^2 & x_5 y_5 & y_5^2 & x_5 & y_5 & 1 \end{vmatrix} = 0.
$$

The fact that five points are sufficient to determine a conic is a consequence of the most fundamental theorem in the theory of conics, *Pascal's theorem*. Consider six points on a conic, arranged as in Figure 14.7. If we connect points as shown, we form six straight lines. Pascal's theorem states that the three intersection points $\mathbf{p}_1, \mathbf{p}_2, \mathbf{p}_3$ are always collinear.

It can be used to *construct* a conic through five points: referring to Figure 14.7 again, let $\mathbf{a}_1, \mathbf{b}_1, \mathbf{c}_1, \mathbf{a}_2, \mathbf{b}_2$ be given (no three of them collinear). Let $\mathbf{p}_1$ be the intersection of the two straight lines through $\mathbf{a}_1, \mathbf{b}_2$ and $\mathbf{a}_2, \mathbf{b}_1$. We may now fix a line $\mathbf{l}$ through $\mathbf{p}_1$, thus obtaining $\mathbf{p}_2$ and $\mathbf{p}_3$. The sixth point on the conic is then determined as the intersection of the two straight lines

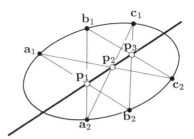

Figure 14.7: Pascal's theorem: the intersection points $\mathbf{p}_1, \mathbf{p}_2, \mathbf{p}_3$ of the indicated pairs of straight lines are collinear.

through $\mathbf{a}_1, \mathbf{p}_2$ and $\mathbf{b}_1, \mathbf{p}_3$. We may construct arbitrarily many points on the conic by letting the straight line l rotate around $\mathbf{p}_1$.

14.6   Two Classic Problems

A large number of methods exist to construct conic sections from given pieces of information, most based on Pascal's theorem. A nice collection is given in the book by R. Liming [314]. An in-depth discussion of those methods is beyond the scope of this book; we restrict ourselves to the solution of two problems.

1. Conic from two points and tangents plus another point. The given data amount to prescribing $\mathbf{b}_0, \mathbf{b}_1, \mathbf{b}_2$. The missing weight w_1 must be determined from the point $\mathbf{p}$, which is assumed to be on the conic. We assume, without loss of generality, that the conic is in standard form ($w_0 = w_2 = 1$).

For the solution, we make use of the implicit form (14.20). We can easily determine the barycentric coordinates τ_0, τ_1, τ_2 of $\mathbf{p}$ with respect to the triangle formed by the three $\mathbf{b}_i$. We can then solve (14.20) for the unknown weight w_1:

$$w_1 = \frac{\tau_1}{2\sqrt{\tau_0 \tau_2}}. \tag{14.21}$$

If $\mathbf{p}$ is inside the triangle formed by $\mathbf{b}_0, \mathbf{b}_1, \mathbf{b}_2$, Eq. (14.21) always has a solution. Otherwise, problems might occur (see Problems). If we do not insist on the conic in standard form, the given point may be given the parameter value $t = 1/2$, in which case it is referred to as a *shoulder point*.

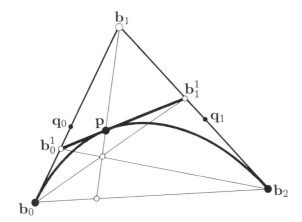

Figure 14.8: Conic constructions: $\mathbf{b}_0, \mathbf{b}_1, \mathbf{b}_2$, and the tangent through $\mathbf{b}_0^1$ and $\mathbf{b}_1^1$ are given.

2. Conic from two points and tangents plus a third tangent.
Again, we are given the Bézier polygon of the conic plus a tangent, which passes through two points that we call $\mathbf{b}_0^1$ and $\mathbf{b}_1^1$. We have to find the interior weight w_1, assuming the conic will be in standard form. The unknown weight w_1 determines the two weight points $\mathbf{q}_0$ and $\mathbf{q}_1$, with $\overline{\mathbf{q}_0, \mathbf{q}_1}$ parallel to $\overline{\mathbf{b}_0, \mathbf{b}_1}$; see Figure 14.8.

We compute the ratios $r_0 = \mathrm{ratio}(\mathbf{b}_0, \mathbf{b}_0^1, \mathbf{b}_1)$ and $r_1 = \mathrm{ratio}(\mathbf{b}_1, \mathbf{b}_1^1, \mathbf{b}_2)$. From the definition of the $\mathbf{q}_i$ in Eq. (14.10), it follows that $\mathrm{ratio}(\mathbf{b}_0, \mathbf{q}_0, \mathbf{b}_1) = w_1$ and $\mathrm{ratio}(\mathbf{b}_1, \mathbf{q}_1, \mathbf{b}_2) = 1/w_1$. The cross ratio property (14.11) now yields

$$\frac{r_0}{w_1} = r_1 w_1, \qquad (14.22)$$

from which we easily determine $w_1 = \sqrt{r_0/r_1}$. The number under the square root must be nonnegative for this to be meaningful (see Problems). Again, if we do not insist on standard form, we may associate the parameter value $t = 1/2$ with the given tangent – it is then called a *shoulder tangent*.

Figure 14.8 also gives a strictly geometric construction: intersect lines $\mathbf{b}_0\mathbf{b}_1^1$ and $\mathbf{b}_2\mathbf{b}_0^1$. Connect the intersection with $\mathbf{b}_1$ and intersect with the given tangent: the intersection is the desired point $\mathbf{p}$.

14.7   Classification

In a projective environment, all conics are equivalent: projective maps map conics to conics. In affine geometry, conics fall into three classes; namely, hyperbolas, parabolas, and ellipses. Thus: ellipses are mapped to ellipses under affine maps, parabolas to parabolas, and hyperbolas to hyperbolas. How can we determine what type a given conic is?

Before we answer that question (following Lee [303]), let us consider the *complementary segment* of a conic. If the conic is in standard form, it is obtained by reversing the sign of w_1. Note that the implicit form (14.20) is not affected by this; hence we still have the same conic, but with a different representation. If $\mathbf{c}(t)$ is a point on the original conic and $\hat{\mathbf{c}}(t)$ is a point on the complementary segment, one easily verifies that $\mathbf{b}_1, \mathbf{c}(t)$, and $\hat{\mathbf{c}}(t)$ are collinear, as shown in Figure 14.9. If we assume that $w_1 > 0$, then the behavior of $\hat{\mathbf{c}}(t)$ determines what type the conic is: if $\hat{\mathbf{c}}(t)$ has no singularities in $[0, 1]$, it is an ellipse; if it has one singularity, it is a parabola; and if it has two singularities, it is a hyperbola.

The singularities, corresponding to points at infinity of $\hat{\mathbf{c}}(t)$, are determined by the real roots of the denominator $\hat{w}(t)$ of $\hat{\mathbf{c}}(t)$. There are at most two real roots, and they are given by

$$t_{1,2} = \frac{1 - w_1 \pm \sqrt{w_1^2 - 1}}{2 - 2w_1}.$$

Thus, a conic is an ellipse if $w_1 < 1$, a parabola if $w_1 = 1$, and a hyperbola if $w_1 > 1$. The three types of conics are shown in Figure 14.10 (see also Figure 14.5).

The circle is one of the more important conic sections; let us now pay some special attention to it. Let our rational quadratic (with $w_1 < 1$) describe an arc of a circle. Because of the symmetry properties of the circle, the control polygon must form an isosceles triangle. If we know the angle $\alpha = \angle(\mathbf{b}_2, \mathbf{b}_0, \mathbf{b}_1)$, we should be able to determine the weight w_1.[2] We may utilize the solution to the second problem in Section 14.6 together with some elementary trigonometry and obtain

$$w_1 = \cos\alpha.$$

---

[2]The actual size of the control polygon does not matter, of course: it can be changed by a scaling to any size we want, and scalings do not affect the weights!

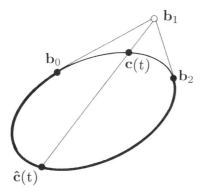

Figure 14.9: The complementary segment: the original conic segment and the complementary segment, both evaluated for all parameter values $t \in [0,1]$, comprise the whole conic section.

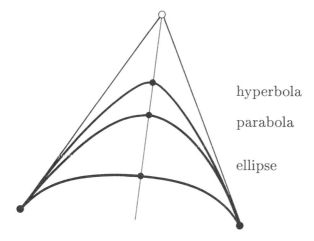

Figure 14.10: Conic classification: the three types of conics are obtained by varying the center weight w_1, assuming $w_0 = w_2 = 1$.

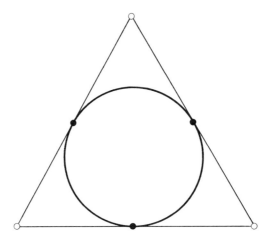

Figure 14.11: Circles: a whole circle may be written as three rational Bézier quadratics.

A whole circle can be represented by piecing several such arcs together. For example, we might choose to represent a circle by three equal arcs, resulting in a configuration like that shown in Figure 14.11. The angles α equal 60 degrees, and so the weights of the inner Bézier points are $1/2$, whereas the junction Bézier points have weights of unity, since each arc is in standard form.

Our representation of the circle is C^1, assuming uniform parameter intervals; see Eq. (14.15). It is not C^2, however! Still we have an *exact* representation of the circle, not an approximation. Thus this particular representation of the circle is an example of a G^2 curve.

We should mention that the parametrization of our circle is not the arc length parametrization as explained in Chapter 11. If uniform traversal of the circle is necessary for some application, one has no choice but to resort to the classical sine and cosine representation. It can be shown (Farouki and Sakkalis [179]) that no rational curve other than the straight line is parametrized with respect to arc length when evaluated at equal increments of its parameter t, and the curve will not be traced out at uniform speed.

14.8 Control Vectors

In principle, any arc of a conic may be written as a rational quadratic curve segment (possibly with negative weights). But what happens for the case where the tangents at $\mathbf{b}_0$ and $\mathbf{b}_2$ become parallel? Intuitively, this would send $\mathbf{b}_1$ to infinity. A little bit of analysis will overcome this problem, as we shall see from the following example.

Let a conic be given by $\mathbf{b}_0 = [-1, 0]^{\mathrm{T}}, \mathbf{b}_2 = [1, 0]^{\mathrm{T}}$, and $\mathbf{b}_1 = [0, \tan \alpha]^{\mathrm{T}}$ and a weight $w_1 = c \cos \alpha$ (we assume standard form). The angle α is formed by $\mathbf{b}_0 \mathbf{b}_1$ and $\mathbf{b}_0 \mathbf{b}_2$ at $\mathbf{b}_0$. Note that for $c = 1$, we obtain a circular arc, as illustrated in Figure 14.12.

The equation of our conic is given by

$$
\mathbf{c}(t) = \frac{(1-t)^2 \begin{bmatrix} -1 \\ 0 \end{bmatrix} + \cos \alpha \cdot 2ct(1-t) \begin{bmatrix} 0 \\ \tan \alpha \end{bmatrix} + t^2 \begin{bmatrix} 1 \\ 0 \end{bmatrix}}{(1-t)^2 + 2ct(1-t) \cos \alpha + t^2}.
$$

What happens as α tends to $\frac{\pi}{2}$? For the limiting conic, we obtain the equation

$$
\mathbf{c}(t) = \frac{(1-t)^2 \begin{bmatrix} -1 \\ 0 \end{bmatrix} + 2t(1-t) \begin{bmatrix} 0 \\ c \end{bmatrix} + t^2 \begin{bmatrix} 1 \\ 0 \end{bmatrix}}{(1-t)^2 + t^2}. \tag{14.23}
$$

The problem of a weight tending to zero and a control point tending to infinity has thus been resolved. For $c = 1$, we obtain a semicircle; other values of c give rise to different conics. For $c = -1$, we obtain the "lower" half of the unit circle.

We have been able to overcome possible problems with parallel end tangents. But there is a price to be paid: if we look at (14.23) closely, we see that it does not constitute a barycentric combination anymore! The factors of $\mathbf{b}_0$ and $\mathbf{b}_2$ sum to one identically, hence $[0, c]^{\mathrm{T}}$ must be interpreted as a *vector*. Thus (14.23) contains both control points and control vectors.[3] An important property of Bézier curves is thus lost; namely, the convex hull property: it is only defined for point sets, not for a potpourri of points and vectors.

The use of control vectors allows a very compact form of writing a semicircle. But two disadvantages argue against its use: first, the loss of the convex hull property. Second: to write the control vector form in the context of "normal" rational quadratics, one will have to resort to a special case treatment. We shall see later (Section 15.6) how to avoid the use of the control vector form. Control points were, under the name "infinite control points," introduced by Piegl [371].

---

[3]In projective geometry, vectors are sometimes called "points at infinity." This has given rise to the name "infinite control points" by L. Piegl [371]. We prefer the term "control vector" since this allows us to distinguish between $[0, c]^{\mathrm{T}}$ and $[0, -c]^{\mathrm{T}}$.

Figure 14.12: Conic arcs: a 168 degree arc of a circle is shown. Note that α is close to 90 degrees.

14.9   Implementation

The following routine solves the first problem in Section 14.6:

```
 float conic_weight(b0,b1,b2,p)
/*
  Input:b0,b1,b2:   conic control polygon vertices
        p:          point on conic
  Output:           weight of b1 (assuming standard form).

  Note:             will crash in "forbidden" situations.
*/
```

14.10   Problems

1. Equation (14.21) does not always have a solution. Identify the "forbidden" regions for the third point **p** on the conic.

2. In the same manner, investigate Eq. (14.22).

3. Prove that the four tangent theorem holds for parabolas.

4. Establish the connection between (14.11) and the four tangent theorem.

5. In Problem 2 from Section 14.6, find a geometric construction for the point on the conic corresponding to the given tangent.

6. Interpret the solutions to the "classical problems" (Section 14.6) as an application of Pascal's theorem.

7. Our discussion of the use of the implicit form of Eq. (14.20) was somewhat academic: in a "real-life" situation, (14.20) will never be satisfied *exactly*. Discuss the tolerance problem that arises here, i.e., how closely does (14.20) have to be satisfied for a point to be within a given tolerance to the conic?

Chapter 15

Rational Bézier and B-spline Curves

Rational B-spline curves[1] are becoming the standard curve and surface description in the field of CAD and graphics. The growing number of articles concerned with them includes those by Vesprille [466], Tiller [456], and Piegl and Tiller [374]. The use of rational curves in CAGD may be traced back to Coons [101], [103], and Forrest [190].

15.1   Rational Bézier Curves

In the previous chapter, we obtained a conic section in $I\!\!E^2$ as the projection of a parabola (a quadratic) in $I\!\!E^3$. Conic sections may be expressed as rational quadratic (Bézier) curves, and their generalization to higher degree rational curves is quite straightforward: a rational Bézier curve of degree n in $I\!\!E^3$ is the projection of an nth degree Bézier curve in $I\!\!E^4$ into the hyperplane $w = 1$. We may view this 4D hyperplane as a copy of $I\!\!E^3$; we assume that a point in $I\!\!E^4$ is given by its coordinates $[\ x\quad y\quad z\quad w\ ]^{\text{T}}$. Proceeding in exactly the same way as we did for conics, we can show that an nth degree rational Bézier curve is given by

$$\mathbf{x}(t) = \frac{w_0\mathbf{b}_0 B_0^n(t) + \cdots + w_n\mathbf{b}_n B_n^n(t)}{w_0 B_0^n(t) + \cdots + w_n B_n^n(t)}; \quad \mathbf{x}(t), \mathbf{b}_i \in I\!\!E^3. \tag{15.1}$$

---

[1]Often called NURBS for *nonuniform rational B-splines*.

The w_i are again called *weights*; the $\mathbf{b}_i$ form the control polygon. It is the projection of the 4D control polygon $[\ w_i\mathbf{b}_i \quad w_i\ ]^{\mathrm{T}}$ of the nonrational 4D preimage of $\mathbf{x}(t)$.

If all weights equal one,[2] we obtain the standard nonrational Bézier curve, in which case, the denominator is identically equal to one. If some w_i are negative, singularities may occur; we will therefore deal only with nonnegative w_i. Rational Bézier curves enjoy all the properties that their nonrational counterparts possess; for example, they are affinely invariant. We can see this by rewriting (15.1) as

$$\mathbf{x}(t) = \sum_{i=0}^{n} \mathbf{b}_i \frac{w_i B_i^n(t)}{\sum_{i=0}^{n} w_i B_i^n(t)}.$$

We see that the basis functions

$$\frac{w_i B_i^n(t)}{\sum_{i=0}^{n} w_i B_i^n(t)}$$

sum to one identically, thus asserting affine invariance. If all w_i are nonnegative, we have the convex hull property. We also have symmetry, invariance under affine parameter transformations, endpoint interpolation, and the variation diminishing property. Obviously, the conic sections from the preceding section are included in the set of all rational Bézier curves, further justifying their increasing popularity.

The w_i are typically used as *shape parameters*. If we increase one w_i, the curve is pulled toward the corresponding $\mathbf{b}_i$, as illustrated in Figure 15.1. Note that the effect of changing a weight is different from that of moving a control vertex, illustrated in Figure 15.1. If we let all weights tend to infinity at the same rate, we do *not* approach the control polygon since a common (if large) factor in the weights does not matter – the rational Bézier curve shape parameters behave differently from γ- or ν-spline shape parameters.

15.2   The de Casteljau Algorithm

A rational Bézier curve may be evaluated by applying the de Casteljau algorithm to both numerator and denominator and finally dividing through. A warning is appropriate: while simple and usually effective, this method is not

---

[2]It is sufficient to set them all equal – a common factor does not matter.

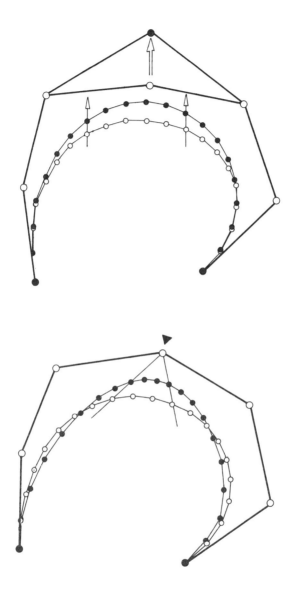

Figure 15.1: Influence of the weights: top, a nonrational curve with a change in one control point; bottom, with one weight changed.

numerically stable for weights that vary significantly in magnitude. If some of the w_i are large, the 3D intermediate points $[w_i \mathbf{b}_i]^r$ (interpreted as points in a given coordinate system) are no longer in the convex hull of the original control polygon $\{\mathbf{b}_i\}$; this may result in a loss of accuracy.[3]

An expensive yet more accurate technique is to project every intermediate de Casteljau point $[\; w_i \mathbf{b}_i \quad w_i \;]^T$; $\mathbf{b}_i \in I\!\!E^3$ into the hyperplane $w = 1$. This yields the rational de Casteljau algorithm (see Farin [156]):

$$\mathbf{b}_i^r(t) = (1-t)\frac{w_i^{r-1}}{w_i^r}\mathbf{b}_i^{r-1} + t\frac{w_{i+1}^{r-1}}{w_i^r}\mathbf{b}_{i+1}^{r-1}, \tag{15.2}$$

with

$$w_i^r(t) = (1-t)w_i^{r-1}(t) + tw_{i+1}^{r-1}(t). \tag{15.3}$$

An explicit form for the intermediate points $\mathbf{b}_i^r$ is given by

$$\mathbf{b}_i^r(t) = \frac{\sum_{j=0}^r w_{i+j}\mathbf{b}_{i+j}B_j^r(t)}{\sum_{j=0}^r w_{i+j}B_j^r(t)}.$$

Note that for positive weights, the $\mathbf{b}_i^r$ are all in the convex hull of the original $\mathbf{b}_i$, thus assuring numerical stability.

The rational de Casteljau algorithm allows a nice geometric interpretation. While the standard de Casteljau algorithm makes use of ratios of three points, this one makes use of the *cross ratio of four points*. Let us define points $\mathbf{q}_i^r(t)$, which are located on the straight lines joining $\mathbf{b}_i^r$ and $\mathbf{b}_{i+1}^r$. Subdividing them in the ratio

$$\text{ratio}(\mathbf{b}_i^r, \mathbf{q}_i^r, \mathbf{b}_{i+1}^r) = \frac{w_{i+1}^r}{w_i^r}.$$

We shall call these points *weight points*, because they indicate the relative magnitude of the weights in a geometric way. Then all of the following cross ratios are equal:

$$\text{cr}(\mathbf{b}_i^r, \mathbf{q}_i^r, \mathbf{b}_i^{r+1}, \mathbf{b}_{i+1}^r) = \frac{1-t}{t} \quad \text{for all } r, i.$$

For $r = 0$, the weight points

$$\mathbf{q}_i = \mathbf{q}_i^0 = \frac{w_i\mathbf{b}_i + w_{i+1}\mathbf{b}_{i+1}}{w_i + w_{i+1}}$$

---

[3]These points are obtained by applying the de Casteljau algorithm to the control points $w_i\mathbf{b}_i$ of the numerator of (15.1). They have no true geometric interpretation, because their location is not invariant under translations of the original control polygon.

We can solve for $\mathbf{x}^{(r)}(t)$:

$$\mathbf{x}^{(r)}(t) = \frac{1}{w(t)}[\mathbf{p}^{(r)} - \sum_{j=1}^{r} \binom{r}{j} w^{(j)}(t)\mathbf{x}^{(r-j)}(t)]. \qquad (15.8)$$

This is a recursive formula for the rth derivative of a rational Bézier curve. It only involves taking derivatives of polynomial curves.

For the first derivative at the endpoint of a rational Bézier curve, we find

$$\dot{\mathbf{x}}(0) = \frac{nw_1}{w_0}\Delta\mathbf{b}_0.$$

Let us now consider two rational Bézier curves, one defined over the interval $[u_0, u_1]$ with control polygon $\mathbf{b}_0, \ldots, \mathbf{b}_n$ and weights $w_0, \ldots, w_n$ and the other defined over the interval $[u_1, u_2]$ with control polygon $\mathbf{b}_n, \ldots, \mathbf{b}_{2n}$ and weights $w_n, \ldots, w_{2n}$. Both segments form a C^1 curve if

$$\frac{w_{n-1}}{\Delta_0}\Delta\mathbf{b}_{n-1} = \frac{w_{n+1}}{\Delta_1}\Delta\mathbf{b}_n, \qquad (15.9)$$

where the appearance of the interval lengths Δ_i is due to the application of the chain rule. This is necessary since we now consider a composite curve with a global parameter u, as explained in Section 7.1. Note that the weight w_n has no influence on differentiability at all! Thus there are C^1 composite rational curves that are the projection of discontinuous composite curves. While the computation of higher order derivatives is quite involved in the case of rational Bézier curves, we note that the computation of curvature or torsion may be simplified by the application of the formulas of Eq. (11.9) or Eq. (11.10) and (11.11).

15.4   Osculatory Interpolation

With rational cubics, it is easy to solve an interesting kind of interpolation problem: given a Bézier polygon $\mathbf{b}_0, \mathbf{b}_1, \mathbf{b}_2, \mathbf{b}_3$ and a curvature value at each endpoint, find a set of weights w_0, w_1, w_2, w_3 such that the corresponding rational cubic assumes the given curvatures at $\mathbf{b}_0$ and $\mathbf{b}_1$. The following method is very similar to one developed by T. Goodman in 1988; see [215].

are directly related to the weights w_i: given the weights, we can find the and given the $\mathbf{q}_i$, we can find the weights w_i.[4] Thus the $\mathbf{q}_i$ may be used *shape parameters*: moving a $\mathbf{q}_i$ along the polygon leg $\mathbf{b}_i, \mathbf{b}_{i+1}$ influences t shape of the curve. It may be preferable to let a designer use these geometr handles rather than requiring him or her to input numbers for the weights.

As in the nonrational case, the de Casteljau algorithm may be used to *subdivide* a curve. The de Casteljau algorithm subdivides the 4D preimage of our 3D rational Bézier curve $\mathbf{x}(t)$, see Section 4.6. The intermediate 4D points $[\, w_i^r \mathbf{b}_i^r \quad w_i^r \,]^\mathrm{T}$; $\mathbf{b}_i^r \in I\!\!E^3$, may be projected into the hyperplane $w = 1$ to provide us with the control polygons for the "left" and "right" curve segment. The control vertices and weights corresponding to the interval $[0, t]$ are given by

$$\mathbf{b}_i^{\mathrm{left}} = \mathbf{b}_0^i(t), \quad w_i^{\mathrm{left}} = w_0^i, \tag{15.4}$$

where the $\mathbf{b}_0^i(t)$ and the w_0^i are computed from (15.2). The control points and weights corresponding to the interval $[t, 1]$ are given by

$$\mathbf{b}_i^{\mathrm{right}} = \mathbf{b}_{n-i}^i(t), \quad w_i^{\mathrm{right}} = w_{n-i}^i. \tag{15.5}$$

15.3 Derivatives

For the first derivative of a rational Bézier curve, we obtain

$$\dot{\mathbf{x}}(t) = \frac{1}{w(t)}[\dot{\mathbf{p}}(t) - \dot{w}(t)\mathbf{x}(t)], \tag{15.6}$$

where we have set

$$\mathbf{p}(t) = w(t)\mathbf{x}(t); \quad \mathbf{p}(t), \mathbf{x}(t) \in I\!\!E^3 \tag{15.7}$$

in complete analogy to the development in Section 14.4. For higher derivatives, we differentiate (15.7) r times:

$$\mathbf{p}^{(r)}(t) = \sum_{j=0}^{r} \binom{r}{j} w^{(j)}(t)\mathbf{x}^{(r-j)}(t).$$

---

[4]To be precise, we can only find them modulo an – immaterial – common factor.
[5]This situation is similar to the way curves are generated using the direct G^2 spline algorithm from Chapter 12 compared to the generation of γ-splines.

We assume without loss of generality that $w_0 = w_3 = 1$.[6] The given curvatures κ_0 and κ_3 are then related to the unknown weights by (11.10):

$$\kappa_0 = \frac{4}{3}\frac{w_2}{w_1^2}c_0, \quad \kappa_3 = \frac{4}{3}\frac{w_1}{w_2^2}c_1, \tag{15.10}$$

where
$$c_0 = \frac{\text{area}[\mathbf{b}_0, \mathbf{b}_1, \mathbf{b}_2]}{\text{dist}^3[\mathbf{b}_0, \mathbf{b}_1]}, \quad c_1 = \frac{\text{area}[\mathbf{b}_1, \mathbf{b}_2, \mathbf{b}_3]}{\text{dist}^3[\mathbf{b}_2, \mathbf{b}_3]}.$$

Equations (15.10) decouple nicely, so that we can determine our unknowns w_1 and w_2:

$$w_1 = \frac{4}{3}[\frac{c_0^2}{\kappa_0^2}\frac{c_1}{\kappa_1}]^{\frac{1}{3}}, \quad w_2 = \frac{4}{3}[\frac{c_0}{\kappa_0}\frac{c_1^2}{\kappa_1^2}]^{\frac{1}{3}}. \tag{15.11}$$

For planar control polygons, the quantities c_0 or c_1 may be negative – this happens when a control polygon is S-shaped. This is meaningful since curvature may be defined as *signed curvature* for 2D curves, as defined in (23.1). Of course, one should then also prescribe the corresponding κ_0 and κ_1 as being negative, so that one ends up with positive weights.

A similar interpolation problem was addressed by Klass [291] and de Boor, Hollig, and Sabin for the nonrational case: they prescribe two points and corresponding tangent directions and curvatures [120]. The solution (when it exists) can only be obtained using an iterative method.

15.5 Reparametrization and Degree Elevation

Arguing exactly as in the conic case (see the end of Section 14.2), we may *reparametrize* a rational Bézier curve by changing the weights according to

$$\hat{w}_i = c^i w_i; \quad i = 0, \ldots, n,$$

where c is any nonzero constant. Figure 15.2 shows how the reparametrization affects the parameter spacing on the curve; note that the curve shape remains the same.

The new weights correspond to new weight points $\hat{\mathbf{q}}_i$. One can show (see Farin and Worsey [171]) that the new and old weight points are strongly related: the cross ratios of any four points $[\mathbf{b}_i, \mathbf{q}_i, \hat{\mathbf{q}}_i, \mathbf{b}_{i+1}]$ are the same for all polygon legs.

---

[6]Goodman [215] assumes that $w_1 = w_2 = 1$.

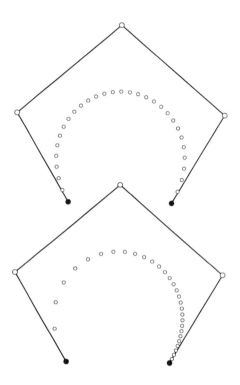

Figure 15.2: Reparametrizations: top, a rational Bézier curve evaluated at parameter values $0, 0.1, 0.2, \ldots, 1$; bottom, the same curve and parameter values but after a reparametrization with $c = 3$.

We may always transform a rational Bézier curve to *standard form* by using the rational linear parameter transformation resulting from the choice

$$c = \sqrt[n]{\frac{w_0}{w_n}}.$$

This results in $\hat{w}_n = w_0$; after dividing all weights through by w_0, we have the standard form $\hat{w}_0 = \hat{w}_n = 1$. Of course, we have to require that the above root exists. A different derivation of this result is in Patterson [356].

How can rational Bézier curves in nonstandard form arise? A common case occurs in connection with rational Bézier surfaces, as discussed in Section 17.7: the end weights of an isoparametric curve will in general not be unity.

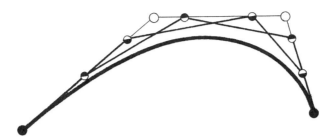

Figure 15.3: Ambiguous curve representations: the two heavy polygons represent the same rational quartic. Also indicated is the rational cubic representation that they were both obtained from.

Such curves are often "extracted" from a surface and then treated as entities in their own right.

We may perform *degree elevation* (in analogy to Section 5.1) by degree elevating the 4D polygon with control vertices $[\; w_i\mathbf{b}_i \quad w_i\;]^{\mathrm{T}}$ and projecting the resulting control vertices into the hyperplane $w = 1$. Let us denote the control vertices of the degree elevated curve by $\mathbf{b}_i^{(1)}$; they are given by

$$\mathbf{b}_i^{(1)} = \frac{w_{i-1}\alpha_i\mathbf{b}_{i-1} + w_i(1 - \alpha_i)\mathbf{b}_i}{w_{i-1}\alpha_i + w_i(1 - \alpha_i)}; \quad i = 0,\ldots,n+1 \qquad (15.12)$$

and $\alpha_i = i/(n+1)$. The weights $w_i^{(1)}$ of the new control vertices are given by

$$w_i^{(1)} = w_{i-1}\alpha_i + w_i(1 - \alpha_i); \quad i = 0,\ldots,n+1.$$

The connection of reparametrization and degree elevation may lead to surprising situations. Consider the following procedure: take any rational Bézier curve in standard form and degree elevate it. Next, take the original curve, reparametrize it, then degree elevate it and bring it to standard form. We end up with two different polygons (and two different sets of standardized weights) that both describe the same rational curve. This situation is very different from the nonrational case! It is illustrated in Figure 15.3.

For the sake of completeness, we should mention that ways other than just by rational linear reparametrizations exist to reparametrize rational curves. For example, the reparametrization $t \leftarrow t(2 - t)$ does not change the curve,

Let a rational quadratic semicircle in control vector form be given by two control points $\mathbf{b}_0, \mathbf{b}_2$ with weights of unity and one control vector $\mathbf{v}_1$, without a weight. Its equation is given by (14.23). After degree elevation, we obtain a rational cubic with four control points:

$$[\mathbf{b}_0, \mathbf{b}_1, \mathbf{b}_2, \mathbf{b}_3] = \left[\begin{bmatrix} -1 \\ 0 \end{bmatrix} \begin{bmatrix} -1 \\ 1 \end{bmatrix} \begin{bmatrix} 1 \\ 1 \end{bmatrix} \begin{bmatrix} 1 \\ 0 \end{bmatrix} \right]$$

and weights

$$[w_0, w_1, w_2, w_3] = [1, \frac{1}{3}, \frac{1}{3}, 1].$$

Example 15.1: Writing a semicircle as a rational cubic.

but it raises the degree from n to $2n$. As long as the reparametrization is of the form $t \leftarrow r(t)$, where $r(t)$ is a rational polynomial, we do not leave the class of rational curves. It is an interesting task to determine if a rational curve has been "degree elevated" in this way. For a solution, consult Sederberg [427].

15.6 Control Vectors

In Section 14.8, we encountered control vectors (also known as infinite control points) as the limiting case of parallel tangents to a conic. The resulting curve representation contained both points and vectors. We can devise a similar form for rational Bézier curves (first suggested by L. Piegl [371]). They will be of the form

$$\mathbf{b}(t) = \frac{\sum_{\text{points}} w_i \mathbf{b}_i B_i^n(t) + \sum_{\text{vectors}} \mathbf{v}_i B_i^n(t)}{\sum_{\text{points}} w_i B_i^n(t)}. \tag{15.13}$$

The control vectors do not have weights in this form; we may multiply each $\mathbf{v}_i$ by a factor, however, and the curve will change accordingly. Note that at least one of the point weights w_i must be nonzero for (15.13) to be meaningful – otherwise a point [namely $\mathbf{b}(t)$] would have to equal a vector!

As in the conic case, we have lost the convex hull property, and evaluation of (15.13) will require special case treatment. However, we can eliminate the control vectors completely – we just have to degree elevate the curve (possibly more than once). Example 15.1 shows how to do this.

15.7 Rational Cubic B-spline Curves

In this section, we take advantage of the special notation from Chapter 7. A 3D rational cubic B-spline curve is the projection through the origin of a 4D nonrational cubic B-spline curve into the hyperplane $w = 1$. The control polygon of the rational B-spline curve is given by vertices $\mathbf{d}_{-1}, \ldots, \mathbf{d}_{L+1}$; each vertex $\mathbf{d}_i \in I\!\!E^3$ has a corresponding weight w_i. The rational B-spline curve has a piecewise rational cubic Bézier representation. It may be obtained by projecting the corresponding 4D Bézier points into the hyperplane $w = 1$. Thus we obtain

$$\mathbf{b}_{3i-2} = \frac{w_{i-1}(1 - \alpha_i)\mathbf{d}_{i-1} + w_i\alpha_i\mathbf{d}_i}{v_{3i-2}}, \qquad (15.14)$$

$$\mathbf{b}_{3i-1} = \frac{w_{i-1}\beta_i\mathbf{d}_{i-1} + w_i(1 - \beta_i)\mathbf{d}_i}{v_{3i-1}}, \qquad (15.15)$$

where all points $\mathbf{b}_j, \mathbf{d}_k$ are in $I\!\!E^3$ and

$$\Delta = \Delta_{i-2} + \Delta_{i-1} + \Delta_i,$$

$$\alpha_i = \frac{\Delta_{i-2}}{\Delta},$$

$$\beta_i = \frac{\Delta_i}{\Delta}.$$

The weights of these Bézier points are given by

$$v_{3i-2} = w_{i-1}(1 - \alpha_i) + w_i\alpha_i, \qquad (15.16)$$

$$v_{3i-1} = w_{i-1}\beta_i + w_i(1 - \beta_i). \qquad (15.17)$$

For the junction points, we obtain

$$\mathbf{b}_{3i} = \frac{\gamma_i v_{3i-1}\mathbf{b}_{3i-1} + (1 - \gamma_i)v_{3i+1}\mathbf{b}_{3i+1}}{v_{3i}}, \qquad (15.18)$$

where

$$\gamma_i = \frac{\Delta_i}{\Delta_{i-1} + \Delta_i}$$

and

$$v_{3i} = \gamma_i v_{3i-1} + (1 - \gamma_i)v_{3i+1}$$

is the weight of the junction point $\mathbf{b}_{3i}$.

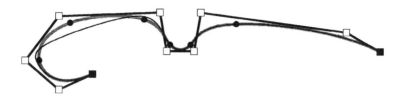

Figure 15.4: Rational B-splines: the weight of the indicated control point is changed. The curve is only affected locally.

Another way to generate the piecewise rational Bézier polygon is by taking the control polygon $[\ w_i\mathbf{d}_i, w_i\ ]^\mathrm{T}$, converting it to Bézier form, and then dividing through by the Bézier weights. This is less geometric, but was still chosen as the basis for the C procedure `ratbspline_to_bezier` at the end of this section simply because it is more efficient.

Designing with rational B-spline curves is not very different from designing with their nonrational counterparts. We now have the added freedom of being able to change weights. A change of only one weight affects a rational B-spline curve only locally, as shown in Figure 15.4.

This development follows the general philosophy of computing with rational curves: we are given 3D points $\mathbf{x}_i$ and their weights w_i. Transform them to 4D points $[\ w_i\mathbf{x}_i\ \ w_i\ ]^\mathrm{T}$ and perform 4D nonrational algorithms (for example, finding the Bézier points of a B-spline curve). The result of these operations will be a set of 4D points $[\ \mathbf{y}_i\ \ v_i\ ]^\mathrm{T}$. From these, obtain 3D points $\mathbf{y}_i/v_i$. The weights of these 3D points are the numbers v_i.

Let us close this section with a result that somewhat limits the use of C^2 rational B-spline curves: *There is no symmetric periodic representation of a circle as a C^2 rational cubic B-spline curve.* If such a representation existed, it would be of the form

$$\mathbf{x}(u) = \sum w_i\mathbf{d}_i N_i^3(u) / \sum w_i N_i^3(u),$$

where all w_i are equal by symmetry. Then the w_i cancel, leaving us with an integral B-spline curve, which is not capable of representing a circle. Note, however, that we can represent any *open* circular arc by C^2 rational cubics.

15.8 Interpolation with Rational Cubics

The interpolation problem in the context of rational B-splines is the following:

Given: 3D data points $\mathbf{x}_0, \ldots, \mathbf{x}_L$, parameter values $u_0, \ldots, u_L$, and weights $w_0, \ldots, w_L$.

Find: a C^2 rational B-spline curve with control vertices $\mathbf{d}_{-1}, \ldots, \mathbf{d}_{L+1}$ and weights $v_{-1}, \ldots, v_{L+1}$ that interpolates to the given data and weights.

For the solution of this problem, we follow the philosophy outlined at the end of the last section: solve a 4D interpolation problem to the data points $[\ w_i \mathbf{x}_i \quad w_i\ ]^T$ and parameter values u_i. All we have to do is to solve the linear system of Eq. (9.7), where input and output is now 4D instead of the usual 3D. We will obtain a 4D control polygon $[\ \mathbf{e}_i \quad v_i\ ]^T$, from which we now obtain the desired $\mathbf{d}_i$ as $\mathbf{d}_i = \mathbf{e}_i / v_i$. The v_i are the weights of the control vertices $\mathbf{d}_i$.

We have not yet addressed the problem of how to choose the weights w_i for the data points $\mathbf{x}_i$. No known algorithms exist for this problem. It seems reasonable to assign high weights in regions where the interpolant is expected to curve sharply. Yet there is a limit to the assignment of weights: if all of them are very high, this will not have a significant effect on the curve since a common factor in all weights will simply cancel. Also, care must be taken to prevent the denominator of the interpolant from being zero. This is not a trivial task – for instance, we might assign a very large weight to one data point while keeping all the others at unity. The resulting weight function $w(t)$ will not be positive everywhere, giving rise to singularities at its zeroes.

Integral cubic spline interpolation has *cubic precision*: if the data points and the parameter values come from one global cubic, the interpolant reproduces that cubic. In the context of rational spline interpolation, an analogous question is that of *conic precision*: if the data points and the parameter values come from one global conic, can we reproduce it? We must also require that the data points have weights assigned to them. With them, we may view the rational spline interpolation problem as an integral spline interpolation problem in $I\!\!E^4$. There, cubic splines have quadratic precision, i.e., we may recapture any parabola. The projection of the parabola yields a conic section; thus if our data – points, parameter values, and weights – were taken from a conic, rational cubic spline interpolation will reproduce the conic.

We should note, however, that this argument is limited to open curves; for closed curves, we have already seen that we cannot represent a circle as a C^2 symmetric periodic B-spline curve.

15.9   Rational B-splines of Arbitrary Degree

The process of generalizing the concept of general B-spline curves to the rational case is now straightforward. A 3D rational B-spline curve is the projection through the origin of a 4D nonrational B-spline curve into the hyperplane $w = 1$. It is thus given by

$$\mathbf{s}(u) = \frac{\sum_{j=0}^{L+n-1} w_i \mathbf{d}_i N_i^n(u)}{\sum_{j=0}^{L+n-1} w_i N_i^n(u)}. \tag{15.19}$$

We have chosen the notation from Chapter 10. Thus (15.19) is the generalization of (10.11) to the rational parametric case.

A rational B-spline curve is given by its knot sequence, its 3D control polygon, and its weight sequence. The control vertices $\mathbf{d}_i$ are the projections of the 4D control vertices $[\ w_i \mathbf{d}_i \quad w_i\ ]^{\mathrm{T}}$.

To evaluate a rational B-spline curve at a parameter value u, we may apply the de Boor algorithm to both numerator and denominator of (15.19) and finally divide through. This corresponds to the evaluation of the 4D nonrational curve with control vertices $[\ w_i \mathbf{d}_i \quad w_i\ ]^{\mathrm{T}}$ and to projecting the result into $I\!\!E^3$. Just as in the case of Bézier curves, this may lead to instabilities, and so we give a rational version of the de Boor algorithm that is more stable but also computationally more involved:

de Boor algorithm, rational: Let $u \in [u_I, u_{I+1}) \subset [u_{n-1}, u_{L+n-1}]$. Define

$$\mathbf{d}_i^k(u) = [(1 - \alpha_i^k)w_{i-1}^{k-1}\mathbf{d}_{i-1}^{k-1}(u) + \alpha_i^k w_i^{k-1}\mathbf{d}_i^{k-1}(u)]/w_i^k \tag{15.20}$$

for $k = 1, \ldots, n - r$, and $i = I - n + k + 1, \ldots, I + 1$, where

$$\alpha_i^k = \frac{u - u_{i-1}}{u_{i+n-k} - u_{i-1}}$$

and

$$w_i^k = (1 - \alpha_i^k)w_{i-1}^{k-1} + \alpha_i^k w_i^{k-1}.$$

Then

$$\mathbf{s}(u) = \mathbf{d}_{I+1}^{n-r}(u) \tag{15.21}$$

is the point on the B-spline curve at parameter value u. Here, r denotes the multiplicity of u in case it was already one of the knots. If it was not, set $r = 0$. As usual, we set $\mathbf{d}_i^0 = \mathbf{d}_i$ and $w_i^0 = w_i$.

The reader is referred to Section 10.3 for the notation.

Knot insertion is, as in the nonrational case, performed by executing just one step of the de Boor algorithm, i.e., by fixing $k = 1$ in the above algorithm. The original polygon vertices $\mathbf{d}_{I-n+2}, \ldots, \mathbf{d}_I$ are replaced by the $\mathbf{d}_{I-n+2}^{(1)}, \ldots, \mathbf{d}_{I+1}^{(1)}$; their weights are the numbers $w_{I-n+2}^{(1)}, \ldots, w_{I+1}^{(1)}$.

A rational B-spline curve, being piecewise rational polynomial, has a piecewise rational Bézier representation. We can find the Bézier points and their weights for each segment by inserting every knot until it has multiplicity n, i.e., by applying the de Boor algorithm to each knot.

It is also possible to *reparametrize* a rational B-spline curve, just as we could do for Bézier curves. For a description, see Lee and Lucian [307].

15.10   Implementation

The following computes a point on a rational Bézier curve:

```
float ratbez(degree,coeff,weight,t)
/*
     uses rational de casteljau to compute
     point on ratbez curve for param. value t.
Input: degree:  degree of curve
       coeff:    control point coordinates
       weight:   weights
       t:        evaluation paramter
*/
```

Reparametrizing a rational Bézier curve:

```
void reparam(wold,degree,s,wnew)
/* reparametrizes ratbez curve: only the weights,
stored in wold, are changed. New weights are in
wnew. Parametrization is determined by shoulder
point s. For s=0.5, nothing changes. Also,
s should be in (0,1).
*/
```

The routine to subdivide a rational Bézier curve at a parameter value t was already given in Section 4.9.

A program that generates the piecewise rational Bézier form from a rational cubic B-spline curve is:

```
void ratbspline_to_bezier(bspl_x,bspl_y,bspl_w,knot,l,bez_x,bez_y,bez_w)
/* converts rational  cubic B-spline polygon into piecewise
rational Bezier polygon
Input: bspl_x, bspl_y: planar B-spline control polygon
       bspl_w:         B-spline weights
       knot:           knot sequence
       l:              no. of intervals

Output: bez_x, bez_y: planar piecewise Bezier polygon
        bez_w:        Bezier weights (not in piecewise standard form!)
*/
```

15.11   Problems

1. In Section 15.8, we investigated rational C^2 cubic spline interpolants. Investigate rational G^2 cubic ν-spline or γ-spline interpolants. Literature: [65], [35], [210].

2. Suppose you are given two coplanar rational quadratic segments that form a C^1 curve, but not a G^2 curve. Can you adjust the weights (not the control polygons!) such that the resulting segments form a G^2 curve? Hint: use (11.9).

3. Define a rational Aitken's algorithm, i.e., one where the data points are assigned weights. Try to adjust those weights in an attempt to reduce the oscillatory behavior of the interpolant.

4. In Section 15.4, we said that signed curvature only makes sense in $I\!\!E^2$. Why not in $I\!\!E^3$?

5. In Section 15.6, we argued that the convex hull property is lost when the control vector form of a rational Bézier curve is used. What can you say about the bounds of such curves?

6. In Section 15.5, we remarked that the cross ratios of any four points $(\mathbf{b}_i, \mathbf{q}_i, \hat{\mathbf{q}}_i, \mathbf{b}_{i+1})$ are the same for all polygon legs. How is this cross ratio related to the reparmetrization constant c?

Chapter 16

Tensor Product Bézier Surfaces

The first person to consider this class of surfaces for design purposes was probably de Casteljau, who investigated them between 1959 and 1963. The popularity of this type of surfaces is, however, due to the work of Bézier only slightly later, as documented in Chapter 1. Initially, Bézier patches were only used to approximate a given surface. It took some time for people to realize that any B-spline surface can also be written in piecewise Bézier form.

We will use the example of Bézier patches to demonstrate the tensor product approach to surface patches. Once that principle is developed, it will be trivial to generalize other curve schemes to tensor product surfaces.

16.1 Bilinear Interpolation

In Section 2.3 we studied linear interpolation in $I\!E^3$ and derived properties of this elementary method that we then used for the development of Bézier curves. In an analogous fashion, one can base the theory of *tensor product Bézier surfaces* on the concept of *bilinear interpolation*. While linear interpolation fits the "simplest" *curve* between two points, bilinear interpolation fits the "simplest" *surface* between four points.

To be more precise: Let $\mathbf{b}_{0,0}, \mathbf{b}_{0,1}, \mathbf{b}_{1,0}, \mathbf{b}_{1,1}$ be four distinct points in $I\!E^3$. The set of all points $\mathbf{x} \in I\!E^3$ of the form

$$\mathbf{x}(u, v) = \sum_{i=0}^{1} \sum_{j=0}^{1} \mathbf{b}_{i,j} B_i^1(u) B_j^1(v) \tag{16.1}$$

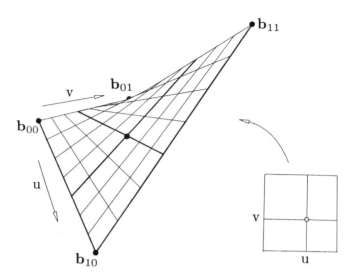

Figure 16.1: Bilinear interpolation: a hyperbolic paraboloid is defined by four points $\mathbf{b}_{i,j}$.

is called a *hyperbolic paraboloid* through the four $\mathbf{b}_{i,j}$. In matrix form:

$$\mathbf{x}(u, v) = \begin{bmatrix} 1 - u & u \end{bmatrix} \begin{bmatrix} \mathbf{b}_{00} & \mathbf{b}_{01} \\ \mathbf{b}_{10} & \mathbf{b}_{11} \end{bmatrix} \begin{bmatrix} 1 - v \\ v \end{bmatrix}. \qquad (16.2)$$

Since (16.1) is linear in both u and v and it interpolates to the input points, the surface $\mathbf{x}$ is called the *bilinear interpolant*. An example is shown in Figure 16.1.

The bilinear interpolant can be viewed as a map of the unit square $0 \leq u, v \leq 1$ in the u, v-plane. We say that the unit square is the *domain* of the interpolant, while the surface $\mathbf{x}$ is its *range*. A line parallel to one of the the axes in the domain corresponds to a curve in the range; it is called an *isoparametric curve*. Every isoparametric curve of the hyperbolic paraboloid (16.1) is a straight line, thus, hyperbolic paraboloids are *ruled surfaces*; see also Sections 20.1 and 22.10. In particular, the isoparametric line $u = 0$ is mapped onto the straight line through $\mathbf{b}_{0,0}$ and $\mathbf{b}_{0,1}$; analogous statements hold for the other three boundary curves.

Instead of evaluating the bilinear interpolant directly, one can apply a two-stage process that we will employ later in the context of tensor product interpolation. We can compute two intermediate points

$$\mathbf{b}_{0,0}^{0,1} = (1 - v)\mathbf{b}_{0,0} + v\mathbf{b}_{0,1}, \tag{16.3}$$

$$\mathbf{b}_{1,0}^{0,1} = (1 - v)\mathbf{b}_{1,0} + v\mathbf{b}_{1,1}, \tag{16.4}$$

and obtain the final result as

$$\mathbf{x}(u, v) = \mathbf{b}_{0,0}^{1,1}(u, v) = (1 - u)\mathbf{b}_{0,0}^{0,1} + u\mathbf{b}_{1,0}^{0,1}.$$

This amounts to computing the coefficients of the isoparametric line $v = const$ first and then evaluating this isoparametric line at u. The reader should verify that the other possibility, computing a $u = const$ isoparametric line first and then evaluating it at v, gives the same result.

Since linear interpolation is an affine map, and since we apply linear interpolation (or affine maps) in both the u- and v-direction, one sometimes sees the term "biaffine map" for bilinear interpolation; see Ramshaw [386].

The term "hyperbolic paraboloid" comes from analytic geometry. We shall justify this name by considering the (nonparametric) surface $z = xy$. It can be interpreted as the bilinear interpolant to the four points

$$\begin{bmatrix} 0 \\ 0 \\ 0 \end{bmatrix}, \begin{bmatrix} 1 \\ 0 \\ 0 \end{bmatrix}, \begin{bmatrix} 0 \\ 1 \\ 0 \end{bmatrix}, \begin{bmatrix} 1 \\ 1 \\ 1 \end{bmatrix}$$

and is shown in Figure 16.2. If we intersect the surface with a plane parallel to the x, y-plane, the resulting curve is a *hyperbola*; if we intersect it with a plane containing the z-axis, the resulting curve is a *parabola*.

16.2   The Direct de Casteljau Algorithm

Bézier curves may be obtained by repeated application of linear interpolation. We shall now obtain surfaces from repeated application of *bilinear interpolation*.

Suppose we are given a rectangular array of points $\mathbf{b}_{i,j}; 0 \leq i, j \leq n$ and parameter values (u, v). The following algorithm generates a point on a surface determined by the array of the $\mathbf{b}_{i,j}$:

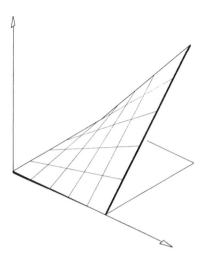

Figure 16.2: Bilinear interpolation: the surface $z = xy$ over the unit square.

Given $\{\mathbf{b}_{i,j}\}_{i,j=0}^{n}$ and $(u,v) \in I\!\!R^2$, set

$$\mathbf{b}_{i,j}^{r,r} = \begin{bmatrix} 1-u & u \end{bmatrix} \begin{bmatrix} \mathbf{b}_{i,j}^{r-1,r-1} & \mathbf{b}_{i,j+1}^{r-1,r-1} \\ \mathbf{b}_{i+1,j}^{r-1,r-1} & \mathbf{b}_{i+1,j+1}^{r-1,r-1} \end{bmatrix} \begin{bmatrix} 1-v \\ v \end{bmatrix} \begin{matrix} \\ {\scriptstyle r=1,\ldots,n} \\ {\scriptstyle i,j=0,\ldots,n-r} \end{matrix}.$$
$$(16.5)$$

and $\mathbf{b}_{i,j}^{0,0} = \mathbf{b}_{i,j}$. Then $\mathbf{b}_{0,0}^{n,n}$ is the point with parameter values (u,v) on the *Bézier surface* $\mathbf{b}^{n,n}$. (The reason for the somewhat clumsy identical superscripts will be explained in the next section.) The net of the $\mathbf{b}_{i,j}$ is called the *Bézier net* or *control net* of the surface $\mathbf{b}^{n,n}$. The $\mathbf{b}_{i,j}$ are called control points or Bézier points, just as in the curve case. Figure 16.3 shows an example for $n = 3$; Example 16.1 shows how to compute the quadratic case. An example of a bicubic ($n = 3$) Bézier patch is shown in Figure 16.4.

We have defined a surface scheme through a constructive algorithm just as we have done in the curve case. We could now continue to derive analytic properties of these surfaces, again as in the curve case. This is possible without much effort; however, we use a different approach in Section 16.3.

In the next section we shall be able to handle surfaces that are of different degrees in u and v. Such surfaces have control nets $\{\mathbf{b}_{i,j}\}$; $i = 0,\ldots,m$,

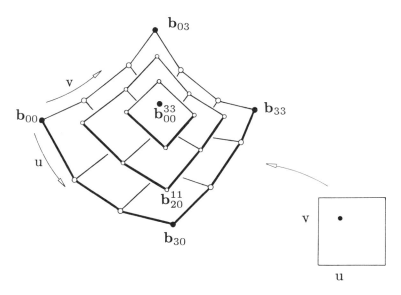

Figure 16.3: The direct de Casteljau algorithm for surfaces: the point on the surface is found from repeated bilinear interpolation.

$j = 0, \ldots, n$. The direct de Casteljau algorithm for such surfaces exists, but it needs a case distinction: consulting Figure 16.5, we see that the direct de Casteljau algorithm cannot be performed until the point of the surface is reached. Instead, after $k = \min(m, n)$, the intermediate $\mathbf{b}_{i,j}^{k,k}$ form a curve control polygon. We now must proceed with the univariate de Casteljau algorithm to obtain a point on the surface. This case distinction is awkward and will not be encountered by the tensor product approach in the next section.

16.3 The Tensor Product Approach

We have seen in the introduction by P. Bézier how stylists in the design shop physically created surfaces: templates were used to scrape material off a rough clay model (see Figure 1.13 in Chapter 1). Different templates are used as more and more of the surface is carved out of the clay. Analyzing this process from a theoretical viewpoint, one arrives at the following intuitive definition

Let a Bézier control net be given by

$$\begin{bmatrix} 0 \\ 0 \\ 0 \end{bmatrix} \begin{bmatrix} 2 \\ 0 \\ 0 \end{bmatrix} \begin{bmatrix} 4 \\ 0 \\ 0 \end{bmatrix}$$

$$\begin{bmatrix} 0 \\ 2 \\ 0 \end{bmatrix} \begin{bmatrix} 2 \\ 2 \\ 0 \end{bmatrix} \begin{bmatrix} 4 \\ 2 \\ 2 \end{bmatrix} .$$

$$\begin{bmatrix} 0 \\ 4 \\ 0 \end{bmatrix} \begin{bmatrix} 2 \\ 4 \\ 4 \end{bmatrix} \begin{bmatrix} 4 \\ 4 \\ 4 \end{bmatrix}$$

After one step of the direct de Casteljau algorithm for $(u, v) = (0.5, 0.5)$, we obtain

$$\begin{bmatrix} 1 \\ 1 \\ 0 \end{bmatrix} \begin{bmatrix} 3 \\ 1 \\ 0.5 \end{bmatrix}$$

$$\begin{bmatrix} 1 \\ 3 \\ 1 \end{bmatrix} \begin{bmatrix} 3 \\ 3 \\ 2.5 \end{bmatrix} .$$

The point on the surface is

$$\begin{bmatrix} 2 \\ 2 \\ 1 \end{bmatrix} .$$

Example 16.1: Computing a point on a Bézier surface using the direct de Casteljau algorithm.

of a surface: *A surface is the locus of a curve that is moving through space and thereby changing its shape.* See Figure 16.6 for an illustration.

We will now formalize this intuitive concept in order to arrive at a mathematical description of a surface. First, we assume that the moving curve is a Bézier curve of constant degree m. (This assumption is made so that the following formulas will work out; it is actually a serious restriction on the class of surfaces that we can represent using the tensor product approach.) At any time, the moving curve is then determined by a set of control points. Each original

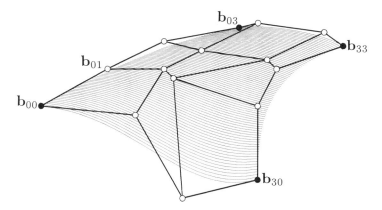

Figure 16.4: Bézier surfaces: a bicubic patch with its defining control net.

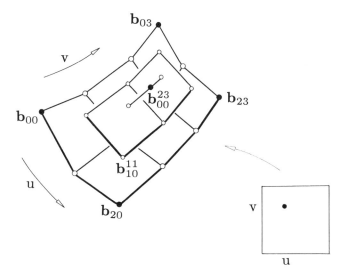

Figure 16.5: The direct de Casteljau algorithm: a surface with $(m, n) = (2, 3)$ proceeds in a univariate manner after no more direct de Casteljau steps can be performed.

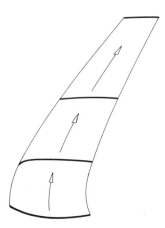

Figure 16.6: Tensor product surfaces: a surface can be thought of as being swept out by a moving and deforming curve.

control point moves through space on a curve. Our next assumption is that this curve is also a Bézier curve, and that the curves on which the control points move are all of the same degree. An example is given in Figure 16.7.

This can be formalized as follows: let the initial curve be a Bézier curve of degree m:

$$\mathbf{b}^m(u) = \sum_{i=0}^{m} \mathbf{b}_i B_i^m(u).$$

Let each $\mathbf{b}_i$ traverse a Bézier curve of degree n:

$$\mathbf{b}_i = \mathbf{b}_i(v) = \sum_{j=0}^{n} \mathbf{b}_{i,j} B_j^n(v).$$

We can now combine these two equations and obtain the point $\mathbf{b}^{m,n}(u, v)$ on the surface $\mathbf{b}^{m,n}$ as

$$\mathbf{b}^{m,n}(u, v) = \sum_{i=0}^{m} \sum_{j=0}^{n} \mathbf{b}_{i,j} B_i^m(u) B_j^n(v). \tag{16.6}$$

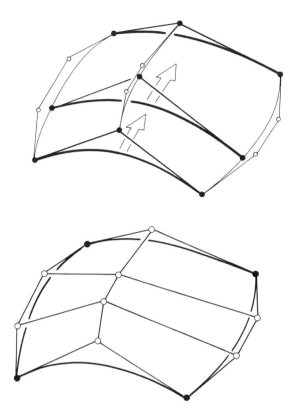

Figure 16.7: Tensor product Bézier surfaces: Top, a surface is obtained by moving the control points of a curve (quadratic) along other Bézier curves (cubic); bottom: the final Bézier net.

With this notation, the original curve $\mathbf{b}^m(u)$ now has Bézier points $\mathbf{b}_{i,0}; i = 0, \ldots, m$.

It is not difficult to prove that the definition of a Bézier surface (16.6) and the definition using the direct de Casteljau algorithm are equivalent (see Problems). Example 16.2 supports this view.

We have described the Bézier surface of Eq. (16.6) as being obtained by moving the isoparametric curve corresponding to $v = 0$. It is an easy exercise to check that the three remaining boundary curves could also have been used as the starting curve.

We can also compute the point on the surface of Example 16.1 by the tensor product method. We then evaluate each row of Bézier points for $u = 1/2$, and obtain the intermediate values

$$\begin{bmatrix} 2 \\ 0 \\ 0 \end{bmatrix}$$
$$\begin{bmatrix} 2 \\ 2 \\ 0.5 \end{bmatrix}.$$
$$\begin{bmatrix} 2 \\ 4 \\ 3 \end{bmatrix}$$

This quadratic control polygon defines the isoparametric curve $\mathbf{b}(\frac{1}{2}, v)$; we evaluate it for $v = 1/2$ obtain the same point as in Example 16.1.

Example 16.2: Computing a point on a Bézier surface using the tensor product method.

An arbitrary isoparametric curve $\hat{v} = const$ of a Bézier surface $\mathbf{b}^{m,n}$ is a Bézier curve of degree m in u, and its $m + 1$ Bézier points are obtained by evaluating all rows of the control net at $v = const$. As a formula:

$$\mathbf{b}_{i,0}^{0n}(\hat{v}) = \sum_{j=0}^{n} \mathbf{b}_{ij} B_j^n(\hat{v}); \quad i = 0, \ldots m.$$

This process of obtaining the Bézier points of an isoparametric line is a second possible interpretation of Figure 16.7. The coefficients of the isoparametric line can be obtained by applying $m + 1$ de Casteljau algorithms. A point on the surface is then obtained by performing one more de Casteljau algorithm.

Isoparametric curves $u = const$ are treated analogously. Note, however, that other straight lines in the domain are mapped to higher degree curves on the patch: they are generally of degree $n + m$. Two special examples of such curves are the two diagonals of the domain rectangle.

16.4 Properties

Most properties of Bézier patches follow in a straightforward way from those of Bézier curves – the reader is referred to Sections 3.3 and 4.2. We give a brief listing:

Affine invariance: The direct de Casteljau algorithm consists of repeated bilinear and possibly subsequent repeated linear interpolation. All these operations are affinely invariant; hence, so is their composition. We can also argue that in order for (16.6) to be a barycentric combination (and therefore affinely invariant), we must have

$$\sum_{j=0}^{n}\sum_{i=0}^{m} B_i^m(u)B_j^n(v) \equiv 1. \tag{16.7}$$

This identity is easily verified algebraically. A warning: there is no projective invariance of Bézier surfaces! In particular, we cannot apply a perspective projection to the control net and then plot the surface that is determined by the resulting image. Such operations will be possible by means of rational Bézier surfaces.

Convex hull property: For $0 \le u, v \le 1$, the terms $B_i^m(u)B_j^n(v)$ are nonnegative. Then, taking (16.7) into account, (16.6) is a convex combination.

Boundary curves: The boundary curves of the patch $\mathbf{b}^{m,n}$ are polynomial curves. Their Bézier polygons are given by the boundary polygons of the control net. In particular, the four corners of the control net all lie on the patch.

Variation diminishing property: This property is *not* inherited from the univariate case. In fact, it is not at all clear what the definition of variation diminution should be in the bivariate case. Counting intersections with straight lines, as we did for curves, would not make Bézier patches variation diminishing; it is easy to visualize a patch that is intersected by a straight line while its control net is not. (Here, we would view the control net as a collection of bilinear patches.) Other attempts at a suitable definition of a bivariate variation diminishing property have been similarly unsuccessful.

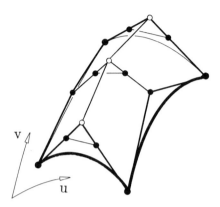

Figure 16.8: Degree elevation: the surface problem can be reduced to a series of univariate problems.

16.5   Degree Elevation

Suppose we want to rewrite a Bézier surface of degree (m, n) as one of degree $(m + 1, n)$. This amounts to finding coefficients $\mathbf{b}_{i,j}^{(1,0)}$ such that

$$\mathbf{b}^{m,n}(u, v) = \sum_{j=0}^{n} \left[\sum_{i=0}^{m+1} \mathbf{b}_{i,j}^{(1,0)} B_i^{m+1}(u) \right] B_j^n(v).$$

The $n + 1$ terms in square brackets represent $n + 1$ univariate degree elevation problems as discussed in Section 5.1. They are solved by a direct application of (5.2):

$$\mathbf{b}_{i,j}^{(1,0)} = (1 - \frac{i}{m+1})\mathbf{b}_{i,j} + \frac{i}{m+1}\mathbf{b}_{i-1,j}; \begin{cases} i = 0, \ldots, m+1 \\ j = 0, \ldots n. \end{cases} \qquad (16.8)$$

A tensor product surface is thus degree elevated in the u-direction by treating all columns of the control net as Bézier polygons of mth degree curves and degree elevating each of them. This is illustrated in Figure 16.8.

Degree elevation in the v-direction works the same way, of course. If we want to degree elevate in both the u- and the v-direction, we can perform the procedure first in the u-direction, then in the v-direction, or we can proceed the other way around. Both approaches yield the same surface of degree $(m+1, n+1)$. Its coefficients $\mathbf{b}_{i,j}^{(1,1)}$ may be found in a one-step method:

$$\mathbf{b}_{i,j}^{(1,1)} = \left[\begin{array}{cc} \frac{i}{m+1} & 1 - \frac{i}{m+1} \end{array}\right] \left[\begin{array}{cc} \mathbf{b}_{i-1,j-1} & \mathbf{b}_{i-1,j} \\ \mathbf{b}_{i,j-1} & \mathbf{b}_{i,j} \end{array}\right] \left[\begin{array}{c} \frac{j}{n+1} \\ 1 - \frac{j}{n+1} \end{array}\right] \quad (16.9)$$

$$i = 0, \ldots, m+1,$$
$$j = 0, \ldots, n+1.$$

The net of the $\mathbf{b}_{i,j}^{(1,1)}$ is obtained by *piecewise bilinear interpolation* from the original control net.

16.6 Derivatives

In the curve case, taking derivatives was accomplished by differencing the control points. The same will be true here. The derivatives that we will consider are *partial derivatives* $\partial/\partial u$ or $\partial/\partial v$. A partial derivative is the tangent vector of an isoparametric curve, and can be found by a straightforward calculation:

$$\frac{\partial}{\partial u}\mathbf{b}^{m,n}(u,v) = \sum_{j=0}^{n}\left[\frac{\partial}{\partial u}\sum_{i=0}^{m}\mathbf{b}_{i,j}B_i^m(u)\right]B_j^n(v).$$

The bracketed terms depend only on u, and we can apply the formula for the derivative of a Bézier curve (4.18):

$$\frac{\partial}{\partial u}\mathbf{b}^{m,n}(u,v) = m\sum_{j=0}^{n}\sum_{i=0}^{m-1}\Delta^{1,0}\mathbf{b}_{i,j}B_i^{m-1}(u)B_j^n(v).$$

Here we have generalized the standard difference operator in the obvious way: the superscript $(1,0)$ means that differencing is performed only on the first subscript: $\Delta^{1,0}\mathbf{b}_{i,j} = \mathbf{b}_{i+1,j} - \mathbf{b}_{i,j}$. If we take v-partials, we employ a difference operator that acts only on the second subscripts: $\Delta^{0,1}\mathbf{b}_{i,j} = \mathbf{b}_{i,j+1} - \mathbf{b}_{i,j}$. We then obtain

$$\frac{\partial}{\partial v}\mathbf{b}^{m,n}(u,v) = n\sum_{i=0}^{m}\sum_{j=0}^{n-1}\Delta^{0,1}\mathbf{b}_{i,j}B_j^{n-1}(v)B_i^m(u).$$

Again, a surface problem can be broken down into several univariate problems: to compute a u-partial, for instance, interpret all columns of the control net as Bézier curves of degree m and compute their derivatives (evaluated at the desired value of u). Then interpret these derivatives as coefficients of another Bézier curve of degree n and compute its value at the desired value of v.

We can write down formulas for higher order partials:

$$\frac{\partial^r}{\partial u^r} \mathbf{b}^{m,n}(u,v) = \frac{m!}{(m-r)!} \sum_{j=0}^{n} \sum_{i=0}^{m-r} \Delta^{r,0} \mathbf{b}_{i,j} B_i^{m-r}(u) B_j^n(v) \qquad (16.10)$$

and

$$\frac{\partial^s}{\partial v^s} \mathbf{b}^{m,n}(u,v) = \frac{n!}{(n-s)!} \sum_{i=0}^{m} \sum_{j=0}^{n-s} \Delta^{0,s} \mathbf{b}_{i,j} B_j^{n-s}(v) B_i^m(u). \qquad (16.11)$$

Here, the difference operators are defined by

$$\Delta^{r,0} \mathbf{b}_{i,j} = \Delta^{r-1,0} \mathbf{b}_{i+1,j} - \Delta^{r-1,0} \mathbf{b}_{i,j}$$

and

$$\Delta^{0,s} \mathbf{b}_{i,j} = \Delta^{0,s-1} \mathbf{b}_{i,j+1} - \Delta^{0,s-1} \mathbf{b}_{i,j}.$$

It is not hard now to write down the most general case, namely *mixed partials* of arbitrary order:

$$\frac{\partial^{r+s}}{\partial u^r \partial v^s} \mathbf{b}^{m,n}(u,v)$$

$$= \frac{m! n!}{(m-r)!(n-s)!} \sum_{i=0}^{m-r} \sum_{j=0}^{n-s} \Delta^{r,s} \mathbf{b}_{i,j} B_i^{m-r}(u) B_j^{n-s}(v). \qquad (16.12)$$

Before we proceed to consider some special cases, the reader should recall that the coefficients $\Delta^{r,s} \mathbf{b}_{i,j}$ are vectors and therefore do not "live" in $\mathbb{E}^3$. See Section 4.3 for more details.

A partial derivative of a point-valued surface is itself a vector-valued surface. We can evaluate it along isoparametric lines, of which the four boundary curves are the ones of most interest. Such a derivative, e.g., $\partial / \partial u \mid_{u=0}$ is called a *cross boundary derivative*. We can thus restrict (16.10) to $u = 0$ and get, with a slight abuse of notation,

$$\frac{\partial^r}{\partial u^r} \mathbf{b}^{m,n}(0,v) = \frac{m!}{(m-r)!} \sum_{j=0}^{n} \Delta^{r,0} \mathbf{b}_{0,j} B_j^n(v). \qquad (16.13)$$

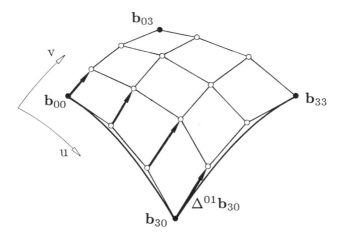

Figure 16.9: Cross boundary derivatives: along the edge $u = 0$, the cross boundary derivative only depends on two rows of control points.

Similar formulas hold for the other three edges. We thus have determined that r^{th} order cross boundary derivatives, evaluated along that boundary, depend only on the $r + 1$ rows (or columns) of Bézier points next to that boundary. This will be important when we formulate conditions for C^r continuity between adjacent patches. The case $r = 1$ is illustrated in Figure 16.9.

16.7   Normal Vectors

The *normal vector* $\mathbf{n}$ of a surface is a normalized vector that is normal to the surface at a given point. It can be computed from the cross product of any two vectors that are tangent to the surface at that point. Since the partials $\partial/\partial u$ and $\partial/\partial v$ are two such vectors, we may set

$$\mathbf{n}(u, v) = \frac{\frac{\partial}{\partial u}\mathbf{b}^{m,n}(u, v) \wedge \frac{\partial}{\partial v}\mathbf{b}^{m,n}(u, v)}{||\frac{\partial}{\partial u}\mathbf{b}^{m,n}(u, v) \wedge \frac{\partial}{\partial v}\mathbf{b}^{m,n}(u, v)||}, \tag{16.14}$$

where $\wedge$ denotes the cross product.

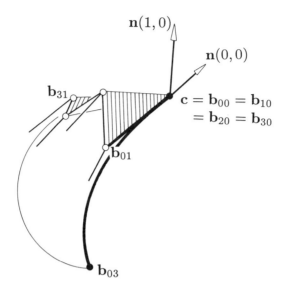

Figure 16.10: Degenerate patches: a "triangular" patch is created by collapsing a whole boundary curve into a point. The normal at that point may be undefined. Normals are shown for $u = 0$ and for $u = 1$.

At the four corners of the patch, the involved partials are simply differences of boundary points, for example,

$$\mathbf{n}(0,0) = \frac{\Delta^{1,0}\mathbf{b}_{0,0} \wedge \Delta^{0,1}\mathbf{b}_{0,0}}{||\Delta^{1,0}\mathbf{b}_{0,0} \wedge \Delta^{0,1}\mathbf{b}_{0,0}||}. \qquad (16.15)$$

The normal at one of the corners (we take $\mathbf{b}_{0,0}$ as an example) is undefined if $\Delta^{1,0}\mathbf{b}_{0,0}$ and $\Delta^{0,1}\mathbf{b}_{0,0}$ are linearly dependent: if that were the case, (16.15) would degenerate into an expression of the form $\frac{\mathbf{0}}{0}$. The corresponding patch corner is then called *degenerate*. Two cases of special interest are illustrated in Figures 16.10, 16.11, and 16.12.

In the first of these, a whole boundary curve is collapsed into a single point. As an example, we could set $\mathbf{b}_{00} = \mathbf{b}_{10} = \cdots = \mathbf{b}_{m0} = \mathbf{c}$. Then the boundary $\mathbf{b}(u,0)$ would degenerate into a single point. In such cases, the normal vector at $v = 0$ may or may not be defined. To examine this in more detail, consider the tangents of the isoparametric lines $u = \hat{u}$, evaluated at $v = 0$. These tangents must be perpendicular to the normal vector, if it exists.

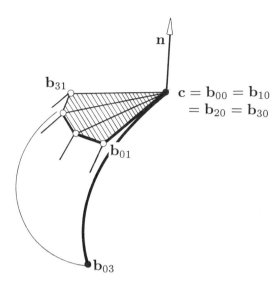

Figure 16.11: Degenerate patches: if all $\mathbf{b}_{i1}$ and $\mathbf{c}$ are coplanar, then the normal vector at $\mathbf{c}$ is perpendicular to that plane.

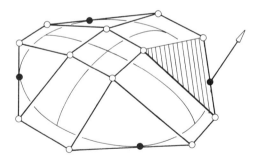

Figure 16.12: Degenerate patches: the normals at all four corners of this patch are determined by the triangles that are formed by the corner subquadrilaterals (one corner highlighted).

So a condition for the existence of the normal vector at c is that all v-partials, evaluated at $v = 0$, are coplanar. But that is equivalent to $b_{01}, b_{11}, \ldots, b_{m1}$ and c being coplanar.

A second possibility in creating degenerate patches is to allow two corner partials to be collinear, for example, $\partial/\partial u$ and $\partial/\partial v$ at $(0,0)$, as shown in Figure 16.12. In that case, b_{10}, b_{01}, and b_{00} are collinear. Then the normal at b_{00} is defined, provided that b_{11} is not collinear with b_{10}, b_{01}, and b_{00}. Recall that b_{00}, b_{10}, b_{01}, and b_{11} form the osculating paraboloid at $(u, v) = (0, 0)$. Then it follows that the tangent plane at b_{00} is the plane through the four coplanar points b_{00}, b_{10}, b_{01}, and b_{11}. The normal at b_{00} is perpendicular to it.

A warning: when we say "the normal is defined" then it should be understood that this is a purely mathematical statement. In any of the above degeneracies, a program using (16.14) will crash. A case distinction is necessary, and then the program can branch into the special cases that we just described. More complex situations are encountered when one also wants to compute curvatures of a degenerate patch. A solution is offered in [473].

16.8   Twists

The *twist* of a surface[1] is its mixed partial $\partial^2/\partial u \partial v$. According to (16.12), the twist surface of $b^{m,n}$ is a Bézier surface of degree $(m - 1, n - 1)$, and its (vector) coefficients have the form $mn\Delta^{1,1}b_{i,j}$. These coefficients have a nice geometric interpretation. For its discussion, we refer to Figure 16.13. The point $p_{i,j}$ in that figure is the fourth point on the parallelogram defined by $b_{i,j}, b_{i+1,j}, b_{i,j+1}$. It is defined by

$$p_{i,j} - b_{i+1,j} = b_{i,j+1} - b_{i,j}. \tag{16.16}$$

Since

$$\Delta^{1,1}b_{i,j} = (b_{i+1,j+1} - b_{i+1,j}) - (b_{i,j+1} - b_{i,j}), \tag{16.17}$$

it follows that

$$\Delta^{1,1}b_{i,j} = b_{i+1,j+1} - p_{i,j}. \tag{16.18}$$

---

[1]In this chapter, we are only dealing with polynomial surfaces. For these, the twist is uniquely defined. For other surfaces, it may depend on the order in which derivatives are taken; see Section 21.1.

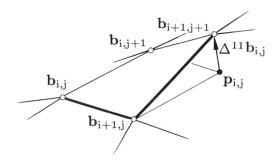

Figure 16.13: Twists: the twist coefficients are proportional to the deviations of the subquadrilaterals from parallelograms.

Thus the terms $\Delta^{1,1}\mathbf{b}_{i,j}$ measure the deviation of each subquadrilateral of the Bézier net from a parallelogram.

The twists at the four patch corners determine the deviation of the respective corner subquadrilaterals of the control net from parallelograms. For example,

$$\frac{\partial^2}{\partial u \partial v}\mathbf{b}^{m,n}(0,0) = mn\Delta^{1,1}\mathbf{b}_{00}. \qquad (16.19)$$

This twist vector is a measure for the deviation of $\mathbf{b}_{11}$ from the tangent plane at $\mathbf{b}_{00}$.

An interesting class of surfaces is obtained if all subquadrilaterals $\mathbf{b}_{i,j}$, $\mathbf{b}_{i+1,j}$, $\mathbf{b}_{i+1,j}$, $\mathbf{b}_{i+1,j+1}$ are parallelograms; in that case the twist vanishes everywhere. Such surfaces are called *translational surfaces* and will be discussed in Section 21.3; an example is shown in Figure 21.4.

16.9 The Matrix Form of a Bézier Patch

In Section 4.8, we formulated a matrix expression for Bézier curves. This approach carries over well to tensor product patches. We can write:

$$\mathbf{b}^{m,n}(u,v) =$$

$$\begin{bmatrix} B_0^m(u) & \cdots & B_m^m(t) \end{bmatrix} \begin{bmatrix} \mathbf{b}_{00} & \cdots & \mathbf{b}_{0n} \\ \vdots & & \vdots \\ \mathbf{b}_{m0} & \cdots & \mathbf{b}_{mn} \end{bmatrix} \begin{bmatrix} B_0^n(v) \\ \vdots \\ B_n^n(v) \end{bmatrix}. \qquad (16.20)$$

The matrix $\{\mathbf{b}_{ij}\}$, defining the control net, is sometimes called the *geometry matrix* of the patch. If we perform a basis transformation and write the Bernstein polynomials in monomial form, we obtain

$$
\mathbf{b}^{m,n}(u,v) = \begin{bmatrix} u^0 & \cdots & u^m \end{bmatrix} M^{\mathrm{T}} \begin{bmatrix} \mathbf{b}_{00} & \cdots & \mathbf{b}_{0n} \\ \vdots & & \vdots \\ \mathbf{b}_{m0} & \cdots & \mathbf{b}_{mn} \end{bmatrix} N \begin{bmatrix} v^0 \\ \vdots \\ v^n \end{bmatrix}. \quad (16.21)
$$

The square matrices M and N are given by

$$
m_{ij} = (-1)^{j-i} \binom{m}{j}\binom{j}{i} \quad (16.22)
$$

and

$$
n_{ij} = (-1)^{j-i} \binom{n}{j}\binom{j}{i}. \quad (16.23)
$$

In the bicubic case, $m = n = 3$, we have

$$
M = N = \begin{bmatrix} 1 & -3 & 3 & -1 \\ 0 & 3 & -6 & 3 \\ 0 & 0 & 3 & -3 \\ 0 & 0 & 0 & 1 \end{bmatrix}.
$$

For reasons of numerical stability, the use of the monomial form (16.21) is not advisable (see the discussion in Section 24.3). It is included here since it is still in widespread use.

16.10   Nonparametric Patches

This section is the bivariate analog of Section 5.5. Having outlined the main ideas there, we can be brief here. A nonparametric surface is of the form $z = f(x,y)$. It has the parametric representation

$$
\mathbf{x}(u,v) = \begin{bmatrix} u \\ v \\ f(u,v) \end{bmatrix},
$$

and we restrict both u and v to between zero and one. We are interested in functions f that are in Bernstein form:

$$
f(x,y) = \sum_{i}^{m} \sum_{j}^{n} b_{ij} B_i^m(x) B_j^n(y).
$$

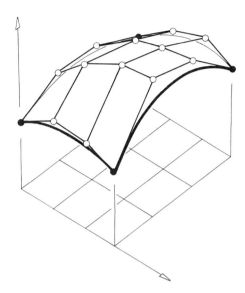

Figure 16.14: Nonparametric patches: the Bézier points are located over a regular partition of the domain rectangle.

Using the identity of Eq. (4.15) for both variables u and v, we see that the Bézier points of $\mathbf{x}$ are given by

$$\mathbf{b}_{ij} = \begin{bmatrix} i/m \\ j/n \\ b_{ij} \end{bmatrix}.$$

The points $(i/m, j/n)$ in the (x, y)-plane are called *Bézier abscissas* of the function f; the b_{ij} are called its *Bézier ordinates*. A nonparametric Bézier function is not constrained to be defined over the unit square; if a point $\mathbf{p}$ and two vectors $\mathbf{v}$ and $\mathbf{w}$ define a parallelogram in the (x, y)-plane, then the Bézier abscissas $\mathbf{a}_{ij} \in I\!E^2$ of a nonparametric Bézier function over this domain are given by $\mathbf{a}_{ij} = \mathbf{p} + i\mathbf{v} + j\mathbf{w}$. Figure 16.14 gives an example.

Integrals also carry over from the univariate case. With a proof analogous to the one in Section 5.7, one can show that

$$\int_0^1 \int_0^1 \sum_i^m \sum_j^n b_{ij} B_i^m(x) B_j^n(x) = \frac{\sum_i^m \sum_j^n b_{ij}}{(m+1)(n+1)}. \qquad (16.24)$$

16.11   Implementation

The following is the header for a program to plot a tensor product Bézier surface, in fact, a rational one. If the polynomial case is desired, just set all weights to unity.

```
void plot_ratsurf(bx,by,bw,degree_u,degree_v,u_points,v_points,
                        scale_x,scale_y)
/* plots v_points isoparametric
        curves of the  rat Bez surface, each with u_points
        points on it.

Input:  bx, by:  arrays with x- and y- coordinates of
            control net.
            degree_u,degree_v: degrees in u- and v- direction
u_points,v_points: plot resolution
scale_x,scale_y:   scale factor for postscript.

Output: postscript file
*/
```

16.12   Problems

1. Draw the hyperbolic paraboloid from Figure 16.2 over the square $(-1, -1)$, $(1, -1)$, $(1, 1)$, $(-1, 1)$. Try to do it manually, i.e., without graphics support.

2. Show that the direct de Casteljau algorithm generates surfaces of the form of Eq. (16.6).

3. Show that Bézier surfaces have bilinear precision: if $\mathbf{b}_{i,j} = \mathbf{x}(\frac{i}{m}, \frac{j}{n})$ and $\mathbf{x}$ is bilinear, then $\mathbf{b}^{m,n}(u, v) = \mathbf{x}(u, v)$ for all u, v and for arbitrary m, n.

4. Give an explicit formula for $\Delta^{r,s}\mathbf{b}_{i,j}$. Hint: use (4.20).

5. If a Bézier surface is given by its control net, we can use the de Casteljau algorithm to compute $\mathbf{b}^{m,n}(u, v)$ in three ways: by the direct form from Section 16.2, or by the two possible tensor product approaches, computing the coefficients of a u (or v) isoparametric line, and then

evaluating that curve at v (or u). While theoretically equivalent, the computation counts for these methods differ. Work out the details.

6. Generalize the routine `degree_elevate` to the tensor product case.

7. Generalize the routine `aitken` to the tensor product case, i.e., program tensor product Lagrange interpolation.

Chapter 17

Composite Surfaces and Spline Interpolation

Tensor product Bézier patches were under development in the early 1960s; at about the same time, people started to think about piecewise surfaces. One of the first publications was de Boor's work on bicubic splines [112] in 1962. Almost simultaneously, and apparently unaware of de Boor's work, J. Ferguson [183] implemented piecewise bicubics at Boeing. His method was used extensively, although it had the serious flaw of using only zero corner twist vectors. An excellent account of the industrial use of piecewise bicubics is the article by G. Peters [362].

17.1 Smoothness and Subdivision

Let $\mathbf{x}(u, v)$ and $\mathbf{y}(u, v)$ be two patches, defined over $[u_{I-1}, u_I] \times [v_J, v_{J+1}]$ and $[u_I, u_{I+1}] \times [v_J, v_{J+1}]$, respectively. They are r times continuously differentiable across their common boundary curve $\mathbf{x}(u_I, v) = \mathbf{y}(u_I, v)$ if all u-partials up to order r agree there:

$$\frac{\partial^r}{\partial u^r} \mathbf{x}(u, v) \Big|_{u=u_I} = \frac{\partial^r}{\partial u^r} \mathbf{y}(u, v) \Big|_{u=u_I}. \tag{17.1}$$

Now suppose both patches are given in Bézier form; let the control net of the "left" patch be $\{\mathbf{b}_{ij}\}; 0 \leq i \leq m, 0 \leq j \leq n$ and $\{\mathbf{b}_{ij}\}; m \leq i \leq 2m, 0 \leq j \leq n$. We can then invoke Eq. (16.13) for the cross boundary derivative of a Bézier patch. That formula is in local coordinates. To make the transition to global

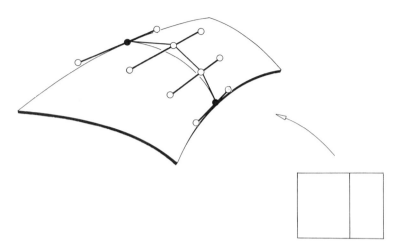

Figure 17.1: C^1 continuous Bézier patches: the shown control points must be collinear and all be in the same ratio.

coordinates (u, v), we must invoke the chain rule, just as we did for composite curves [Eq. (7.6)]:

$$(\frac{1}{\Delta_{I-1}})^r \sum_{j=0}^{n} \Delta^{r,0} \mathbf{b}_{m-r,j} B_j^n(v) = (\frac{1}{\Delta_I})^r \sum_{j=0}^{n} \Delta^{r,0} \mathbf{b}_{m,j} B_j^n(v), \qquad (17.2)$$

where $\Delta_I = u_{I+1} - u_I$. Since the $B_j^n(v)$ are linearly independent, we can compare coefficients:

$$(\frac{1}{\Delta_{I-1}})^r \Delta^{r,0} \mathbf{b}_{m-r,j} = (\frac{1}{\Delta_I})^r \Delta^{r,0} \mathbf{b}_{m,j}; \quad j = 0, \ldots, n.$$

This is the C^r condition of Eq. (7.6) for Bézier curves, applied to all $n + 1$ rows of the composite Bézier net. We thus have the C^r condition for composite Bézier surfaces: *two adjacent patches are C^r across their common boundary if and only if all rows of their control net vertices can be interpreted as polygons of C^r piecewise Bézier curves.* We have again succeeded in reducing a surface problem to several curve problems. The smoothness conditions apply analogously to the v-direction.

The case $r = 1$ is illustrated in Figure 17.1. The C^1 condition states that for every j, the polygon formed by $\mathbf{b}_{0,j}, \ldots, \mathbf{b}_{2m,j}$ is the control polygon of a C^1 piecewise Bézier curve. For this to be the case, the three points $\mathbf{b}_{m-1,j}, \mathbf{b}_{m,j}, \mathbf{b}_{m+1,j}$ must be collinear and in the ratio $\Delta_I : \Delta_{I+1}$. This ratio must be the same for all j. Simple collinearity is *not* sufficient: composite surfaces that have $\mathbf{b}_{m-1,j}, \mathbf{b}_{m,j}, \mathbf{b}_{m+1,j}$ collinear for each j but not in the same ratio will in general not be C^1. Moreover, they will not even have a continuous tangent plane. The rigidity of the C^1 condition can be a serious obstacle in the design of surfaces that consist of a network of Bézier patches (or of piecewise polynomial patches in other representations).

In the univariate case, the de Casteljau algorithm not only provided the point on the curve, but could also be used to *subdivide* the curve. A similar result is true for the surface case. Suppose the domain rectangle of a Bézier patch is subdivided into two subrectangles by a straight line $u = \hat{u}$. That line maps to an isoparametric curve on the patch, which is thus subdivided into two subpatches. We wish to find the control nets for each patch. These two patches, being part of one global surface, meet with C^n continuity. Therefore, all their rows of control points must be control polygons of C^n piecewise n^{th} degree curves. Those curves are related to each other by the univariate subdivision process from Section 4.6.

We now have the following subdivision algorithm: interpret all rows of the control net as control polygons of Bézier curves. Subdivide each of these

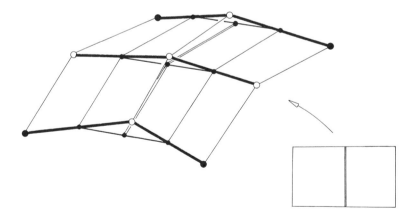

Figure 17.2: Subdivision of a Bézier patch: all rows are subdivided using the de Casteljau algorithm.

curves at $u = \hat{u}$. The resulting control points form the two desired control nets. For an example, see Figure 17.2.

Subdivision along an isoparametric line $v = \hat{v}$ is treated analogously. If we want to subdivide a patch into four subpatches that are generated by two isoparametric lines $u = \hat{u}$ and $v = \hat{v}$, we apply the subdivision procedure twice. It does not matter in which direction we subdivide first.

17.2 Bicubic B-spline Surfaces

B-spline surfaces (both rational and nonrational) play an important role in current surface design methods and will be discussed here in more detail. Using the notation from Chapter 10, a parametric tensor product B-spline surface may be written as

$$\mathbf{x}(u, v) = \sum_i \sum_j \mathbf{d}_{ij} N_i^3(u) N_j^3(v), \tag{17.3}$$

where we assume that one knot sequence in the u-direction and one in the v-direction are given. A typical control net, corresponding to triple end knots[1] and consisting of 5×3 bicubic patches, is shown in Figure 17.3.

Two more bicubic B-spline surfaces, again with triple end knots, are shown in Plates IV, IX, X, and XI. For curves, triple end knots meant that the first and last two B-spline control points were also Bézier control points; the same is true here. The B-spline control points $\mathbf{d}_{ij}$ for which i or j equal 0 or 1, are also control vertices of the piecewise Bézier net of the surface. Thus they determine the boundary curves and the cross boundary derivatives.

This raises a general question: since a bicubic B-spline surface is a collection of bicubic patches, how can we find the Bézier net of each patch? The answer to this question may be useful for the conversion of a B-spline data format to the piecewise Bézier form. It is also relevant if we decide to evaluate a B-spline surface by first breaking it down into bicubics. The solution arises, as usual for tensor products, from the breakdown of this surface problem into a series of curve problems. If we rewrite (17.3) as

$$\mathbf{x}(u, v) = \sum_i N_i^3(v) [\sum_j \mathbf{d}_{ij} N_j^3(u)],$$

---

[1]This is the notation from Chapter 10. The notation from Chapter 7 is implicitly based on triple end knots.

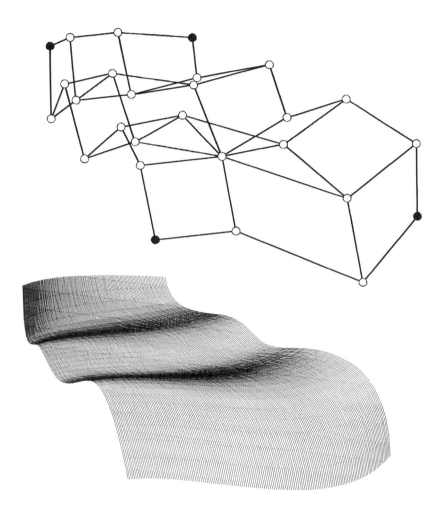

Figure 17.3: Bicubic B-spline surfaces: top, a control net for a bicubic B-spline surface consisting of 5 × 3 patches; bottom, the corresponding surface.

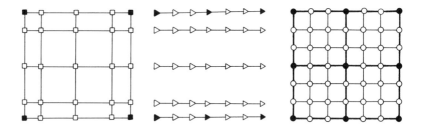

Figure 17.4: Bringing a bicubic B-spline surface into piecewise bicubic Bézier form: we first perform B-spline–Bézier curve conversion row by row, then column by column.

we see that for each i the sum in square brackets describes a B-spline curve in the variable u. We may convert it to Bézier form by using the univariate methods described in Chapter 7 or 10. This corresponds to interpreting the B-spline control net row by row as univariate B-spline polygons and then converting them to piecewise Bézier form. The Bézier points thus obtained – column by column – may be interpreted as B-spline polygons, which we may again transform to Bézier form one by one. This final family of Bézier polygons constitutes the piecewise Bézier net of the surface, as illustrated in Figure 17.4. Plate XI shows the piecewise Bézier net obtained from the B-spline net of the surface in Plates IX and X.

Needless to say, we could have started the B-spline–Bézier conversion process column by column. From the Bézier form, we may now transform to any other piecewise polynomial form, such as the piecewise monomial or the piecewise Hermite form.

17.3 Twist Estimation

Suppose that we are given a rectangular network of points $\mathbf{x}_{IJ}$; $0 \leq I \leq M$, $0 \leq J \leq N$ and two sets of parameter values u_I and v_J. We want a C^1 piecewise cubic surface $\mathbf{x}(u, v)$ that interpolates to the data points:

$$\mathbf{x}(u_I, v_J) = \mathbf{x}_{IJ}.$$

For a solution, we utilize curve methods wherever possible. We will first fit piecewise cubics to all rows and columns of data points using methods

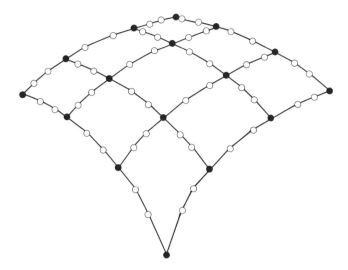

Figure 17.5: Piecewise bicubic interpolation: after a network of curves has been created, one still must determine four more coefficients per patch. A network of 3×3 patches is shown.

that were developed in Chapter 8. We must keep in mind, however, that all curves in the u-direction have the same parametrization, given by the u_I; the v-curves are all defined over the v_J.

Creating a network of C^1 (or C^2) piecewise cubics through the data points is only the first step toward a surface, however. Our aim is a C^1 piecewise bicubic surface, and so far we have only constructed the boundary curves for each patch. This constitutes 12 data out of the 16 needed for each patch. Figure 17.5 illustrates the situation. In Bézier form, we are still missing four interior Bézier points per patch, namely, $\mathbf{b}_{11}, \mathbf{b}_{21}, \mathbf{b}_{12}, \mathbf{b}_{22}$; in terms of derivatives, we must still determine the *corner twists* of each patch.

We now list a few methods to determine the missing twists.

Zero twists: Historically, this is the first twist estimation "method." It appears, hidden in a set of formulas in pseudo-code, in the paper by Ferguson [183]. Ferguson did not comment on the effects that this choice of twist vectors might have.

As we have seen above, surfaces exist that have identically vanishing twists – these are translational surfaces (see Figure 21.4). If the boundary

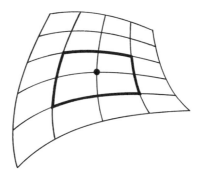

Figure 17.6: Adini's twist: the outer boundary curves of four adjacent patches define a Coons surface; its twist at the "middle" point is Adini's twist.

curves of a patch are pairwise related by translations, then the assignment of zero twists is a good idea, but not otherwise. In these other cases, the boundary curves are *not* the generating curves of a translational surface. If zero twists are assigned, the generated patch *will* locally behave like a translational surface, giving rise to the infamous "flat spots" of zero twists. The effects of zero twists will be illustrated in Chapter 23.

If a network of patches has to be created, this choice of twists automatically guarantees C^1 continuity of the overall surface. Thus it is mathematically "safe," but does not guarantee "nice" shapes.

Adini's twist: This method has been introduced into the CAGD literature through the paper by Barnhill, Brown, and Klucewicz [22], based on a scheme ("Adini's rectangle") from the finite element literature. The basic idea is this: the four cubic boundary curves define a bilinearly blended Coons patch (see Chapter 20), which happens to be a bicubic patch itself. Take the corner twists of that patch to be the desired twist vectors.

If a network of patches has to be generated, the above Adini's twists would not guarantee a C^1 surface. A simple modification is necessary: let four patches meet at a point, as in Figure 17.6. The four outer boundary curves of the four patches again define a bilinearly blended Coons patch. This Coons patch (consisting of four bicubics) has a well-defined twist at the parameter

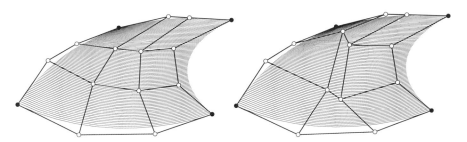

Figure 17.7: Twist estimation: the four interior Bézier points are computed to yield zero corner twists (left), and then according to Adini's method (right).

value where the four bicubics meet. Take that twist to be the desired twist. It is given by

$$
\begin{aligned}
\mathbf{x}_{uv}(u_i, v_j) \\
= \frac{\mathbf{x}_v(u_{i+1}, v_j) - \mathbf{x}_v(u_{i-1}, v_j)}{u_{i+1} - u_{i-1}} \\
+ \frac{\mathbf{x}_u(u_i, v_{j+1}) - \mathbf{x}_u(u_i, v_{j-1})}{v_{j+1} - v_{j-1}} \\
- \frac{\mathbf{x}(u_{i+1}, v_{j+1}) - \mathbf{x}(u_{i-1}, v_{j+1}) - \mathbf{x}(u_{i+1}, v_{j-1}) + \mathbf{x}(u_{i-1}, v_{j-1})}{(u_{i+1} - u_{i-1})(v_{j+1} - v_{j-1})}.
\end{aligned}
$$

It is easy to check that Adini's method, applied to patch boundaries of a translational surface, yields zero twists, which is desirable for that situation. Adini's twist is a reasonable choice, because, considered as an interpolant, it reproduces all bivariate polynomials of the form $uv^j, u^iv; i, j \in \{0, 1, 2, 3\}$, which is a surprisingly large set.[2]

Figure 17.7 compares zero and Adini's twist if only one patch is used. The zero twists give rise to undesirable distortions.

Bessel twists: This method estimates the twist at $\mathbf{x}(u_I, v_J)$ to be the twist of the biquadratic interpolant to the nine points $\mathbf{x}(u_{I+r}, v_{J+s}); r, s \in \{-1, 0, 1\}$. Since a biquadratic patch has a bilinear twist, Bessel's twist is the

---

[2]This is why this twist is called, in the context of finite elements, a "serendipity element."

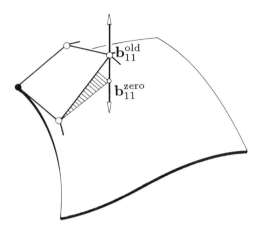

Figure 17.8: Brunet's twist: a convex combination is taken between the given interior Bézier point and the "Adini" Bézier point.

bilinear interpolant to the twists of the four bilinear patches formed by the nine points. Those twists are given by

$$\mathbf{q}_{I,J} = \frac{\Delta^{1,1}\mathbf{x}(u_I, v_J)}{\Delta_I \Delta_J},$$

and Bessel's twist can now be written

$$\mathbf{x}_{uv}(u_I, v_J) = \begin{bmatrix} 1 - \alpha_I & \alpha_I \end{bmatrix} \begin{bmatrix} \mathbf{q}_{I-1,J-1} & \mathbf{q}_{I-1,J} \\ \mathbf{q}_{I,J-1} & \mathbf{q}_{I,J} \end{bmatrix} \begin{bmatrix} 1 - \beta_J \\ \beta_J \end{bmatrix},$$

where

$$\alpha_I = \frac{\Delta_{I-1}}{u_{I+1} - u_{I-1}}, \quad \beta_J = \frac{\Delta_{J-1}}{v_{J+1} - v_{J-1}}.$$

Brunet's twist: (Brunet [79]) has developed a method different from the previous ones in that it does not start from a network of patch boundary curves, but rather from a C^1 network of bicubic patches. It then tries to modify the existing twist vectors in order to achieve a surface that "wiggles" less than the original surface. In that sense, this method is a surface fairing method rather than a pure twist estimation procedure.

Brunet makes the assumption that the bilinearly blended Coons patch (see Chapter 20) to the given patch boundaries would be "optimally smooth."

(Note that by our remarks on Adini's twist that surface is the one obtained from applying Adini's method to the curve network.) He then modifies the given surface to come closer to the "Adini surface." Brunet's method, in Bézier terms, is illustrated in Figure 17.8: a convex combination is taken between the interior Bézier points of the given surface and those corresponding to zero twists. Brunet states that

$$\mathbf{b}_{11}^{\text{new}} = .25\mathbf{b}_{11}^{\text{zero}} + .75\mathbf{b}_{11}^{\text{old}} \tag{17.4}$$

yields surfaces that exhibit fewer undulations than the original surface.

Other methods for twist estimation exist, including Selesnick [445], Hagen and Schulze [248], and Farin and Hagen [166].

17.4 Tensor Product Interpolants

In Chapter 16, we saw one example of a tensor product patch. We shall now exploit the tensor product method in order to obtain interpolating rectangular surfaces. Let

$$\mathbf{x}(u,v) = \sum_{i=0}^{m}\sum_{j=0}^{n}\mathbf{c}_{ij}A_i(u)B_j(v)$$

be a tensor product surface. The functions A_i and B_j are arbitrary at this point, but for the sake of concreteness, one may think of them as being Bernstein polynomials or B-spline basis functions.[3] This equation may be written in matrix form as

$$\mathbf{x}(u,v) = \begin{bmatrix} A_0(u) & \cdots & A_m(u) \end{bmatrix} \begin{bmatrix} \mathbf{c}_{00} & \cdots & \mathbf{c}_{0n} \\ \vdots & & \vdots \\ \mathbf{c}_{m0} & \cdots & \mathbf{c}_{mn} \end{bmatrix} \begin{bmatrix} B_0(v) \\ \vdots \\ B_n(v) \end{bmatrix}. \tag{17.5}$$

Suppose now that we are given an $(m+1) \times (n+1)$ array of data points $\mathbf{x}_{ij}$; $0 \le i \le m$, $0 \le j \le n$. If the surface of Eq. (17.5) is to interpolate to them, Eq. (17.5) must be true for each pair (u_i, v_j). We thus obtain $(n+1) \times (m+1)$ equations, which we may write concisely as

$$\mathbf{X} = ACB, \tag{17.6}$$

---

[3]They may be different basis functions; e.g., B-splines in u and Lagrange polynomials in v.

---

where

$$\mathbf{X} = \begin{bmatrix} \mathbf{x}_{00} & \cdots & \mathbf{x}_{0n} \\ \vdots & & \vdots \\ \mathbf{x}_{m0} & \cdots & \mathbf{x}_{mn} \end{bmatrix},$$

$$A = \begin{bmatrix} A_0(u_0) & \cdots & A_0(u_m) \\ \vdots & & \vdots \\ A_m(u_0) & \cdots & A_m(u_m) \end{bmatrix},$$

$$\mathbf{C} = \begin{bmatrix} \mathbf{c}_{00} & \cdots & \mathbf{c}_{0n} \\ \vdots & & \vdots \\ \mathbf{c}_{m0} & \cdots & \mathbf{c}_{mn} \end{bmatrix},$$

and

$$B = \begin{bmatrix} B_0(v_0) & \cdots & B_0(v_n) \\ \vdots & & \vdots \\ B_n(v_0) & \cdots & B_n(v_n) \end{bmatrix}.$$

Matrices A and B already appeared in Section 6.3; they are *Vandermonde matrices.*

In an interpolation context, the $\mathbf{x}_{ij}$ are known and the coefficients $\mathbf{c}_{ij}$ are unknown. They are found from (17.6) by setting

$$\mathbf{C} = A^{-1}\mathbf{X}B^{-1}. \tag{17.7}$$

The inverse matrices in (17.7) exist provided that the functions A_i and B_j are linearly independent (which they are in all cases of practical interest).

Equation (17.7) shows how a solution to the interpolation problem *could* be found, but one should not try to invert the matrices A and B explicitly! To solve and understand better the tensor product interpolation problem, we rewrite (17.6) as

$$\mathbf{X} = \mathbf{D}B, \tag{17.8}$$

where we have set

$$\mathbf{D} = A\mathbf{C}. \tag{17.9}$$

Note that $\mathbf{D}$ consists of $(m+1)$ rows and $(n+1)$ columns. Equation (17.8) can be interpreted as a family of $(m+1)$ univariate interpolation problems – one for each row of $\mathbf{X}$ and $\mathbf{D}$, where $\mathbf{D}$ contains the unknowns. Having solved

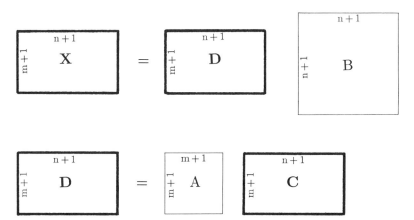

Figure 17.9: Tensor product interpolation: the dimensions of the involved matrices.

all $(n+1)$ problems (all having the same coefficient matrix B!), we can attack (17.9), since we have just computed D. Equation (17.9) may be interpreted as a family of $(n+1)$ univariate interpolation problems, all having the same coefficient matrix A. Figure 17.9 shows the dimensions of the involved matrices.

We thus see how the tensor product form allows a significant "compactification" of the interpolation process. Without the tensor product structure, we would have to solve a linear system of order $(m+1)(n+1) \times (m+1)(n+1)$. That is an order of magnitude more complex than solving $m+1$ problems with the same $(n+1) \times (n+1)$ matrix and then solving $n+1$ problems with the same $(m+1) \times (m+1)$ matrix. If $m = n$, the naive approach would thus need $O(m^6)$ computations, whereas the tensor product approach just needs $O(m^4)$.

Let us look at one specific example, namely bicubic B-spline interpolation. Suppose we have $(K+1) \times (L+1)$ data points x_{IJ} and two knot sequences $u_0, \ldots, u_K$ and $v_0, \ldots, v_L$. Our development is illustrated in Figure 17.10. We use the notation from Chapter 9. For each row of data points, we prescribe two end conditions (e.g., by specifying tangent vectors or Bézier points) and solve the univariate B-spline interpolation problem as described in Section 9.1. This produces the elements of the matrix D, marked by triangles in Figure 17.10. We now take every row of D and perform univariate B-spline interpolation on

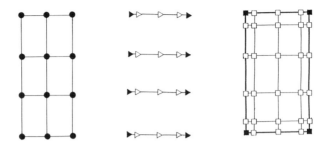

Figure 17.10: Tensor product bicubic spline interpolation: the solution is obtained in a two-step process.

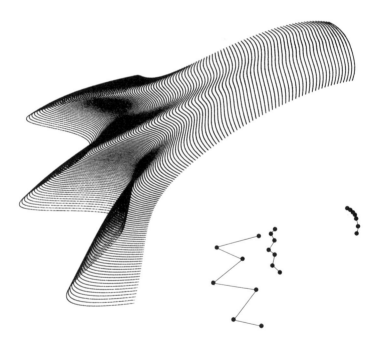

Figure 17.11: Tensor product bicubic spline interpolation: the given data, lower right (scaled down), and the solution, using Bessel end conditions and uniform parametrizations.

it, again by prescribing end conditions such as end tangents.[4] Note that at the four corners, this amounts to the prescription of twist vectors. An example is shown in Figure 17.11. In it, the data points are connected in the u-direction – this is just to highlight the structure of the data.

Although mathematically equivalent, the two processes – first row by row, then column by column; or first column by column, then row by row – do not yield the same computation count.

17.5 The Parametrization

While tensor product interpolation is very elegant, its use is limited to cases where the data points possess a rectangular structure. When the data points deviate from a "nice" grid, we have to face a problem that is associated with the process of bicubic spline interpolation, in fact, for most kinds of tensor product interpolation: the problem of finding an appropriate parametrization. In the curve case (Section 9.4), we were able to devise several methods that assigned parameter values to the given data points. So why not take those methods and apply them to the tensor product case in much the same way in which we generalized curve methods to their tensor product counterparts? The problem is that we have to produce *one* set of parameter values for *all* isoparametric curves in the u-direction; the same holds for the v-direction.

We may endow each isoparametric curve (in the u-direction, say) with a parametrization from Section 9.4. To arrive at *one* parametrization for all of them, we may then carry out some averaging process. Such an approach will only produce acceptable results if all our isoparametric curves have the same shape characteristics, i.e., if they essentially yield the same parametrization. This is, however, not always the case, as Figures 17.12, 17.13, 17.14 illustrate.[5]

Are there ways out of the dilemma? Not if one has unevenly distributed data and insists on bicubic spline interpolation. If one is willing to go to higher degrees and to replace C^1 or C^2 continuity by G^1 continuity (see Chapter 19), then several methods exist – see the literature cited in Chapter 19.

---

[4]Note that **D** has two more rows than **X**. This is due to the end conditions; to resolve the apparent discrepancy, we may think of **X** as having two additional rows that constitute the end condition data.

[5]Another interesting phenomenon may be observed here: note how the first and the third of this set of surfaces have varying densities in their plots. The reason is that each cubic isoparametric curve was plotted in 90 increments. With very unequal parameter spacing, this generates abruptly varying spacing on the curves.

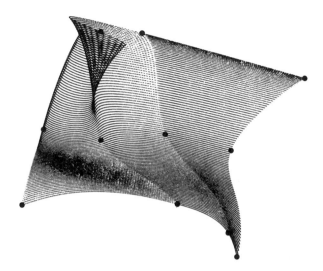

Figure 17.12: Parametrizing a bicubic spline interpolant: all "horizontal" isoparametric curves have the parametrization $u_i = [0, 4, 5.5, 6.0]$. Note that the bottom curve has a reasonable shape.

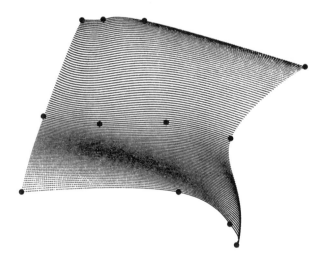

Figure 17.13: Parametrizing a bicubic spline interpolant: all "horizontal" isoparametric curves have the parametrization $u_i = [0, 1, 2, 3]$.

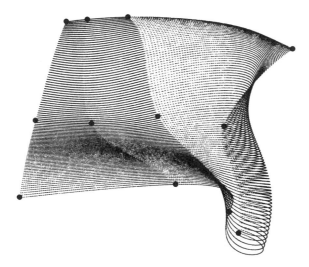

Figure 17.14: Parametrizing a bicubic spline interpolant: all "horizontal" isoparametric curves have the parametrization $u_i = [0, 0.5, 1.5, 6.5]$. Now the top curve has a good shape.

17.6 Bicubic Hermite Patches

We emphasize the Bézier and B-spline form for composite surfaces, although one can also form the tensor product of the cubic Hermite interpolation scheme (see Section 6.5). The input parameters to this patch representation are points, partials, and mixed partials. A bicubic patch in Hermite form is given by

$$\mathbf{x}(u, v) = \sum_{i=0}^{3} \sum_{j=0}^{3} \mathbf{h}_{i,j} H_i^3(u) H_j^3(v); \quad 0 \le u, v \le 1, \tag{17.10}$$

where the H_i^3 are the cubic Hermite functions from Section 6.5 and the $\mathbf{h}_{i,j}$ are given by

$$[\mathbf{h}_{i,j}] = \begin{bmatrix} \mathbf{x}(0,0) & \mathbf{x}_v(0,0) & \mathbf{x}_v(0,1) & \mathbf{x}(0,1) \\ \mathbf{x}_u(0,0) & \mathbf{x}_{uv}(0,0) & \mathbf{x}_{uv}(0,1) & \mathbf{x}_u(0,1) \\ \mathbf{x}_u(1,0) & \mathbf{x}_{uv}(1,0) & \mathbf{x}_{uv}(1,1) & \mathbf{x}_u(1,1) \\ \mathbf{x}(1,0) & \mathbf{x}_v(1,0) & \mathbf{x}_v(1,1) & \mathbf{x}(1,1) \end{bmatrix}. \tag{17.11}$$

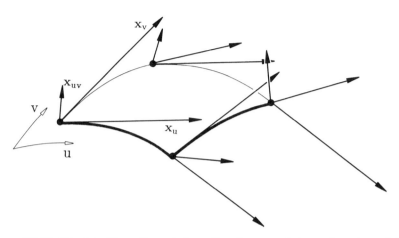

Figure 17.15: Bicubic Hermite patches: the shown points and vectors define a patch over the unit square.

The coefficients of this form are shown in Figure 17.15. Note how the coefficients in the matrix are grouped into four 2×2 partitions, each holding the data pertaining to one corner.[6]

In the context of composite surfaces, one is given a network of bicubic patches. The equation of patch (I, J) in the network is given by

$$\mathbf{x}(u, v) = \sum_{i=0}^{3} \sum_{j=0}^{3} \mathbf{h}_{i,j} H_i^3(s) H_j^3(t); \quad 0 \le s, t \le 1. \tag{17.12}$$

Here, s and t are local coordinates of the intervals $[u_I, u_{I+1}]$ and $[v_J, v_{J+1}]$. The coefficient matrix now changes:

$$[\mathbf{h}_{i,j}] = \begin{bmatrix} \mathbf{x}(0,0) & \Delta_J \mathbf{x}_v(0,0) & \Delta_J \mathbf{x}_v(0,1) & \mathbf{x}(0,1) \\ \Delta_I \mathbf{x}_u(0,0) & \Delta_I \Delta_J \mathbf{x}_{uv}(0,0) & \Delta_I \Delta_J \mathbf{x}_{uv}(0,1) & \Delta_I \mathbf{x}_u(0,1) \\ \Delta_I \mathbf{x}_u(1,0) & \Delta_I \Delta_J \mathbf{x}_{uv}(1,0) & \Delta_I \Delta_J \mathbf{x}_{uv}(1,1) & \Delta_I \mathbf{x}_u(1,1) \\ \mathbf{x}(1,0) & \Delta_J \mathbf{x}_v(1,0) & \Delta_J \mathbf{x}_v(1,1) & \mathbf{x}(1,1) \end{bmatrix}. \tag{17.13}$$

---

[6]This is a deviation from standard notation. Standard notation groups by order of derivatives, i.e., there is a group of four positions, four u-partials, and so on. The form of Eq. (17.11) was chosen since it groups coefficients according to their geometric location. Also, it is closer to the Bézier form of a bicubic patch.

A composite surface, expressed in piecewise bicubic Hermite form, is C^1 if adjacent patches have coefficient matrices that coincide in the two rows or columns parallel to the common boundary. Thus, storage requirements may be reduced.

Bicubic patches in Hermite form may also be utilized for bicubic spline interpolation as described above for the B-spline form. One solves $(m + 1)$ univariate interpolation problems as outlined in Section 9.2 in order to obtain the v-partials at the data points. Next, one solves $(n + 1)$ interpolation problems to obtain the u-partials. Now one takes the u-partials and solves row by row to obtain the twists at the data points. (Or one takes the v-partials and solves column by column to obtain the twists.) This is a three-stage process, and does not compare favorably with the two-stage B-spline solution to the same problem.

17.7   Rational Bézier and B-spline Surfaces

We can generalize Bézier and B-spline surfaces to their rational counterparts in much the same way as we did for the curve cases. In other words, we define a rational Bézier or B-spline surface as the projection of a 4D tensor product Bézier or B-spline surface. Thus, the rational Bézier patch takes the form

$$\mathbf{x}(u, v) = \frac{\sum_i \sum_j w_{i,j} \mathbf{b}_{i,j} B_i^m(u) B_j^n(v)}{\sum_i \sum_j w_{i,j} B_i^m(u) B_j^n(v)}, \tag{17.14}$$

and a rational B-spline surface is written as

$$\mathbf{s}(u, v) = \frac{\sum_i \sum_j w_{i,j} \mathbf{d}_{i,j} N_i^m(u) N_j^n(v)}{\sum_i \sum_j w_{i,j} N_i^m(u) N_j^n(v)}. \tag{17.15}$$

Figure 17.16 shows an example of a rational B-spline surface. It was obtained from the same control net as the surface in Figure 17.3, but with weights as shown in the figure. Note how the "dip" became more pronounced, as well as the "vertical ridge."

Rational surfaces are obtained as the projections of tensor product patches – but they are not tensor product patches themselves. Recall that a tensor product surface is of the form $\mathbf{x}(u, v) = \sum_i \sum_j \mathbf{c}_{i,j} F_{i,j}(u, v)$, where the basis

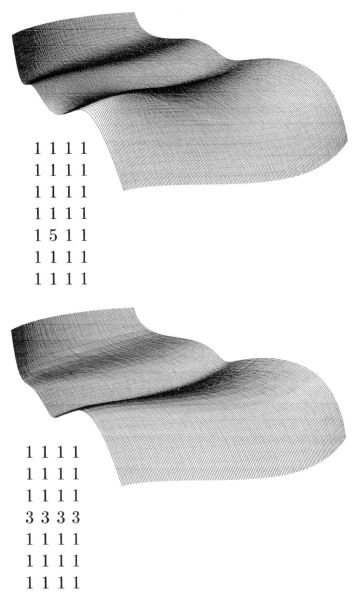

```
1 1 1 1
1 1 1 1
1 1 1 1
1 1 1 1
1 5 1 1
1 1 1 1
1 1 1 1
```

```
1 1 1 1
1 1 1 1
1 1 1 1
3 3 3 3
1 1 1 1
1 1 1 1
1 1 1 1
```

Figure 17.16: Rational B-spline surfaces: a surface together with the set of weights used to generate it.

functions $F_{i,j}$ may be expressed as products $F_{i,j}(u,v) = A_i(u)B_j(v)$. The basis functions for (17.15) are of the form

$$F_{i,j}(u,v) = \frac{w_{i,j}N_i^m(u)N_j^n(v)}{\sum_i \sum_j w_{i,j}N_i^m(u)N_j^n(v)}.$$

Because of the structure of the denominator, this may in general not be factored into the required form $F_{i,j}(u,v) = A_i(u)B_j(v)$.

But even though rational surfaces do not possess a tensor product structure, we may utilize many tensor product algorithms for their manipulation. Consider, for example, the problem of finding the piecewise rational bicubic Bézier form of a rational bicubic B-spline surface. All we have to do is to convert each row of the B-spline control net into piecewise rational Bézier cubics (according to Section 15.7). Then we repeat this process for each column of the resulting net (and the resulting weights!), simply following the principle outlined in Figure 17.4.

As another example, consider the problem of extracting an isoparametric curve from a rational Bézier surface. Suppose the curve corresponds to $v = \hat{v}$. We simply interpret all columns of the control net as control polygons and evaluate each at $\hat{v}$, using the rational de Casteljau algorithm, for example. Keep in mind that we also have to compute a weight for each control polygon. We can now interpret all obtained points together with their weights as the Bézier control polygon of the desired isoparametric curve. In general, its end weights will not be unity, i.e., the curve will not be in standard form (as described in Section 15.5). This situation may be remedied by the use of the reparametrization algorithm, which is also described in that section.

17.8 Surfaces of Revolution

Currently, rational B-spline surfaces are used for two reasons: they allow the exact representation of surfaces of revolution and of quadric surfaces. We will briefly describe surfaces of revolution in rational B-spline form here – quadric surfaces will be treated in Section 18.11.

A surface of revolution is given by

$$\mathbf{x}(u,v) = \begin{bmatrix} r(v)\cos u \\ r(v)\sin u \\ z(v) \end{bmatrix}.$$

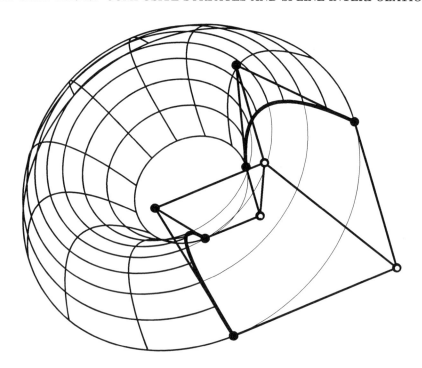

Figure 17.17: Surfaces of revolution: the surface is represented as four rational biquadratic patches. The solid control points (shown for one patch only) have weight 1; the open points have weight 0.71. Graphics courtesy of MCS, system ANVIL-5000©.

For fixed v, an isoparametric line $v = const$ traces out a circle of radius $r(v)$, called a *meridian*. Since a circle may be exactly represented by rational quadratic arcs, we may find an exact rational representation of a surface of revolution provided we can represent $r(v), z(v)$ in rational form.

The most convenient way to define a surface of revolution is to prescribe the (planar) generating curve, or generatrix, given by

$$\mathbf{g}(v) = [r(v), 0, z(v)]^{\mathrm{T}}$$

and by the axis of revolution, in the same plane as $\mathbf{g}$. Suppose $\mathbf{g}$ is given by its control polygon, knot sequence, and weight sequence. We can construct a surface of revolution such that each meridian consists of three rational quadratic arcs, as shown in Figure 14.11. For each vertex of the generating polygon, construct an equilateral triangle (perpendicular to the axis of revolution) as in Figure 14.11. Assign the given weights of the generatrix to the three polygons corresponding to the triangle edge midpoints; assign half those weights to the three control polygons corresponding to the triangle vertices. In this way, we represent *exactly* "classical" surfaces such as cylinders, spheres, or tori.

Instead of breaking down each meridian into three arcs, we might have used four. The resulting four biquadratic control nets then form three concentric squares in the projection into the $z = 0$ plane. The control points at the squares' midpoints are copies of the generatrix control points; their weights are those of the generatrix. The remaining weights, corresponding to the squares' corners, are multiplied by $\cos(45^{o}) = \sqrt{2}/2$. Figure 17.17 gives an example of a parabola that sweeps out a surface of revolution.

Note that although the generatrix may be defined over a knot sequence $\{v_j\}$ with only simple knots, this is not possible for the knots of the meridian circles; we have to use double knots, thereby essentially reducing it to the piecewise Bézier form.

17.9 Volume Deformations

Sometimes local control of a surface, nice as it may be, is not what is needed. A typical design request is "stretch this surface in that direction," or "bend that surface like so." These are *global* shape deformations, and the usual tweaking of control polygon vertices is somewhat cumbersome for this task. P. Bézier devised a method to deform a Bézier patch in a manner that would satisfy this global deformation principle. We shall see that it is also applicable to B-spline surfaces. For literature, see Bézier [53], [56], [57]. A more graphics-oriented version of this principle was presented by Sederberg and Parry [434].

To illustrate the principle, let us consider the 2D case first. Let $\mathbf{x}(t)$ be a planar curve (Bézier, B-spline, rational B-spline, etc.), which is, without loss of generality, located within the unit square. Next, let us cover the square with a regular grid of points $\mathbf{b}_{i,j} = [i/m, j/n]^{\mathrm{T}}; i = 0, \ldots, m; j = 0, \ldots, n$. We can now write every point (u, v) as

$$(u, v) = \sum_{i=0}^{m} \sum_{j=0}^{n} \mathbf{b}_{i,j} B_i^m(u) B_j^n(v);$$

this follows from the linear precision property of Bernstein polynomials (4.15).

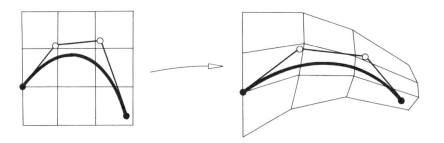

Figure 17.18: Global curve distortions: a Bézier polygon is distorted into another polygon, resulting in a deformation of the initial curve.

If we now distort the grid of $\mathbf{b}_{i,j}$ into a grid $\hat{\mathbf{b}}_{i,j}$, the point (u, v) will be mapped to a point $(\hat{u}, \hat{v})$:

$$(\hat{u}, \hat{v}) = \sum_{i=0}^{m} \sum_{j=0}^{n} \hat{\mathbf{b}}_{i,j} B_i^m(u) B_j^n(v). \qquad (17.16)$$

In other words, we are dealing with a mapping of $I\!\!E^2$ to $I\!\!E^2$.

In particular, the control vertices of the curve $\mathbf{x}(t)$ will be mapped to new control vertices, which in turn determine a new curve $\mathbf{y}(t)$. Note that $\mathbf{y}$ is only an approximation to the image of $\mathbf{x}$ under (17.16).[7] This is highlighted by the fact that the image of $\mathbf{x}$'s control polygon under (17.16) would be a collection of curve arcs, not another piecewise linear polygon.

We now have an *indirect* method for curve design: changing the $\mathbf{b}_{i,j}$ will produce globally deformed curves. This technique may facilitate certain design tasks that are otherwise tedious to perform. Figure 17.18 gives an example of the use of this global design technique.

This technique may be generalized. For instance, we may replace the Bézier distortion (17.16) by an analogous tensor product B-spline distortion. This would reintroduce some form of local control into our design scheme.

The next level of generalization is to $I\!\!E^3$: we introduce a *trivariate* Bézier patch by

$$(\hat{u}, \hat{v}, \hat{w}) = \sum_{i=0}^{m} \sum_{j=0}^{n} \sum_{k=0}^{l} \hat{\mathbf{b}}_{i,j,k} B_i^m(u) B_j^n(v) B_k^l(w), \qquad (17.17)$$

---

[7]An exact procedure is described be T. DeRose [131].

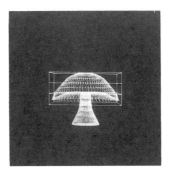

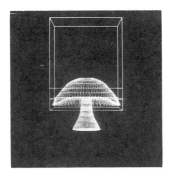

Figure 17.19: Global surface distortions: part of a surface is embedded in a
Bézier volume (top). That volume is distorted (middle), leading to a distorted
final object (bottom).

which constitutes a deformation of 3D space $I\!\!E^3$. We may use (17.17) to deform the control net of a surface embedded in the unit cube. Again, the use of a Bézier patch for the distortion is immaterial; we might have used trivariate B-splines, etc., in order to introduce some degree of locality into the method.

An example is shown in Figure 17.19. Part of the mushroom-shaped surface is embedded in a trivariate Bézier volume that is cubic in the vertical direction and linear in the other two. The top layer of control points is moved upward, leading to a C^2 distortion of the initial object.

Why use deformation methods instead of just manipulating control vertices interactively? Volume deformation methods allow a designer to modify whole assemblies of surfaces at once, in a way that spreads out the changes in each part of the assembly in a very harmonic way. By tweaking control vertices one by one, a similarly balanced modification cannot be the result.

17.10   Trimmed Surfaces

Tensor product patches (or surfaces) are very handy when they can be applied, and they can be applied to data that are inherently rectilinear. Not all situations lend themselves to a rectangular arrangement of patches or surfaces, and we should anticipate problems for such situations.

An example for which rectangular topology is not appropriate is given in Plate III. Toward the lower right quadrant of that figure, we see a small "patch" surface that blends the central part of the hood to the part over the fender. (Such surfaces, by the way, are extremely tedious to design.) If you take a close look at Plate III, you will see that the surfaces covered by the patch surface are not drawn where the patch surface is drawn. In fact, they are not defined there. The parts that are occupied by the patch surface are not part of the "regular" surfaces.

Such modification of tensor product surfaces is achieved by a technique known as "trimming." A trimmed tensor product surface is essentially a regular tensor product surface, but with certain areas marked as invalid or invisible. These areas are usually defined by (fairly dense) polygons in the u, v-domain of the patch, as shown in Figure 17.20. Trimmed surfaces should be viewed as an "engineering" extension of tensor product patches. That is to say, they are not a panacea to all surface problems either. Consider, for example, the problem of joining two trimmed surfaces together in a smooth

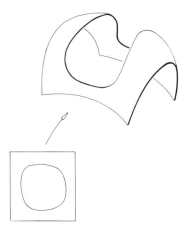

Figure 17.20: Trimmed surfaces: certain parts of a tensor product surface are marked as "invalid."

way. If they are to join along trim curves, there is no way of ensuring exact tangent plane continuity between them, as was the case for standard tensor patches. Such smoothness questions must be dealt with on a case by case basis, which is clearly not very desirable. Just consider the problem of fitting the above-mentioned blend surface from Plate III between its neighbors!

Literature on trimmed surfaces: Farouki and Hinds [176], Shantz and Chang [446], Casale and Bobrow [83], Miller [335].

17.11 Implementation

The routines in this section are written for rational surfaces. By setting all weights equal to one, the standard piecewise polynomial case is recovered.

The routine that converts a rational bicubic B-spline control net into the piecewise bicubic Bézier form:

```
void ratbspl_to_bez_surf(bspl_x,bspl_y,bspl_w,lu,lv,knot_u,
                         knot_v,bez_x,bez_y,bez_w,aux_x,aux_y,aux_w)
/*    Converts B-spline control net into piecewise
      Bezier control net (bicubic).
```

```
Input:       bspl_x,bspl_y:    B-spline control net (one coordinate only
             bspl_w:           B-spline weights
             lu,lv:            no. of intervals in u- and v-direction
             knot_u, knot_v:   knot vectors in u- and v-direction
Output:      bez_x,bez_y:      piecewise bicubic Bezier net.
             bez_w:            Bezier weights.
Work space:aux_x,aux_y,aux_w: needed to store intermediate results.

Remark:      The piecewise Bezier net only stores each control point onc
             i.e., neighboring patches share the same boundary.
             Knots are simple (but, in the language of Chapter 10, the
             boundary knots have multiplicity three).
*/
```

Once the piecewise rational Bézier representation of a bicubic spline surface is achieved, the following routine plots the whole surface:

```
void
plot_ratbez_surfaces(bez_x,bez_y,bez_w,lu,lv,u_points,v_points,
                     scale_x,scale_y,value)
/* Plots piecewise cubic surface, i.e., generates postscript output
Input: bez_x, bez_y:    control nets
lu,lv:                  no. of segments in u- and v- direction
u_points,v_p oints: per patch: v_points many
                        isoparametric curves with u_points
                        points on each
value:    minmax box of all control nets.
scale_x,scale_y:  scale factors for  postscript
*/
```

Tensor product spline interpolation (bicubic) is carried out by the following routine. It utilizes Bessel end conditions.

```
void spline_surf_int(data_x,data_y,bspl_x,bspl_y,lu,lv,knot_u,
                     knot_v,aux_x,aux_y)
/*      Interpolates to an array of size [0,lu+2]x[0,lv+2]
Input:     data_x, data_y:   data array (one coordinate only)
           lu,lv:            no. of intervals in u- and v-direction
           knot_u, knot_v:   knot vectors in u- and v-direction
Output:    bspl_x,bspl_y:    B-spline control net.

Work space: aux_x, aux_y.
```

Remark: On input, it is assumed that data_x and data_y have rows
           1 and lu+1 and columns 1 and lv+1 empty, i.e., they are
           not filled with data points. Example for lu=4, lv=7:

```
x0xxxxxx0x
0000000000
x0xxxxxx0x          x=data coordinate,
x0xxxxxx0x          0=unused  input array
x0xxxxxx0x            element.
0000000000          The 0's will be filled with
x0xxxxxx0x          'tangent Bezier points'.
```

   This approach makes it easy to feed in clamped end conditions
   if so desired: put in values in the 0's and delete the calls
   to bessel_ends below.
*/

Next is the header of a program that plots the control net of a Bézier
surface or of a composite surface.

```
void psplot_net(lu,lv,bx,by,step_u,step_v,scale_x,scale_y,value)
/* plots  control net      into postscript-file.
Input: lu,lv: dimensions of net
bx,by: net vertices
step_u,step_v: subnet sizes (e.g. both=3 for pw bicubic net)
scale_x,scale_y:scale factors for ps
value: window size in world coords
Output: written into postscript file
*/
```

17.12 Problems

1. Justify that in tensor product interpolation (Section 17.4), it does not
   matter if one starts with the row interpolation process or with the column interpolation process. Give computation counts for both strategies.
   (They are not equal!)

2. Generalize quintic Hermite and Lagrange interpolation to the tensor
   product case.

3. Generalize the B-spline knot insertion algorithm to the tensor product
   case.

4. Show that if two polynomial surfaces are C^1 across a common boundary, then they are also twist continuous across that boundary.

5. We defined a direct de Casteljau algorithm for tensor product Bézier surfaces in Section 16.2. Define a direct de Boor algorithm for tensor product B-spline surfaces.

6. Suppose we want to find a parametrization $\{u_i\}$ for a tensor product interpolant. We may parametrize all rows of data points and then form the averages of the parametrizations thus obtained. Or we could average all rows of data points, e.g., by setting $\mathbf{p}_i = \sum_j \frac{i}{n}\mathbf{x}_{i,j}$ and we could then parametrize the $\mathbf{p}_i$. Do we get the same result? Discuss both methods.

7. We mentioned trivariate B-splines in Section 17.9. Give an exact definition of these "hypersurfaces."

8. Problem 6 in Chapter 9 aimed at the introduction of a linear segment into a spline curve. Modify the surface interpolation programs to incorporate this curve scheme.

Chapter 18

Bézier Triangles

When de Casteljau invented Bézier curves in 1959, he realized the need for the extension of the curve ideas to surfaces. Interestingly enough, the first surface type that he considered was what we now call Bézier triangles. This historical "first" of triangular patches is reflected by the mathematical statement that they are a more "natural" generalization of Bézier curves than are tensor product patches. We should note that while de Casteljau's work was never published, Bézier's was, therefore the corresponding field now bears Bézier's name. For the placement of triangular Bernstein-Bézier surfaces in the field of CAGD, see Barnhill [18].

While de Casteljau's work (established in two internal Citroën technical reports [121] and [122]) remained unknown until its discovery by W. Boehm around 1975, other researchers realized the need for triangular patches. M. Sabin [401] worked on triangular patches in terms of Bernstein polynomials, unaware of de Casteljau's work. Among the people concerned with the development of triangular patches we name P. J. Davis [111], L. Frederickson [195], P. Sablonnière [407], and D. Stancu [452]. All of their Bézier-type approaches relied on the fact that piecewise surfaces were defined over regular triangulations; arbitrary triangulations were considered by Farin [152]. Two surveys on the field of triangular Bézier patches are Farin [160] and de Boor [115].

18.1 Barycentric Coordinates and Linear Interpolation

Barycentric coordinates were discussed in Section 2.3, where they were used in connection with straight lines. Now we will use them as coordinate systems when dealing with the plane. Barycentric coordinates are at the origin of affine geometry – they were first introduced by F. Moebius in 1827; see his collected works [337].

Consider a triangle with vertices $\mathbf{a}$, $\mathbf{b}$, $\mathbf{c}$ and a fourth point $\mathbf{p}$, all in $I\!\!E^2$. It is always possible to write $\mathbf{p}$ as a barycentric combination of $\mathbf{a}$, $\mathbf{b}$, $\mathbf{c}$:

$$\mathbf{p} = u\mathbf{a} + v\mathbf{b} + w\mathbf{c}. \tag{18.1}$$

A reminder: if Eq. (18.1) is to be a barycentric combination (and hence geometrically meaningful), we require that

$$u + v + w = 1. \tag{18.2}$$

The coefficients $\mathbf{u} := (u, v, w)$ are called *barycentric coordinates* of $\mathbf{p}$ with respect to $\mathbf{a}$, $\mathbf{b}$, $\mathbf{c}$. We will often drop the distinction between the barycentric coordinates of a point and the point itself; we then speak of "the point $\mathbf{u}$."

If the four points $\mathbf{a}$, $\mathbf{b}$, $\mathbf{c}$, and $\mathbf{p}$ are given, we can always determine $\mathbf{p}$'s barycentric coordinates u, v, w: Eqs. (18.1) and (18.2) can be viewed as a linear system of three equations [recall that (18.1) is shorthand for two scalar equations] in three unknowns u, v, w. The solution is obtained by an application of Cramer's rule:

$$u = \frac{\text{area}(\mathbf{p}, \mathbf{b}, \mathbf{c})}{\text{area}(\mathbf{a}, \mathbf{b}, \mathbf{c})}, \quad v = \frac{\text{area}(\mathbf{a}, \mathbf{p}, \mathbf{c})}{\text{area}(\mathbf{a}, \mathbf{b}, \mathbf{c})}, \quad w = \frac{\text{area}(\mathbf{a}, \mathbf{b}, \mathbf{p})}{\text{area}(\mathbf{a}, \mathbf{b}, \mathbf{c})}. \tag{18.3}$$

Actually, Cramer's rule makes use of determinants; they are related to areas by the identity

$$\text{area}(\mathbf{a}, \mathbf{b}, \mathbf{c}) = \frac{1}{2} \begin{vmatrix} a_x & b_x & c_x \\ a_y & b_y & c_y \\ 1 & 1 & 1 \end{vmatrix}. \tag{18.4}$$

We note that in order for (18.3) to be well defined, we must have $\text{area}(\mathbf{a}, \mathbf{b}, \mathbf{c}) \neq 0$, which means that $\mathbf{a}$, $\mathbf{b}$, $\mathbf{c}$ must not lie on a straight line.

Due to their connection with barycentric combinations, barycentric coordinates are *affinely invariant*: let $\mathbf{p}$ have barycentric coordinates u, v, w with respect to $\mathbf{a}$, $\mathbf{b}$, $\mathbf{c}$. Now map all four points to another set of four points by an affine map Φ. Then $\Phi\mathbf{p}$ has the same barycentric coordinates u, v, w with respect to $\Phi\mathbf{a}$, $\Phi\mathbf{b}$, $\Phi\mathbf{c}$.

Figure 18.1 illustrates more of the geometric properties of barycentric coordinates. An immediate consequence of Figure 18.1 is known as *Ceva's theorem:*

$$\text{ratio}(\mathbf{a}, \mathbf{p}_c, \mathbf{b}) \cdot \text{ratio}(\mathbf{b}, \mathbf{p}_a, \mathbf{c}) \cdot \text{ratio}(\mathbf{c}, \mathbf{p}_b, \mathbf{a}) = 1.$$

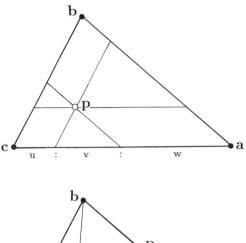

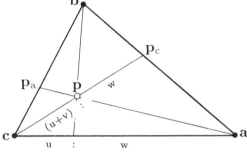

Figure 18.1: Barycentric coordinates: let $\mathbf{p} = u\mathbf{a} + v\mathbf{b} + w\mathbf{c}$. The two figures show some of the ratios generated by certain straight lines through $\mathbf{p}$.

More details on this and related theorems can be found in most geometry books, e.g., Gans [199] or Berger [46].

Any three noncollinear points $\mathbf{a}$, $\mathbf{b}$, $\mathbf{c}$ define a barycentric coordinate system in the plane. The points inside the triangle $\mathbf{a}$, $\mathbf{b}$, $\mathbf{c}$ have positive barycentric coordinates, while the remaining ones have (some) negative barycentric coordinates. Figure 18.2 shows more.

We may use barycentric coordinates to define *linear interpolation*. Suppose we are given three points $\mathbf{p}_1, \mathbf{p}_2, \mathbf{p}_3 \in I\!\!E^3$. Then any point of the form

$$\mathbf{p} = \mathbf{p}(\mathbf{u}) = \mathbf{p}(u, v, w) = u\mathbf{p}_1 + v\mathbf{p}_2 + w\mathbf{p}_3 \qquad (18.5)$$

with $u + v + w = 1$ lies in the plane spanned by $\mathbf{p}_1, \mathbf{p}_2, \mathbf{p}_3$. This map from $I\!\!E^2$ to $I\!\!E^3$ is called *linear interpolation*. Since $u + v + w = 1$, we may interpret

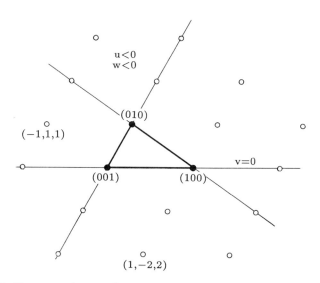

Figure 18.2: Barycentric coordinates: a triangle defines a coordinate system in the plane.

u, v, w as barycentric coordinates of $\mathbf{p}$ relative to $\mathbf{p}_1, \mathbf{p}_2, \mathbf{p}_3$. We may also interpret u, v, w as barycentric coordinates of a point in $I\!E^2$ relative to some triangle $\mathbf{a}, \mathbf{b}, \mathbf{c} \in I\!E^2$. Then (18.5) may be interpreted as a map of the triangle $\mathbf{a}, \mathbf{b}, \mathbf{c} \in I\!E^2$ onto the triangle $\mathbf{p}_1, \mathbf{p}_2, \mathbf{p}_3 \in I\!E^3$. We call the triangle $\mathbf{a}, \mathbf{b}, \mathbf{c}$ the *domain triangle*. Note that the actual location or shape of the domain triangle is totally irrelevant to the definition of linear interpolation. (Of course, we must demand that it be nondegenerate.) Since we can interpret u, v, w as barycentric coordinates in both two and three dimensions, it follows that linear interpolation (18.5) is an affine map.

18.2   The de Casteljau Algorithm

The de Casteljau algorithm for triangular patches is a direct generalization of the corresponding algorithm for curves. The curve algorithm uses repeated linear interpolation, and that process is also the key ingredient in the triangle case. The "triangular" de Casteljau algorithm is completely analogous to the univariate one, the main difference being notation. The control net is

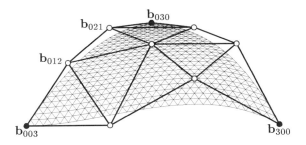

Figure 18.3: Bézier triangles: a cubic patch with its control net.

now of a triangular structure; in the quartic case, the control net consists of vertices

$$\mathbf{b}_{040}$$
$$\mathbf{b}_{031}\mathbf{b}_{130}$$
$$\mathbf{b}_{022}\mathbf{b}_{121}\mathbf{b}_{220}$$
$$\mathbf{b}_{013}\mathbf{b}_{112}\mathbf{b}_{211}\mathbf{b}_{310}$$
$$\mathbf{b}_{004}\mathbf{b}_{103}\mathbf{b}_{202}\mathbf{b}_{301}\mathbf{b}_{400}$$

Note that all subscripts sum to 4. In general, the control net consists of $\frac{1}{2}(n+1)(n+2)$ vertices. The numbers $\frac{1}{2}(n+1)(n+2)$ are called *triangle numbers*. Figure 18.3 gives an example of a cubic patch with its control net.

Some notation: we denote the point $\mathbf{b}_{ijk}$ by $\mathbf{b_i}$. Also, we use the abbreviations $\mathbf{e1} = (1,0,0)$, $\mathbf{e2} = (0,1,0)$, $\mathbf{e3} = (0,0,1)$, and $|\mathbf{i}| = i + j + k$. When we say $|\mathbf{i}| = n$, we mean $i + j + k = n$, always assuming $i, j, k \geq 0$.

The de Casteljau algorithm follows.

de Casteljau algorithm:

Given: a triangular array of points $\mathbf{b_i} \in I\!\!E^3$; $|\mathbf{i}| = n$ and a point in $I\!\!E^2$ with barycentric coordinates $\mathbf{u}$.

Set:

$$\mathbf{b_i^r}(\mathbf{u}) = u\mathbf{b}_{\mathbf{i+e1}}^{r-1}(\mathbf{u}) + v\mathbf{b}_{\mathbf{i+e2}}^{r-1}(\mathbf{u}) + w\mathbf{b}_{\mathbf{i+e3}}^{r-1}(\mathbf{u}), \qquad (18.6)$$

where

$$r = 1, \ldots, n \quad \text{and} \quad |\mathbf{i}| = n - r$$

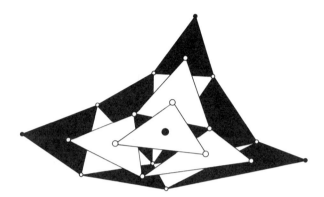

Figure 18.4: The "triangular" de Casteljau algorithm: a point is constructed by repeated linear interpolation.

and $\mathbf{b}_\mathbf{i}^0(\mathbf{u}) = \mathbf{b}_\mathbf{i}$. Then $\mathbf{b}_\mathbf{0}^n(\mathbf{u})$ is the point with parameter value $\mathbf{u}$ on the *Bézier triangle*[1] $\mathbf{b}^n$.

Figure 18.4 illustrates the construction of a point on a cubic Bézier triangle. We give a simple example: for $n = 3, r = 1$, and $\mathbf{i} = (2, 0, 0)$, we would obtain $\mathbf{b}_{200}^1 = u\mathbf{b}_{300} + v\mathbf{b}_{210} + w\mathbf{b}_{201}$. A complete numerical example is given in Example 18.1.

At this point, the reader should compare the "triangular" de Casteljau algorithm with the univariate one, and also have a look at the barycentric form of Bézier curves (see Section 5.9).

Based on the de Casteljau algorithm, we can state many properties of Bézier triangles:

Affine invariance: This property follows since linear interpolation is an affine map and since the de Casteljau algorithm makes use of linear interpolation only.

Invariance under affine parameter transformations: This property is guaranteed since such a reparametrization amounts to choosing a new domain triangle, but we have not even specified any particular domain triangle. More precisely, a point $\mathbf{u}$ will have the same barycentric coordinates $\mathbf{u}$ after an affine transformation of the domain triangle.

---

[1]More precisely, a triangular Bézier patch.

Let the coefficients $\mathbf{b_i}$ of a quadratic patch be given by

$$\begin{bmatrix} 0 \\ 6 \\ 0 \end{bmatrix}$$

$$\begin{bmatrix} 0 \\ 3 \\ 0 \end{bmatrix} \begin{bmatrix} 3 \\ 3 \\ 6 \end{bmatrix}$$

$$\begin{bmatrix} 0 \\ 0 \\ 0 \end{bmatrix} \begin{bmatrix} 3 \\ 0 \\ 0 \end{bmatrix} \begin{bmatrix} 6 \\ 0 \\ 9 \end{bmatrix},$$

and let $\mathbf{u} = (\frac{1}{3}, \frac{1}{3}, \frac{1}{3})$. Also, we make the assumption that $\mathbf{b_{300}} = [6, 0, 9]^{\mathrm{T}}$ is the image of $\mathbf{e1}$, and $\mathbf{b_{030}} = [0, 6, 0]^{\mathrm{T}}$ is the image of $\mathbf{e2}$.
The de Casteljau steps are as follows: for $r = 1$, the $\mathbf{b_i^1}$ are given by

$$\begin{bmatrix} 1 \\ 4 \\ 2 \end{bmatrix}$$

$$\begin{bmatrix} 1 \\ 1 \\ 0 \end{bmatrix} \begin{bmatrix} 4 \\ 1 \\ 5 \end{bmatrix}$$

The result $\mathbf{b_0^2}$ is

$$\begin{bmatrix} 2 \\ 2 \\ 7/3 \end{bmatrix}.$$

Example 18.1: Computing a point with the de Casteljau algorithm.

The convex hull property: Guaranteed since for $0 \le u, v, w \le 1$, each $\mathbf{b_i^r}$ is a convex combination of the previous $\mathbf{b_i^{r-1}}$.

The boundary curves: For a triangular patch, these curves are determined by the boundary control vertices (having at least one zero as a subscript). For example, a point on the boundary curve $\mathbf{b}^n(u, 0, w)$ is generated by

$$\mathbf{b}_i^r(u, 0, w) = u\mathbf{b}_{i+\mathbf{e1}}^{r-1} + w\mathbf{b}_{i+\mathbf{e3}}^{r-1}; \quad u + w = 1,$$

which is the univariate de Casteljau algorithm for Bézier curves.

18.3 Triangular Blossoms

The blossoming principle was introduced in Section 3.4 and also proves useful here. We will develop blossoms for triangular patches very much like we did in Section 5.9 for curves; the reader is assumed to be familiar with that material.

Our development follows a familiar flavor: we feed different arguments into the de Casteljau algorithm. At level r of the algorithm, we will use $\mathbf{u}_r$ as its argument, arriving finally at level n, with the blossom value $\mathbf{b}[\mathbf{u}_1, \ldots, \mathbf{u}_n]$. Note that all arguments are triples of numbers, because they represent points in the domain plane. The multivariate polynomial $\mathbf{b}[\mathbf{u}_1, \ldots, \mathbf{u}_n]$ is called the *blossom* of the triangular patch $\mathbf{b}(\mathbf{u})$. This blossom has all the properties that we encountered earlier: it agrees with the patch if all arguments are equal: $\mathbf{b}(\mathbf{u}) = \mathbf{b}[\mathbf{u}^{<n>}]$ (recall that $\mathbf{u}^{<n>}$ is short for n-fold repetition of $\mathbf{u}$), it is multiaffine, and it is symmetric.

Let us consider a special case, namely, that of fixing one argument and letting the remaining ones be equal, similar to the developments of polars in Section 4.7. So consider $\mathbf{b}[\mathbf{e1}, \mathbf{u}^{<n-1>}]$. We have to carry out one de Casteljau step with respect to $\mathbf{e1}$, and then continue as in the standard algorithm. Since a step with respect to $\mathbf{e1}$ yields

$$\mathbf{b}_i^1(\mathbf{e1}) = \mathbf{b}_{i+1,j,k}; \quad |\mathbf{i}| = n - 1,$$

we end up with a triangular patch of degree $n - 1$ that has all vertices of the form $\mathbf{b}_{0,j,k}$ "peeled away." We may continue this experiment: if we next use $\mathbf{e2}$, we peel another layer of coefficients, and so on. Let us utilize $\mathbf{e1}$ i times, $\mathbf{e2}$ j times, and $\mathbf{e3}$ k times. We are then left with a single control point:

$$\mathbf{b_i} = \mathbf{b}[\mathbf{e1}^{<i>}, \mathbf{e2}^{<j>}, \mathbf{e3}^{<k>}] \quad |\mathbf{i}| = n. \tag{18.7}$$

So again the Bézier control points are obtainable as special blossom values!

We may also write the intermediate points of the de Casteljau algorithm as special blossom values:

$$\mathbf{b}_i^r(\mathbf{u}) = \mathbf{b}[\mathbf{u}^{<r>}, \mathbf{e1}^{<i>}, \mathbf{e2}^{<j>}, \mathbf{e3}^{<k>}]; \quad i + j + k + r = n. \tag{18.8}$$

If we are interested in the control vertices $\mathbf{c_i}$ with respect to a different triangle, with vertices $\mathbf{f1, f2, f3}$, say, all we have to do is to evaluate the blossom:

$$\mathbf{c_i} = \mathbf{b}[\mathbf{f1}^{<i>}, \mathbf{f2}^{<j>}, \mathbf{f3}^{<k>}]. \tag{18.9}$$

This relationship, without use of the blossoming principle, is discussed by Goldman [206] and by Boehm and Farin [72]. See also Seidel [438].

18.4 Bernstein Polynomials

Univariate Bernstein polynomials are the terms of the binomial expansion of $[t + (1 - t)]^n$. In the bivariate case, Bernstein polynomials $B_{\mathbf{i}}^n$ are defined by[2]

$$B_{\mathbf{i}}^n(\mathbf{u}) = \binom{n}{\mathbf{i}} u^i v^j w^k = \frac{n!}{i!j!k!} u^i v^j w^k; \quad |\mathbf{i}| = n. \tag{18.10}$$

We define $B_{\mathbf{i}}^n(\mathbf{u}) = 0$ if some of the (i, j, k) are negative. Some of the quartic Bernstein polynomials are shown in Figure 18.5.

As an example, the quartic Bernstein polynomials could be arranged in the following triangular scheme (corresponding to the control point arrangement in the de Casteljau algorithm):

$$
\begin{array}{ccccccccc}
& & & & v^4 & & & & \\
& & & 4v^3w & & 4uv^3 & & & \\
& & 6v^2w^2 & & 12uv^2w & & 6u^2v^2 & & \\
& 4vw^3 & & 12uvw^2 & & 12u^2vw & & 4u^3v & \\
w^4 & & 4uw^3 & & 6u^2w^2 & & 4u^3w & & u^4
\end{array}
$$

Bernstein polynomials satisfy the following recursion:

$$B_{\mathbf{i}}^n(\mathbf{u}) = u B_{\mathbf{i-e1}}^{n-1}(\mathbf{u}) + v B_{\mathbf{i-e2}}^{n-1}(\mathbf{u}) + w B_{\mathbf{i-e3}}^{n-1}(\mathbf{u}); \quad |\mathbf{i}| = n. \tag{18.11}$$

This follows from their definition, as given in Eq. (18.10), and the use of the identity

$$\binom{n}{\mathbf{i}} = \binom{n}{i}\binom{n-i}{j}.$$

---

[2]Keep in mind that although $B_{\mathbf{i}}^n(u, v, w)$ *looks* trivariate, it is not, since $u + v + w = 1$.

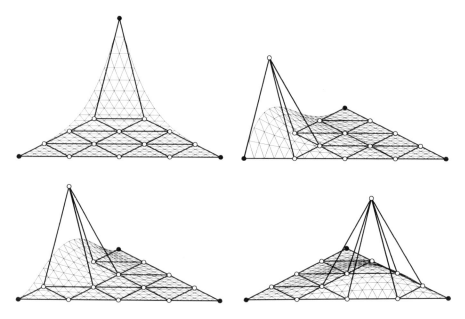

Figure 18.5: Bernstein polynomials: four quartic basis functions are shown; the shapes of the remaining ones follow from symmetry.

The intermediate points $\mathbf{b_i^r}$ in the de Casteljau algorithm can be expressed in terms of Bernstein polynomials:

$$\mathbf{b_i^r(u)} = \sum_{|\mathbf{j}|=r} \mathbf{b_{i+j}} B_{\mathbf{j}}^r(\mathbf{u}); \quad |\mathbf{i}| = n - r. \tag{18.12}$$

Setting $r = n$, we see that a triangular Bézier patch can be written in terms of Bernstein polynomials:

$$\mathbf{b^n(u)} = \mathbf{b_0^n(u)} = \sum_{|\mathbf{j}|=n} \mathbf{b_j} B_{\mathbf{j}}^n(\mathbf{u}). \tag{18.13}$$

We still need to prove (18.12). We use induction and the recursive definition of Bernstein polynomials:

$$\begin{aligned} \mathbf{b_i^{r+1}} &= u\mathbf{b_{i+e1}^r} + v\mathbf{b_{i+e2}^r} + w\mathbf{b_{i+e3}^r} \\ &= u\sum_{|\mathbf{j}|=r} \mathbf{b_{i+j+e1}} B_{\mathbf{j}}^r + v\sum_{|\mathbf{j}|=r} \mathbf{b_{i+j+e2}} B_{\mathbf{j}}^r + w\sum_{|\mathbf{j}|=r} \mathbf{b_{i+j+e3}} B_{\mathbf{j}}^r \end{aligned}$$

$$= \sum_{|\mathbf{j}|=r+1} u\mathbf{b_{i+j}}B^r_{\mathbf{j}-\mathbf{e1}} + v\mathbf{b_{i+j}}B^r_{\mathbf{j}-\mathbf{e2}} + w\mathbf{b_{i+j}}B^r_{\mathbf{j}-\mathbf{e3}}$$

$$= \sum_{|\mathbf{j}|=r+1} \mathbf{b_{i+j}}B^{r+1}_{\mathbf{j}}.$$

Compare with the similar proof for the univariate case of Eq. (4.6)!

We can generalize (18.13) just as we could in the univariate case:

$$\mathbf{b}^n(\mathbf{u}) = \sum_{|\mathbf{j}|=n-r} \mathbf{b}^r_{\mathbf{j}}(\mathbf{u})B^{n-r}_{\mathbf{j}}(\mathbf{u}); \quad 0 \le r \le n. \tag{18.14}$$

18.5 Derivatives

When we discussed derivatives for tensor product patches (Section 16.6), we considered *partials* because they are easily computed for those surfaces. The situation is different for triangular patches; the appropriate derivative here is the *directional derivative*. Let $\mathbf{u}_1$ and $\mathbf{u}_2$ be two points in the domain. Their difference $\mathbf{d} = \mathbf{u}_2 - \mathbf{u}_1$ defines a vector.[3] The directional derivative of a surface at $\mathbf{x}(\mathbf{u})$ with respect to $\mathbf{d}$ is given by

$$\mathbf{D_d}\mathbf{x}(\mathbf{u}) = d\mathbf{x}_u(\mathbf{u}) + e\mathbf{x}_v(\mathbf{u}) + f\mathbf{x}_w(\mathbf{u}).$$

A geometric interpretation: in the domain, draw the straight line through $\mathbf{u}$ that is parallel to $\mathbf{d}$. This straight line will be mapped to a curve on the patch. Its tangent vector at $\mathbf{x}(\mathbf{u})$ is the desired directional derivative.

The partials of a Bézier patch are not hard to compute; we have, for the u-partial:

$$\frac{\partial}{\partial u}\mathbf{b}(\mathbf{u}) = \frac{\partial}{\partial u} \sum_{|\mathbf{i}|=n} \frac{n!}{i!j!k!} u^i v^j w^k \mathbf{b_i}$$

$$= n \sum_{|\mathbf{i}|=n} \frac{(n-1)!}{(i-1)!j!k!} u^{i-1} v^j w^k \mathbf{b_i}$$

$$= n \sum_{|\mathbf{i}|=n-1} \frac{(n-1)!}{i!j!k!} u^i v^j w^k \mathbf{b}_{i+1,j,k}$$

$$= n \sum_{|\mathbf{i}|=n-1} \mathbf{b_{i+e1}}B^{n-1}_{\mathbf{i}}(\mathbf{u}).$$

---

[3]In barycentric coordinates, a point $\mathbf{u}$ is characterized by $u + v + w = 1$, while a vector $\mathbf{d} = (d, e, f)$ is characterized by $d + e + f = 0$.

Working out similar expressions for the v- and w-partials, we have found our directional derivative:

$$D_{\mathbf{d}}\mathbf{b}(\mathbf{u}) = n \sum_{|\mathbf{i}|=n-1} [d\mathbf{b}_{\mathbf{i}+\mathbf{e1}} + e\mathbf{b}_{\mathbf{i}+\mathbf{e2}} + f\mathbf{b}_{\mathbf{i}+\mathbf{e3}}] B_{\mathbf{i}}^{n-1}(\mathbf{u}). \qquad (18.15)$$

A closer look at the terms in square brackets brings to mind the de Casteljau algorithm, and we may write (18.15) as

$$D_{\mathbf{d}}\mathbf{b}(\mathbf{u}) = n \sum_{|\mathbf{i}|=n-1} \mathbf{b}_{\mathbf{i}}^1(\mathbf{d}) B_{\mathbf{i}}^{n-1}(\mathbf{u}). \qquad (18.16)$$

Thus a directional derivative is obtained by performing one step of the de Casteljau algorithm with respect to the direction vector $\mathbf{d}$, and $n-1$ more with respect to the position $\mathbf{u}$. Such configurations can be succinctly expressed in blossom form:

$$D_{\mathbf{d}}\mathbf{b}(\mathbf{u}) = n\mathbf{b}[\mathbf{d}, \mathbf{u}^{<n-1>}]. \qquad (18.17)$$

We may continue taking derivatives, arriving at:

$$D_{\mathbf{d}}^r\mathbf{b}(\mathbf{u}) = \frac{n!}{(n-r)!}\mathbf{b}[\mathbf{d}^{<r>}, \mathbf{u}^{<n-r>}]. \qquad (18.18)$$

Thus the rth directional derivative at $\mathbf{b}(\mathbf{u})$ is found by performing r steps of the de Casteljau algorithm with respect to $\mathbf{d}$, and then by performing $n-r$ more steps with respect to $\mathbf{u}$. Of course it is irrelevant in which order we take these steps (first noted in [151]).

In the same way, we may also compute mixed directional derivatives: if $\mathbf{d}_1$ and $\mathbf{d}_2$ are two vectors in the domain, then their mixed directional derivatives are

$$D_{\mathbf{d}_1,\mathbf{d}_2}^{r,s}\mathbf{b}(\mathbf{u}) = \frac{n!}{(n-r-s)!}\mathbf{b}[\mathbf{d}_1^{<r>}, \mathbf{d}_2^{<s>}, \mathbf{u}^{<n-r-s>}]. \qquad (18.19)$$

The blossom notation may be expressed in terms of Bernstein polynomials. Taking $n-r$ steps of the de Casteljau algorithm with respect to $\mathbf{u}$, and then r more with respect to $\mathbf{d}$ gives

$$D_{\mathbf{d}}^r\mathbf{b}^n(\mathbf{u}) = \frac{n!}{(n-r)!} \sum_{|\mathbf{j}|=r} \mathbf{b}_{\mathbf{j}}^{n-r}(\mathbf{u}) B_{\mathbf{j}}^r(\mathbf{d}). \qquad (18.20)$$

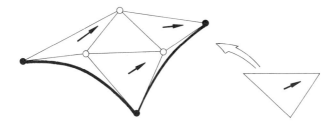

Figure 18.6: Directional derivatives: the coefficients of the directional derivative of a triangular patch are the vectors $\mathbf{b}_{\mathbf{j}}^1(\mathbf{d})$.

Or, we might have taken r steps with respect to $\mathbf{d}$ first, and then $n - r$ ones with respect to $\mathbf{u}$. This gives:

$$D_{\mathbf{d}}^r \mathbf{b}^n(\mathbf{u}) = \frac{n!}{(n-r)!} \sum_{|\mathbf{j}|=n-r} \mathbf{b}_{\mathbf{j}}^r(\mathbf{d}) B_{\mathbf{j}}^{n-r}(\mathbf{u}). \qquad (18.21)$$

Let us now spend some time interpreting our results. First we note that (18.20) is the analog of (4.25) in the univariate case. This sounds surprising at first, since (18.20) does not contain differences. Recall, however, that some of the components of $\mathbf{d}$ must be negative (since $d + e + f = 0$). Then the $B_{\mathbf{j}}^r(\mathbf{d})$ yield positive and negative values. We may therefore view terms involving $B_{\mathbf{j}}^r(\mathbf{d})$ as generalized differences. (In the univariate case, this is easily verified using the barycentric form of a Bézier curve.) Similarly, (18.21) may be viewed as a generalization of the univariate (4.21).

For $r = 1$, the terms $\mathbf{b}_{\mathbf{j}}^1(\mathbf{d})$ in (18.21) have a simple geometric interpretation: since $\mathbf{b}_{\mathbf{j}}^1(\mathbf{d}) = d\mathbf{b}_{\mathbf{j+e1}} + e\mathbf{b}_{\mathbf{j+e2}} + f\mathbf{b}_{\mathbf{j+e3}}$ and $|\mathbf{j}| = n-1$, they denote the affine map of the vector $\mathbf{d} \in I\!\!E^2$ to the triangle formed by $\mathbf{b}_{\mathbf{j+e1}}, \mathbf{b}_{\mathbf{j+e2}}, \mathbf{b}_{\mathbf{j+e3}}$. The directional derivative of $\mathbf{b}^n$ is thus a triangular patch whose coefficients are the images of $\mathbf{d}$ on each subtriangle in the control net (see Figure 18.6). For computational purposes, we would compute the net of the $n\mathbf{b}_{\mathbf{j}}^1(\mathbf{d})$ and use them as the input for a de Casteljau algorithm of an $(n-1)^{\text{st}}$ degree Bézier patch.

Similarly, let us set $r = 1$ in (18.20). Then,

$$\begin{aligned} D_{\mathbf{d}} \mathbf{b}^n(\mathbf{u}) &= n \sum_{|\mathbf{j}|=1} \mathbf{b}_{\mathbf{j}}^{n-1}(\mathbf{u}) B_{\mathbf{j}}^1(\mathbf{d}) \\ &= n(d\mathbf{b}_{\mathbf{e1}}^{n-1} + e\mathbf{b}_{\mathbf{e2}}^{n-1} + f\mathbf{b}_{\mathbf{e3}}^{n-1}). \end{aligned}$$

Let the coefficients $\mathbf{b_i}$ of a quadratic patch be those from Example 18.1. Let us pick the direction $\mathbf{d} = (1, 0, -1)$. We can then compute the $\mathbf{b_i^1}(\mathbf{d})$:

$$\begin{bmatrix} 3 \\ 0 \\ 6 \end{bmatrix}$$

$$\begin{bmatrix} 3 \\ 0 \\ 0 \end{bmatrix} \begin{bmatrix} 3 \\ 0 \\ 9 \end{bmatrix}$$

If we now evaluate at $\mathbf{u} = (\frac{1}{3}, \frac{1}{3}, \frac{1}{3})$, we obtain the vector

$$\begin{bmatrix} 3 \\ 0 \\ 5 \end{bmatrix},$$

which must still be multiplied by a factor of $n-1 = 2$ to obtain the directional derivative $D_{(1,0,-1)}\mathbf{b}^2(\frac{1}{3}, \frac{1}{3}, \frac{1}{3})$.

Alternatively, we might have taken the $\mathbf{b_i^1}$ from Example 18.1 and evaluated them at $\mathbf{d}$. The result is the same!

Example 18.2: Computing a directional derivative.

Since this is true for all directions $\mathbf{d} \in I\!E^2$, it follows that $\mathbf{b_{e1}^{n-1}}, \mathbf{b_{e2}^{n-1}}, \mathbf{b_{e3}^{n-1}}$ define the *tangent plane* at $\mathbf{b}^n(\mathbf{u})$. This is the direct generalization of the corresponding univariate result. In particular, the three vertices $\mathbf{b}_{0,n,0}, \mathbf{b}_{0,n-1,1}, \mathbf{b}_{1,n-1,0}$ span the tangent plane at $\mathbf{b}_{0,n,0}$ with analogous results for the remaining two corners. Again, we see that the de Casteljau algorithm produces derivative information as a by-product of the evaluation process; see Ex. 18.2 for a numerical example.

We next discuss *cross boundary derivatives* of Bézier triangles. Consider the edge $u = 0$ and a direction $\mathbf{d}$ not parallel to it. The directional derivative with respect to $\mathbf{d}$, evaluated along $u = 0$, is the desired cross boundary derivative. It is given by

$$D_{\mathbf{d}}\mathbf{b}^n(\mathbf{u})\Big|_{u=0} = \frac{n!}{(n-r)!} \sum_{|\mathbf{i_0}|=n-r} \mathbf{b_{i_0}^r}(\mathbf{d}) B_{\mathbf{i_0}}^{n-r}(\mathbf{u})\Big|_{u=0}, \qquad (18.22)$$

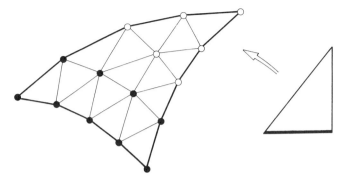

Figure 18.7: Cross boundary derivatives: any first-order cross boundary directional derivative of a quartic, evaluated along the indicated edge, depends only on the two indicated rows of Bézier points.

where $i_0 = (0, j, k)$. Since $\mathbf{u} = (0, v, w) = (0, v, 1 - v)$, this is a univariate expression, a Bézier curve in barycentric form as discussed in Section 5.9. Note that it depends only on the $r + 1$ rows of Bézier points closest to the boundary under consideration. Analogous results hold for the other two boundaries; see Figure 18.7. This result is the straightforward generalization of the corresponding univariate result. We will use it for the construction of composite surfaces, just as we did for curves.

18.6 Subdivision

We will later study surfaces that consist of several triangular patches forming a C^r overall surface. Now, we start with a surface consisting of just two triangular patches. Let their domain triangles be defined by points $\mathbf{a}, \mathbf{b}, \mathbf{c}, \hat{\mathbf{a}}$, as shown in Figure 18.8. If the common boundary is through $\mathbf{b}$ and $\mathbf{c}$, then the (domain!) point $\hat{\mathbf{a}}$ can be expressed in terms of barycentric coordinates of $\mathbf{a}, \mathbf{b}, \mathbf{c}$:

$$\hat{\mathbf{a}} = v_1\mathbf{a} + v_2\mathbf{b} + v_3\mathbf{c}.$$

Suppose now that a Bézier triangle $\mathbf{b}^n$ is given that has the triangle $\mathbf{a}, \mathbf{b}, \mathbf{c}$ as its domain, such that we have barycentric coordinates $\mathbf{a} = \mathbf{e1}, \mathbf{b} = \mathbf{e2}, \mathbf{c} = \mathbf{e3}$. Of course the patch is defined over the whole plane, in particular over $\hat{\mathbf{a}}, \mathbf{c}, \mathbf{b}$. What are the Bézier points $\mathbf{c_i}$ of $\mathbf{b}^n$ if we consider only the part of it that is defined over $\hat{\mathbf{a}}, \mathbf{c}, \mathbf{b}$?

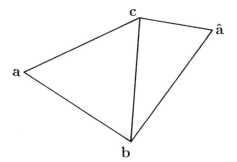

Figure 18.8: Subdivision: the domain geometry.

The answer was already given with (18.9):

$$\mathbf{c_i} = \mathbf{b}[\hat{\mathbf{a}}^{<i>}, \mathbf{e2}^{<j>}, \mathbf{e3}^{<k>}]; \quad i + j + k = n.$$

So all we have to do is compute the point $\mathbf{b}(\hat{\mathbf{a}})$ using the de Casteljau algorithm, and the intermediate points are the desired patch control vertices!

Some of the $\mathbf{c_i}$ deserve special attention: the common boundary of the two patches is characterized by $u = 0$. The Bézier points of the corresponding boundary curve must be same for both patches; we have

$$\mathbf{c}_{0,j,k} = \mathbf{b}[\mathbf{e2}^{<j>}, \mathbf{e3}^{<k>}] = \mathbf{b}_{0,j,k}; \quad j + k = n.$$

We also have

$$\mathbf{c}_{n00} = \mathbf{b}[\hat{\mathbf{a}}^{<n>}],$$

thus asserting, as expected, that we find $\mathbf{c}_{n00}$ as a point on the surface, evaluated at $\hat{\mathbf{a}}$. A numerical example is given in Example 18.3.

We thus have an algorithm that allows us to construct the Bézier points of the "extension" of $\mathbf{b}^n$ to an adjacent patch. It should be noted that this algorithm does not use convex combinations (when $\hat{\mathbf{a}}$ is outside $\mathbf{a}, \mathbf{b}, \mathbf{c}$). It performs piecewise linear extrapolation, and should therefore not be expected to be very stable.

If $\hat{\mathbf{a}}$ is inside $\mathbf{a}, \mathbf{b}, \mathbf{c}$, then we do use convex combinations only, and (18.6) provides a *subdivision algorithm*. Just as $\hat{\mathbf{a}}$ subdivides the triangle $\mathbf{a}, \mathbf{b}, \mathbf{c}$ into three subtriangles, the point $\mathbf{b}^n(\mathbf{v})$ subdivides the triangular patch into three subpatches. Equation (18.6) provides the Bézier points for each of them. Figure 18.9 gives an illustration.

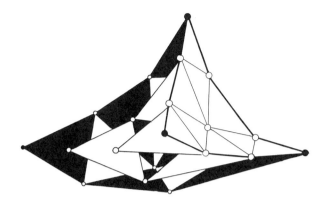

Figure 18.9: Subdivision: the intermediate points from the de Casteljau algorithm form the three subpatch control nets.

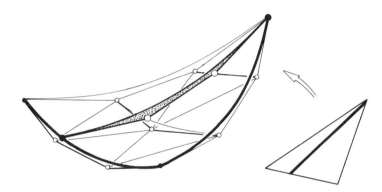

Figure 18.10: Subdivision: the Bézier points of a surface curve that is the image of a straight line through one of the domain triangle vertices.

Just as in the curve case, if we insert a dense sequence of points into the domain triangle, the resulting sequence of control nets will converge to the surface. This fact may be used in rendering techniques or in intersection algorithms.

A special case arises if $\mathbf{v}$ is on one of the edges of the domain triangle $\mathbf{a}, \mathbf{b}, \mathbf{c}$. Then (18.6) generates the Bézier points of the surface curve through $\mathbf{b}^n(\mathbf{v})$ and the opposite patch corner; see Figure 18.10. Such curves, joining a vertex to a point on the opposite edge, are called "radial lines."

18.7 Differentiability

The concept of differentiability is straightforward for tensor product patches: we compare cross boundary partials. As we have seen, partials are not an adequate tool when dealing with triangular patches. We will use the following generalization: consider two triangular patches that are maps of two adjacent domain triangles. Any straight line in the domain that crosses the common edge is mapped onto a composite curve in $I\!E^3$, having one segment in each patch. If all composite curves that can be obtained in this way are C^r curves, then we say that the two patches are C^r continuous.

Equation (18.6) gives a condition by which two adjacent patches can be part of one global polynomial surface. If we do not let r vary from 0 to n, but from 0 to some $s \leq n$, we have a condition for C^s continuity between adjacent patches:

$$\hat{\mathbf{b}}_{(r,j,k)} = \mathbf{b}_{\mathbf{j}_0}^r(\mathbf{v}); r = 0, \ldots, s. \tag{18.23}$$

Equation (18.23) is a necessary and sufficient condition for the C^s continuity of two adjacent patches. We can make that claim since cross boundary derivatives up to order s depend only on the $s+1$ rows of control points "parallel" to the considered boundary.

For $s = 0$, (18.23) states that the two patches must share a common boundary control polygon. The case $s = 1$ is more interesting because (18.23) becomes

$$\hat{\mathbf{b}}_{(1,j,k)} = v_1\mathbf{b}_{1,j,k} + v_2\mathbf{b}_{0,j+1,k} + v_3\mathbf{b}_{0,j,k+1}.$$

Thus each $\hat{\mathbf{b}}_{(1,j,k)}$ is obtained as a barycentric combination of the vertices of a boundary subtriangle of the control net of $\mathbf{b}^n$. Moreover, for all $j + k = n - 1$, these barycentric combinations are identical. Thus all pairs of subtriangles shown in Figure 18.11 are coplanar, and each pair is an affine map of the pair of domain triangles of the two patches.[4] We call the pairs of coplanar subtriangles that satisfy this condition *affine pairs*. Example 18.3 gives more details.

Figure 18.12 shows a composite C^1 surface that consists of several Bézier triangles. The (wire frame) plot of the surface does not look very smooth. This is due to the different spacing of isoparametric lines in the plot, not to discontinuities in the surface. The generation of planar slices of the surfaces shows that it is in fact C^1.

---

[4]It is *not* sufficient that the pairs are coplanar – this does not even guarantee a continuous tangent plane.

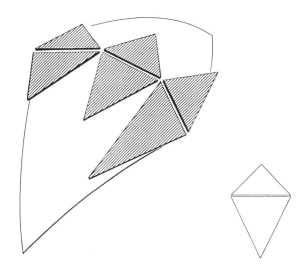

Figure 18.11: C^1 continuity: the shaded pairs of triangles must be coplanar and be an affine map of the two domain triangles.

18.8 Degree Elevation

It is possible to write $\mathbf{b}^n$ as a Bézier triangle of degree $n + 1$:

$$\sum_{|\mathbf{i}|=n} \mathbf{b_i} B_{\mathbf{i}}^n(\mathbf{u}) = \sum_{|\mathbf{i}|=n+1} \mathbf{b_i}^{(1)} B_{\mathbf{i}}^{n+1}(\mathbf{u}). \qquad (18.24)$$

The control points $\mathbf{b_i}^{(1)}$ are obtained from

$$\mathbf{b_i}^{(1)} = \frac{1}{n+1}[i\mathbf{b_{i-e1}} + j\mathbf{b_{i-e2}} + k\mathbf{b_{i-e3}}]. \qquad (18.25)$$

For a proof, we multiply the left-hand side of (18.24) by $u + v + w$ and compare coefficients of like powers. Figure 18.13 illustrates the case $n = 2$. Degree elevation is performed by piecewise linear interpolation of the original control net. Therefore, the degree elevated control net lies in the convex hull of the original one.

As in the univariate case, degree elevation may be repeated. That process generates a sequence of control nets that have the surface patch as their limit (Farin [151]). More details are in Farin [160].

We refer to Example 18.1. Let us define a second domain triangle that shares the edge $w = 0$ with the given domain triangle and that has a third vertex $\hat{\mathbf{a}}$. Let $\hat{\mathbf{a}}$'s barycentric coordinates with respect to the initial domain triangle be $(1,1,-1)$. We can perform one de Casteljau algorithm step and obtain for the $\mathbf{b}_\mathbf{i}^1(\hat{\mathbf{a}})$:

$$\begin{bmatrix} 3 \\ 6 \\ 6 \end{bmatrix}$$

$$\begin{bmatrix} 3 \\ 3 \\ 0 \end{bmatrix} \begin{bmatrix} 6 \\ 3 \\ 15 \end{bmatrix}.$$

A quadratic patch that is C^1 with the given one along its edge $w = 0$ is then given by the control net

$$\begin{bmatrix} 0 \\ 6 \\ 0 \end{bmatrix}$$

$$\begin{bmatrix} 3 \\ 6 \\ 6 \end{bmatrix} \begin{bmatrix} 3 \\ 3 \\ 6 \end{bmatrix}$$

$$\begin{bmatrix} \bullet \\ \bullet \\ \bullet \end{bmatrix} \begin{bmatrix} 6 \\ 3 \\ 15 \end{bmatrix} \begin{bmatrix} 6 \\ 0 \\ 9 \end{bmatrix},$$

where the $\bullet$-entries could be any numbers; they have no influence on C^1 continuity between the two patches.

Example 18.3: Computing a C^1 patch extension.

18.9   Nonparametric Patches

In an analogy to the univariate case, we may write the function

$$z = \sum_{|\mathbf{i}|=n} b_\mathbf{i} B_\mathbf{i}^n(\mathbf{u}) \tag{18.26}$$

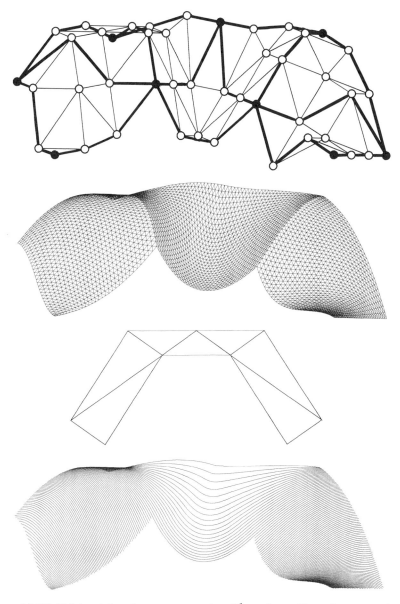

Figure 18.12: Bézier triangles: a composite C^1 surface. Top: the control net; next: the piecewise cubic surface; next: the domain triangles; next: planar slices through the surface.

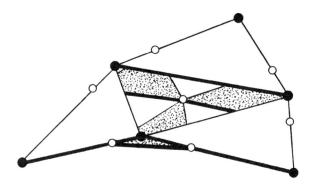

Figure 18.13: Degree elevation: a quadratic control net together with the equivalent cubic net.

as a surface

$$
\begin{bmatrix} u \\ v \\ w \\ z \end{bmatrix} = \sum_{|\mathbf{i}|=n} \begin{bmatrix} i/n \\ j/n \\ k/n \\ b_{\mathbf{i}} \end{bmatrix} B_{\mathbf{i}}^n(\mathbf{u}).
$$

Thus the abscissa values of the control polygon of a nonparametric patch are given by the triples $\mathbf{i}/n$, as illustrated in Figure 18.14. The last equation holds because of the *linear precision* property of the Bernstein polynomials $B_{\mathbf{i}}^n$,

$$
u = \sum_{|\mathbf{i}|=n} \frac{i}{n} B_{\mathbf{i}}^n(\mathbf{u}),
$$

and analogous formulas for v and w. The proof is by degree elevation from one to n of the linear function u. Example 18.4 shows a nonparametric patch.

Nonparametric Bézier triangles play an important role in the investigation of spaces of piecewise polynomials, as studied in approximation theory. Their use has facilitated the investigation of one of the main open questions in that field: what is the dimension of those function spaces? (See, for instance, Alfeld and Schumaker [6]). They have also been useful in defining nonparametric piecewise polynomial interpolants; see, for example, Barnhill and Farin [23], Farin [157], Petersen [366], or Sablonnière [410].

Nonparametric Bézier triangles may be generalized to *Bézier tetrahedra* by introducing barycentric coordinates in tetrahedra; see Problem 5 at the end of the chapter.

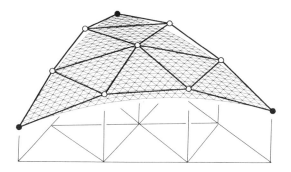

Figure 18.14: Nonparametric patches: the abscissas of the control net are the n-partition points of the domain triangle.

18.10 Rational Bézier Triangles

Following the familiar theme of generating rational curve and surface schemes, we define a rational Bézier triangle to be the projection of a nonrational 4D Bézier triangle. We thus have:

$$\mathbf{b}^n(\mathbf{u}) = \mathbf{b}_0^n(\mathbf{u}) = \frac{\sum_{|\mathbf{i}|=n} w_{\mathbf{i}} \mathbf{b}_{\mathbf{i}} B_{\mathbf{i}}^n(\mathbf{u})}{\sum_{|\mathbf{i}|=n} w_{\mathbf{i}} B_{\mathbf{i}}^n(\mathbf{u})}, \tag{18.27}$$

where, as usual, the $w_{\mathbf{i}}$ are the weights associated with the control vertices $\mathbf{b}_{\mathbf{i}}$. Needless to say, for positive weights we have the convex hull property, and we have affine and projective invariance.

Rational Bézier triangles may be evaluated by a de Casteljau algorithm in a not too surprising way:

Rational de Casteljau algorithm:

Given: a triangular array of points $\mathbf{b}_{\mathbf{i}} \in I\!\!E^3$; $|\mathbf{i}| = n$, corresponding weights $w_{\mathbf{i}}$, and a point in a domain triangle with barycentric coordinates $\mathbf{u}$.

Set:

$$\mathbf{b}_{\mathbf{i}}^r(\mathbf{u}) = \frac{u w_{\mathbf{i}+\mathbf{e1}}^{r-1} \mathbf{b}_{\mathbf{i}+\mathbf{e1}}^{r-1} + v w_{\mathbf{i}+\mathbf{e2}}^{r-1} \mathbf{b}_{\mathbf{i}+\mathbf{e2}}^{r-1} + w w_{\mathbf{i}+\mathbf{e3}}^{r-1} \mathbf{b}_{\mathbf{i}+\mathbf{e3}}^{r-1}}{w_{\mathbf{i}}^r}, \tag{18.28}$$

The bivariate function $z = x^2 + y^2$ may be written as a quadratic nonparametric Bézier patch over the triangle $(0,0), (2,0), (0,2)$. It coefficients are:

$$\begin{bmatrix} 0 \\ 2 \\ 4 \end{bmatrix}$$

$$\begin{bmatrix} 0 \\ 1 \\ 0 \end{bmatrix} \quad \begin{bmatrix} 1 \\ 1 \\ 0 \end{bmatrix}$$

$$\begin{bmatrix} 0 \\ 0 \\ 0 \end{bmatrix} \quad \begin{bmatrix} 1 \\ 0 \\ 0 \end{bmatrix} \quad \begin{bmatrix} 2 \\ 0 \\ 4 \end{bmatrix}$$

Example 18.4: A nonparametric quadratic patch.

where

$$w_{\mathbf{i}}^r = w_{\mathbf{i}}^r(\mathbf{u}) = u w_{\mathbf{i}+\mathbf{e1}}^{r-1}(\mathbf{u}) + v w_{\mathbf{i}+\mathbf{e2}}^{r-1}(\mathbf{u}) + w w_{\mathbf{i}+\mathbf{e3}}^{r-1}(\mathbf{u})$$

and

$$r = 1, \ldots, n \quad \text{and} \quad |\mathbf{i}| = n - r$$

and $\mathbf{b}_{\mathbf{i}}^0(\mathbf{u}) = \mathbf{b}_{\mathbf{i}}, w_{\mathbf{i}}^0 = w_{\mathbf{i}}$. Then $\mathbf{b}_{\mathbf{0}}^n(\mathbf{u})$ is the point with parameter value $\mathbf{u}$ on the rational Bézier triangle $\mathbf{b}^n$.

This algorithm works since we can interpret each intermediate $\mathbf{b}_{\mathbf{i}}^r$ as the projection of the corresponding point in the de Casteljau algorithm of the nonrational 4D preimage of our patch.

Something surprising happens now. Everything thus far was yet another exercise in generating rational schemes. In the case of rational Bézier curves, the initial weights could be used to define weight point $\mathbf{q}_i$, as described in Section 15.2. In the triangle case we can also define weight points $\mathbf{q}_{\mathbf{i}}$ by setting

$$\mathbf{q}_{\mathbf{i}} = \frac{w_{\mathbf{i}+\mathbf{e1}}\mathbf{b}_{\mathbf{i}+\mathbf{e1}} + w_{\mathbf{i}+\mathbf{e2}}\mathbf{b}_{\mathbf{i}+\mathbf{e2}} + w_{\mathbf{i}+\mathbf{e3}}\mathbf{b}_{\mathbf{i}+\mathbf{e3}}}{w_{\mathbf{i}+\mathbf{e1}} + w_{\mathbf{i}+\mathbf{e2}} + w_{\mathbf{i}+\mathbf{e3}}}; \quad |\mathbf{i}| = n - 1.$$

The usefulness of the $\mathbf{q}_i$ in the curve case stemmed from the fact that they could be used as a design handle: we could define points $\mathbf{q}_i$ and then retrieve

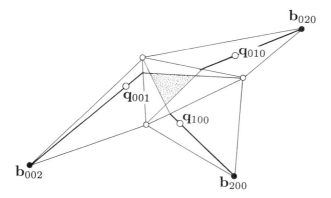

Figure 18.15: Weight points: an arbitrary choice of the three weight points $\mathbf{q}_{e1}, \mathbf{q}_{e2}, \mathbf{q}_{e3}$ would overdetermine the location of $\mathbf{p}$.

the weights w_i. Now, in the triangle case, this is no longer possible (first noted by Ramshaw [386]). We can see why just by considering the quadratic case $n = 2$, illustrated in Figure 18.15.

If we were given a set of weights $w_\mathbf{i}$; $|\mathbf{i}| = 2$, we would not only generate the weight points $\mathbf{q_i}$, but also the point

$$\mathbf{p} = \frac{w_{011}\mathbf{b}_{011} + w_{101}\mathbf{b}_{101} + w_{110}\mathbf{b}_{110}}{w_{011} + w_{101} + w_{110}}.$$

The point $\mathbf{p}$ is then at the intersection of the three straight lines $\overline{\mathbf{q}_{001}, \ \mathbf{b}_{110}}$, etc. If we prescribed the three $\mathbf{q_i}$ arbitrarily (as shown in Figure 18.15), those straight lines would not intersect in one point any more, thus leaving $\mathbf{p}$ overdetermined. Since the existence of a set of weights implies the existence of $\mathbf{p}$, the nonexistence of $\mathbf{p}$ implies that we cannot find a set of weights if we prescribe the $\mathbf{q_i}$ arbitrarily.

For higher degrees, the situation is analogous. So far, no geometric means is known that could define weights similar to the weight point approach for curves (see Problems).

We now give a formula for the *directional derivative* of a rational Bézier triangle. Just as in Section 18.5, let $\mathbf{d}$ denote a direction in the domain triangle, expressed in barycentric coordinates. We are interested in the directional derivative $D_{\mathbf{d}}$ of a rational triangular Bézier patch $\mathbf{b}^n(\mathbf{u})$. Proceeding exactly as in the curve case (see Section 14.4), we obtain:

$$D_{\mathbf{d}}\mathbf{b}^n(\mathbf{u}) = \frac{1}{w(\mathbf{u})}[\dot{\mathbf{p}}(\mathbf{u}) - D_{\mathbf{d}}(\mathbf{u})\mathbf{b}^n(\mathbf{u})],$$

where we have set

$$\mathbf{p}(\mathbf{u}) = w(\mathbf{u})\mathbf{b}^n(\mathbf{u}) = \sum_{|\mathbf{i}|=n} w_{\mathbf{i}}\mathbf{b}_{\mathbf{i}}B_{\mathbf{i}}^n(\mathbf{u}).$$

Higher derivatives follow the pattern outlined in Section 14.4, i.e.,

$$D_{\mathbf{d}}^r\mathbf{b}^n(\mathbf{u}) = \frac{1}{w(\mathbf{u})}[D_{\mathbf{d}}^r\mathbf{p}(\mathbf{u}) - \sum_{j=1}^{r} D_{\mathbf{d}}^j w(\mathbf{u})D_{\mathbf{d}}^{r-j}(\mathbf{u})].$$

18.11   Quadrics

There were (at least) two motivations for the use of rational Bézier curves: they are projectively invariant, and they allow us to represent conics in the form of rational quadratics. While the first argument holds trivially for rational Bézier triangles, the second one does not carry over immediately.

The proper generalization of a conic to the case of surfaces is a *quadric surface*, quadric for short. A conic (curve) has the implicit equation $q(x, y) = 0$, where q is a quadratic polynomial in x and y (see Section 14.5). Similarly, a quadric has the implicit equation $q(x, y, z) = 0$, where q is quadratic in x, y, z.

Quadrics are of importance in almost all solid modeling systems – these systems rely heavily on the ability to decide quickly if a given point is inside or outside a given object. If that object is bounded by simple implicit surfaces, such a decision is simple and reliable. If the object's boundaries are made up from, say, bicubics, the same decision is much more time-consuming and error-prone.

Every finite arc of a conic could be written as a quadratic rational Bézier curve – can we also write every triangle-shaped region on a quadric as a rational quadratic Bézier triangle? Let us try an octant of a sphere. In rational quadratic form, it would have six control net coefficients and six associated weights. Since a rational quadratic has no interior Bézier points, we only have to concentrate on the boundary curve representation. They are quarters of circles, and their representation is given in Section 14.7; see also Figure 18.16.

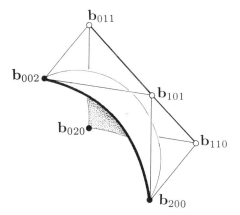

Figure 18.16: The octant of a sphere: an attempt to write it as a quadratic rational Bézier triangle.The weights of the solid control points are unity; the others have weight 1/2.

Thus we should be done, but actually, we are stuck: if we try to evaluate our rational quadratic at $\mathbf{u} = [\frac{1}{3}, \frac{1}{3}, \frac{1}{3}]$, we do not end up on the sphere! Thus *not every triangular quadric patch can be represented as a triangular rational quadratic patch.*

We have seen that rational quadratic triangular patches and quadric triangular patches are far from being in a one-to-one relationship. Luckily, this does not mean that we cannot express quadric patches as rational Bézier triangles – we just have to use higher degrees. In fact, rational *quartic* Bézier triangles are always sufficient for this purpose. Our initial example, the octant of the sphere, has the following rational quartic representation (Farin *et al.* [168]). The control net (for the octant of the *unit sphere*) is given by

$$[0,0,1]$$
$$[\alpha,0,1] \quad [0,\alpha,1]$$
$$[\beta,0,\beta] \quad [\gamma,\gamma,1] \quad [0,\beta,\beta]$$
$$[1,0,\alpha] \quad [1,\gamma,\gamma] \quad [\gamma,1,\gamma] \quad [0,1,\alpha]$$
$$[1,0,0] \quad [1,\alpha,0] \quad [\beta,\beta,0] \quad [\alpha,1,0] \quad [0,1,0]$$

where

$$\alpha = (\sqrt{3}-1)/\sqrt{3}, \quad \beta = (\sqrt{3}+1)/2\sqrt{3}, \quad \gamma = 1 - (5-\sqrt{2})(7-\sqrt{3})/46.$$

The weights are:

$$w_{040} = 4\sqrt{3}(\sqrt{3} - 1), w_{031} = 3\sqrt{2}, w_{202} = 4, w_{121} = \frac{\sqrt{2}}{\sqrt{3}}(3 + 2\sqrt{2} - \sqrt{3}),$$

the other ones following by symmetry.

To represent the whole sphere, we would assemble eight copies of this octant patch. Other representations are also possible: each octant may be written as a rational biquadratic patch (introducing singularities at the north and south poles); see [374]. A representation of the whole sphere as two rational bicubics (Piegl [369]) turned out to be incorrect (see Cobb [97]). Quite a different way of representing the sphere is also due to J. Cobb: he covers it with six rational bicubics having a cube-like connectivity [96].

Let us now discuss the following question: given a rational quadratic Bézier triangle, what are the conditions under which it represents a quadric patch? The following will be useful for that purpose.

We defined a conic section as the projective map of a parabola, the only conic allowing a polynomial parametrization. For surfaces, we again consider those quadrics that permit polynomial parametrizations: these are the paraboloids, consisting of elliptic and hyperbolic paraboloids and of the parabolic cylinders. Every quadric surface may be defined as a projective image of one of these paraboloids.[5]

A paraboloid may be represented by a parametric polynomial surface of degree two. However, as we have seen, not every parametric quadratic is a paraboloid. We need an extra condition, which is easily formulated if we write the quadratic surface in triangular Bézier form: *A quadratic Bézier triangle is an elliptic or hyperbolic paraboloid if and only if the second derivative vectors of the three boundary curves are parallel to each other. It is a parabolic cylinder if those three vectors are only coplanar.* This statement is due to W. Boehm.

For a proof, we observe that nonparametric or functional quadratic polynomials [i.e., of the form $z = f(x, y)$] include all three types of paraboloids, and all three satisfy the conditions of the theorem. Next, we observe that every paraboloid may be obtained as an affine map of a paraboloid of the same type. Thus every paraboloid may be obtained as an affine map of a functional quadratic surface. Consequently, the control net of any paraboloid must be an affine image of the control net of a functional quadratic Bézier triangle.

---

[5]W. Boehm (1990), private communication.

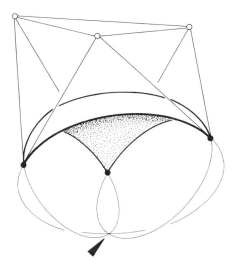

Figure 18.17: Quadrics as rational quadratics: the defining condition is that the extensions of the three boundary curves meet in one point (marked by an arrow) and have coplanar tangents there.

In a projective setting, we would say that the boundary curves of functional paraboloids intersect in one point (this may be the point at infinity) and have coplanar tangents there. Since all quadrics may be obtained from the functional ones by projective maps, we obtain the following characterization of quadric surfaces: *A rational quadratic Bézier triangle is a quadric if and only if all three boundary curves meet in a common point and have coplanar tangents there.* Figure 18.17 gives an illustration; for more details, see Boehm and Hansford [74].

A quadric is determined by nine points. If nine points (x_1, y_1, z_1), ..., (x_9, y_9, z_9) are given, then their interpolating quadric may be written in implicit form as follows:

$$q(x, y, z) = \begin{vmatrix} x^2 & y^2 & z^2 & xy & xz & yz & x & y & z & 1 \\ x_1^2 & y_1^2 & z_1^2 & x_1 y_1 & x_1 z_1 & y_1 z_1 & x_1 & y_1 & z_1 & 1 \\ \vdots & \vdots & \vdots & \vdots & \vdots & \vdots & \vdots & \vdots & \vdots & \vdots \\ x_9^2 & y_9^2 & z_9^2 & x_9 y_9 & x_9 z_9 & y_9 z_9 & x_9 & y_9 & z_9 & 1 \end{vmatrix} = 0. \quad (18.29)$$

18.12   Implementation

We include a function for the evaluation of a Bézier triangle. It uses a linear
array for the coefficients $\mathbf{b_i}$ in order to avoid a waste of storage by putting
them into a square matrix.

```
tri_decas(bpts, tri_num, ndeg, u, b, patch_pt )
/*
 Function: Triangular de Casteljau algorithm for an n^th
           degree triangular Bezier patch.
           Algorithm is applied once for a given (u,v,w) and works on o
           coordinate only.

    Input:    bpts[i]  ............ Bezier points (of one coordinate)
                                    as a linear array (see below).
                                    i=0...tri_num
              tri_num  ............ Based on the degree of the patch.
                                    (n+1)(n+2)/2
              ndeg ............... Degree (n) of the patch.
              u[i] ............... Barycentric coordinates (u,v,w) of
                                    evaluation point.  i=0,2
              b[i]................ A working array with dimension >=
                                    to bpts[].

    Output:   patch_pt  ........... One coordinate of the point on
                                    the patch evaluated at (u,v,w).
              b[]  ............... Contents have been changed.

 Linear array structure: It is assumed that the usual (i,k,j) structure
                         of the Bezier net has been put into a linear
                         array in the following manner.
                         (E.g., for n=3)
                         b_(300) --> bpts[0]  (u=1)
                         b_(030) --> bpts[6]  (v=1)
                         b_(003) --> bpts[9]  (w=1)
*/
```

18.13   Problems

1. Barycentric coordinates are affinely invariant, being the ratio of two
   triangle areas in the plane. Prove the more general statement: the ratio
   of the areas of any two coplanar regions is invariant under affine maps.

2. Find the equation of an edge bisector of a triangle in terms of barycentric coordinates.

3. Work out exactly how terms involving $B_j^r(\mathbf{d})$ generalize the univariate difference operator.

4. Work out a degree reduction procedure for Bézier triangles. Literature: Petersen [366].

5. In Section 17.9, we saw how to modify surfaces by embedding them in the unit cube and then how to distort it using trivariate tensor product schemes. Experiment with the following: instead of embedding a surface in a cube, embed it in a tetrahedron and distort it using a *Bézier tetrahedron*:

$$\mathbf{b}(\mathbf{u}) = \sum_{|\mathbf{i}|=n} \mathbf{b_i} B_{\mathbf{i}}^n(\mathbf{u}),$$

where now $\mathbf{u} = (u_1, u_2, u_3, u_4)$ are barycentric coordinates in a tetrahedron. Also try to embed the surface into a collection of tetrahedra for more local control. More literature on Bézier tetrahedra: de Boor [115], Farin [160], Hoschek and Lasser [272], Goldman [206], Lasser [300], and Worsey and Farin [474].

6. After completing Problem 5, experiment with *rational* Bézier tetrahedra.

7. Suppose that in Figure 18.15 we prescribed $\mathbf{p}$ instead of the three $\mathbf{q_i}$. We could then find several sets of weights. Can that idea be generalized to higher degrees?

8. Problem 2 of Chapter 6 addressed the similarity between the Aitken's and de Casteljau's algorithms. Generalize to the triangular patch case and write a program that modifies tri_decas accordingly.

Chapter 19

Geometric Continuity for Surfaces

The concept of geometric continuity is not restricted to curves. Surfaces in this regard are much more complex to deal with, so we will restrict ourselves to the case of first-order geometric continuity. Here are some pointers to the literature: Boehm [69], Charrot and Gregory [91], DeRose [133], Farin [153], Gregory [235], Hahn [249], Herron [257], Hoschek and Liu [316], Kahmann [284], Jensen [277], Jones [283], Nielson [351], Piper [375], Sarraga [418], [419], Shirman and Séquin [448], van Wijk [460], Vinacua and Brunet [467], and Veron *et al.* [465].

19.1 Introduction

Let us take a look at Figure 19.1. It shows the (potential) boundary curves of two cubic triangular Bézier patches. In each, the interior control point $\mathbf{b}_{111}$ is missing. Can we determine these two missing points such that the resulting two patches form a C^1 surface? We must thus produce two control nets that satisfy the C^1 condition as illustrated in Figure 18.11. But this is not possible in our example: the two shaded pairs of triangles in Figure 19.1 are not affine pairs in the sense of Section 18.7, i.e., they cannot be obtained as an affine map of *one* pair of domain triangles.

We have a better chance of solving the problem if we relax the requirement of C^1 continuity to that of G^1 continuity: *two patches with a common boundary curve are called G^1 if they have a continuously varying tangent plane along that boundary curve.* The concept of G^1 continuity is a genuine

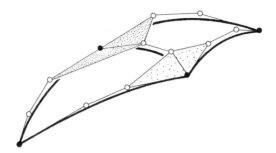

Figure 19.1: G^1 continuity: the shown cubic curves cannot be the boundary curves of two C^1 cubic Bézier triangles since no suitable pair of domain triangles can be found.

generalization of C^1 continuity: all (nondegenerate) C^1 surfaces are G^1, but not the other way around.

An example (if somewhat simplistic) of a G^1 yet non-C^1 surface is easily constructed: take two triangles formed by the diagonal of a square in the x, y-plane and interpret them as two linear Bézier triangular patches. They are clearly G^1, but they are not C^1 if we pick as their domain the two adjacent triangles with vertices $(0,0), (1,0), (0,1)$ and $(0,0), (1,0), (-1,0)$.

One important aspect of G^1 continuity is that it is completely independent of the domains of the two involved surface patches. For C^1 continuity, the interplay between range and domain geometry was crucial, but now the domains are only needed so that we can evaluate each patch.

We will next discuss the different configurations of G^1 continuity between triangular and rectangular patches.

19.2   Triangle-Triangle

In this section, we shall construct a (sufficient) condition for two adjacent triangular Bézier patches to be G^1. We only have to consider the control polygon of the common boundary curve and the two "parallel" rows of control points in each patch. The situation is illustrated in Figure 19.2, where some suitable abbreviations are introduced.

Let $\mathbf{x}(t)$ be a point on the common boundary curve of the two patches. It may be constructed using the de Casteljau algorithm from either patch since

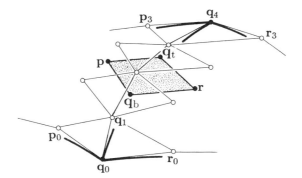

Figure 19.2: G^1 continuity: the G^1 constraints may be expressed in terms of the de Casteljau algorithm. The quartic case is shown.

the de Casteljau algorithm yields the tangent plane at a point (see Section 18.5), being spanned by the $\mathbf{b}_i^{n-1}$. These points are labeled $\mathbf{p}(t), \mathbf{q}_b(t), \mathbf{q}_t(t)$, and $\mathbf{r}(t)$ in Figure 19.2. The two patches are G^1 if these four points are coplanar for all t, i.e., if the lines $\mathbf{pr}$ and $\mathbf{q}_b\mathbf{q}_t$ always intersect.[1] This means that numbers λ and μ exist such that

$$\lambda\mathbf{p}(t) + (1 - \lambda)\mathbf{r}(t) = \mu\mathbf{q}_b(t) + (1 - \mu)\mathbf{q}_t(t) \qquad (19.1)$$

for all t. Of course, $\mu = \mu(t)$ and $\lambda = \lambda(t)$.

We can write explicit expressions for $\mathbf{p}(t), \mathbf{q}_b(t), \mathbf{q}_t(t)$, and $\mathbf{r}(t)$:

$$\mathbf{p}(t) = \sum_{i=0}^{n-1} \mathbf{p}_i B_i^{n-1}(t),$$

$$\mathbf{q}_b(t) = \sum_{i=0}^{n-1} \mathbf{q}_i B_i^{n-1}(t),$$

$$\mathbf{q}_t(t) = \sum_{i=0}^{n-1} \mathbf{q}_{i+1} B_i^{n-1}(t),$$

$$\mathbf{r}(t) = \sum_{i=0}^{n-1} \mathbf{r}_i B_i^{n-1}(t).$$

---

[1] To be precise, we must also require that $\mathbf{p}$ and $\mathbf{r}$ be on *different sides* of $\mathbf{q}_t$ and $\mathbf{q}_b$; otherwise we might generate surfaces that double back on themselves along the common boundary curve.

We now attempt to simplify our task of formulating G^1 conditions: we make the assumption that $\lambda(t)$ and $\mu(t)$ are both *linear*, i.e.,

$$\lambda(t) = \lambda_0 + \lambda_1 t, \quad \mu(t) = \mu_0 + \mu_1 t.$$

Our G^1 condition of Eq. (19.1) can then be written

$$(\lambda_0 + \lambda_1 t) \sum_{i=0}^{n-1} \mathbf{p}_i B_i^{n-1}(t) + (1 - \lambda_0 - \lambda_1 t) \sum_{i=0}^{n-1} \mathbf{r}_i B_i^{n-1}(t)$$

$$= (\mu_0 + \mu_1 t) \sum_{i=0}^{n-1} \mathbf{q}_i B_i^{n-1}(t) + (1 - \mu_0 - \mu_1 t) \sum_{i=0}^{n-1} \mathbf{q}_{i+1} B_i^{n-1}(t).$$

Rearranging:

$$\sum_{i=0}^{n-1} [\lambda_0 \mathbf{p}_i + (1 - \lambda_0) \mathbf{r}_i] B_i^{n-1} + \lambda_1 \sum_{i=0}^{n-1} [\mathbf{p}_i - \mathbf{r}_i] \frac{i+1}{n} B_{i+1}^n$$

$$= \sum_{i=0}^{n-1} [\mu_0 \mathbf{q}_i + (1 - \mu_0) \mathbf{q}_{i+1}] B_i^{n-1} + \mu_1 \sum_{i=0}^{n-1} [\mathbf{q}_i - \mathbf{q}_{i+1}] \frac{i+1}{n} B_{i+1}^n.$$

The terms B_{i+1}^n were obtained from Eq. (5.29). We next perform degree elevation on the B_i^{n-1} terms and compare coefficients of the resulting B_i^n, which yields

$$\frac{i}{n}[(\lambda_0 + \lambda_1)\mathbf{p}_{i-1} + (1 - \lambda_0 - \lambda_1)\mathbf{r}_{i-1}] + (1 - \frac{i}{n})[\lambda_0 \mathbf{p}_i + (1 - \lambda_0)\mathbf{r}_i]$$

$$= \frac{i}{n}[(\mu_0 + \mu_1)\mathbf{q}_{i-1} + (1 - \mu_0 - \mu_1)\mathbf{q}_i] + (1 - \frac{i}{n})[\mu_0 \mathbf{q}_i + (1 - \mu_0)\mathbf{q}_{i+1}].$$

We may simplify this by setting

$$\alpha_1 = \lambda_0 + \lambda_1, \quad \beta_1 = \mu_0 + \mu_1, \quad \alpha_0 = \lambda_0, \quad \beta_0 = \mu_0$$

and rearranging:

$$\frac{i}{n}\Big[[\alpha_1 \mathbf{p}_{i-1} + (1 - \alpha_1)\mathbf{r}_{i-1}] - [\beta_1 \mathbf{q}_{i-1} + (1 - \beta_1)\mathbf{q}_i]\Big]$$
$$= -(1 - \frac{i}{n})\Big[[\alpha_0 \mathbf{p}_i + (1 - \alpha_0)\mathbf{r}_i] - [\beta_0 \mathbf{q}_i + (1 - \beta_0)\mathbf{q}_{i+1}]\Big]. \tag{19.2}$$

This is our desired G^1 condition. If it holds for all $i = 0, \ldots, n$, the two patches have a continuous tangent plane all along their common boundary.

To arrive at a geometric interpretation of (19.2), we first consider the special cases $i = 0$ and $i = n$. For $i = 0$, we obtain

$$\alpha_0 \mathbf{p}_0 + (1 - \alpha_0)\mathbf{r}_0 = \beta_0 \mathbf{q}_0 + (1 - \beta_0)\mathbf{q}_1, \tag{19.3}$$

and $i = n$ yields

$$\alpha_1 \mathbf{p}_{n-1} + (1 - \alpha_1)\mathbf{r}_{n-1} = \beta_1 \mathbf{q}_{n-1} + (1 - \beta_1)\mathbf{q}_n. \tag{19.4}$$

These equations describe the geometry of the planar quadrilaterals formed by $\mathbf{p}_0, \mathbf{q}_0, \mathbf{r}_0, \mathbf{q}_1$ and $\mathbf{p}_{n-1}, \mathbf{q}_{n-1}, \mathbf{r}_{n-1}, \mathbf{q}_n$, respectively. If we were given the two quadrilaterals in some application, we could then readily determine $\alpha_0, \alpha_1, \beta_0, \beta_1$ by interpreting (19.3) and (19.4) as two overdetermined linear systems for those quantities. If the involved quadrilaterals are planar, unique solutions will exist.

Let us concentrate on $\mathbf{p}_0, \mathbf{q}_0, \mathbf{r}_0, \mathbf{q}_1$. The two numbers α_0 and β_0 tell us how to compute the intersection of the two diagonals: either as $\alpha_0 \mathbf{p}_0 + (1 - \alpha_0)\mathbf{r}_0$ or as $\beta_0 \mathbf{q}_0 + (1 - \beta_0)\mathbf{q}_1$. In any other quadrilateral, formed by $\mathbf{p}_i, \mathbf{r}_i, \mathbf{q}_i, \mathbf{q}_{i+1}$, the analogous two expressions $\alpha_0 \mathbf{p}_i + (1 - \alpha_0)\mathbf{r}_i$ and $\beta_0 \mathbf{q}_i + (1 - \beta_0)\mathbf{q}_{i+1}$ will in general yield two different points. The difference vector of these two points is an indication of how much the shape of the arbitrary quadrilateral differs from that of $\mathbf{p}_0, \mathbf{q}_0, \mathbf{r}_0, \mathbf{q}_1$. This difference vector may be written as

$$\mathbf{d}_{i,0} = [\alpha_0 \mathbf{p}_i + (1 - \alpha_0)\mathbf{r}_i] - [\beta_0 \mathbf{q}_i + (1 - \beta_0)\mathbf{q}_{i+1}].$$

Similarly, we may measure how the quadrilateral $\mathbf{p}_i, \mathbf{r}_i, \mathbf{q}_i, \mathbf{q}_{i+1}$ differs from $\mathbf{p}_{n-1}, \mathbf{r}_{n-1}, \mathbf{q}_{n-1}, \mathbf{q}_n$:

$$\mathbf{d}_{i,n} = [\alpha_1 \mathbf{p}_{i-1} + (1 - \alpha_1)\mathbf{r}_{i-1}] - [\beta_1 \mathbf{q}_{i-1} + (1 - \beta_1)\mathbf{q}_i].$$

Now our G^1 condition takes on the simple form

$$\frac{i}{n}\mathbf{d}_{i,n} + (1 - \frac{i}{n})\mathbf{d}_{i,0} = \mathbf{0}; \quad i = 0, \ldots, n. \tag{19.5}$$

The quadratic case is illustrated in Figure 19.3. This case is of special interest: our G^1 condition is only sufficient, not necessary, in general. However, for quadratics, it is both sufficient *and* necessary.[2]

---

[2]This was observed by T. DeRose (1989), private communication.

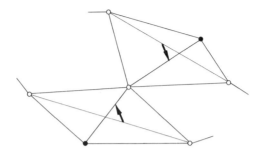

Figure 19.3: G^1 continuity: for two quadratics, the shape difference vectors $\mathbf{d}_{1,2}$ and $\mathbf{d}_{1,0}$ are shown. Note that they sum to $\mathbf{0}$.

19.3   Rectangle-Rectangle

We now consider two tensor product Bézier patches with a common boundary curve of degree n, illustrated in Figure 19.4. Consider any point $\mathbf{x}(t)$ on this curve. It may be constructed using the (univariate) de Casteljau algorithm. The de Casteljau algorithm yields the tangent of a curve as a by-product (Section 4.5), namely, as the difference of the two intermediate points $\mathbf{b}_1^{n-1}$ and $\mathbf{b}_0^{n-1}$. In Figure 19.4, those two points are labeled $\mathbf{q}_b$ and $\mathbf{q}_t$. It now follows that the tangent plane of the left patch in Figure 19.4 is spanned by the points $\mathbf{p}, \mathbf{q}_b, \mathbf{q}_t$ and that of the right patch is spanned by $\mathbf{q}_b, \mathbf{q}_t, \mathbf{r}$, where

$$\mathbf{p}(t) = \sum_{i=0}^{n} \mathbf{p}_i B_i^n(t)$$

and

$$\mathbf{q}(t) = \sum_{i=0}^{n} \mathbf{r}_i B_i^n(t).$$

We could now follow a similar development, as in the previous section, but a little trick will give us our desired G^1 condition easily. Let us simply degree elevate the common boundary curve from degree n to $n+1$. Call the degree elevated control points $\hat{\mathbf{q}}_0, \ldots, \hat{\mathbf{q}}_{n+1}$ (see Section 5.1 for the degree elevation procedure). Now we are in the situation of the previous section! Namely, we have $n+1$ control points $\mathbf{p}_i$, $n+2$ control points $\hat{\mathbf{q}}_i$, and $n+1$ control points $\mathbf{r}_i$. Our situation is equivalent to that of a G^1 join between two triangular patches of degree $n+1$.

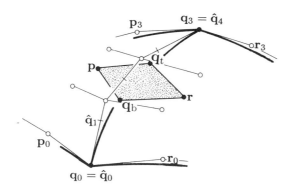

Figure 19.4: G^1 continuity for rectangular patches: the shaded quadrilateral must be planar as it varies along the common boundary. The case of a cubic boundary curve is shown.

The desired G^1 condition is therefore given by

$$\frac{i}{n+1}\Big[[\alpha_1\mathbf{p}_{i-1} + (1 - \alpha_1)\mathbf{r}_{i-1}] - [\beta_1\hat{\mathbf{q}}_{i-1} + (1 - \beta_1)\hat{\mathbf{q}}_i]\Big]$$
$$= -(1 - \tfrac{i}{n+1})\Big[[\alpha_0\mathbf{p}_i + (1 - \alpha_0)\mathbf{r}_i] - [\beta_0\hat{\mathbf{q}}_i + (1 - \beta_0)\hat{\mathbf{q}}_{i+1}]\Big],\tag{19.6}$$

where i varies from 0 to $n + 1$.

The geometric interpretation is analogous to that of the preceding section.

19.4 Rectangle-Triangle

This situation, illustrated in Figure 19.5, is now treated in a completely analogous way. We assume that both patches have a common boundary curve of degree n. If we degree elevate the triangular patch (Section 18.8), we have the control point rows $\hat{\mathbf{p}}_0, \ldots, \hat{\mathbf{p}}_n$ (from the tensor product patch), $\hat{\mathbf{q}}_0, \ldots, \hat{\mathbf{q}}_{n+1}$ (the degree elevated common boundary curve), and $\mathbf{r}_0, \ldots, \mathbf{r}_n$ (from the degree elevated triangular patch). Thus the G^1 condition is

$$\frac{i}{n+1}\Big[[\alpha_1\hat{\mathbf{p}}_{i-1} + (1 - \alpha_1)\mathbf{r}_{i-1}] - [\beta_1\hat{\mathbf{q}}_{i-1} + (1 - \beta_1)\hat{\mathbf{q}}_i]\Big]$$
$$= -(1 - \tfrac{i}{n+1})\Big[[\alpha_0\hat{\mathbf{p}}_i + (1 - \alpha_0)\mathbf{r}_i] - [\beta_0\hat{\mathbf{q}}_i + (1 - \beta_0)\hat{\mathbf{q}}_{i+1}]\Big],\tag{19.7}$$

again with i ranging from 0 to n+1.

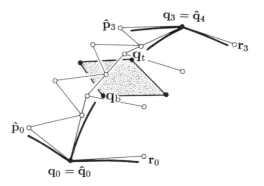

Figure 19.5: G^1 continuity for triangular and rectangular patches: the shaded quadrilateral must be planar as it varies along the common boundary. The case of a cubic boundary curve is shown.

19.5 "Filling in" Rectangular Patches

Suppose that we were given the boundary curves of two bicubic patches, as shown in Figure 19.4. At the endpoints of the common boundary curve, we assume that the three curves meeting there have coplanar tangents. Can we find interior Bézier points $\mathbf{p}_1, \mathbf{p}_2$ and $\mathbf{r}_1, \mathbf{r}_2$ such that the two patches will have a continuously varying tangent plane along the common boundary? We shall employ the G^1 condition of Eq. (19.6) for this purpose. As a first step, we determine α_0 and β_0 from $\mathbf{p}_0, \hat{\mathbf{q}}_0, \mathbf{r}_0, \hat{\mathbf{q}}_1$ and α_1, β_1 from $\mathbf{p}_3, \hat{\mathbf{q}}_3, \mathbf{r}_3, \hat{\mathbf{q}}_4$.

There are three equations in (19.6) that involve our four unknowns $\mathbf{p}_1, \mathbf{p}_2$ and $\mathbf{r}_1, \mathbf{r}_2$. After some suitable modification, they are of the form

$$
\begin{aligned}
\alpha_0\mathbf{p}_1 + (1 - \alpha_0)\mathbf{r}_1 &= \mathbf{rhs1}, \\
\alpha_1\mathbf{p}_1 + (1 - \alpha_1)\mathbf{r}_1 + \alpha_0\mathbf{p}_2 + (1 - \alpha_0)\mathbf{r}_2 &= \mathbf{rhs2}, \\
\alpha_1\mathbf{p}_2 + (1 - \alpha_1)\mathbf{r}_2 &= \mathbf{rhs3}.
\end{aligned}
$$

We have abbreviated the right-hand sides of the equations somewhat.

The above may be written in matrix form:

$$
A\mathbf{x} = \mathbf{b}, \tag{19.8}
$$

with

$$A = \begin{bmatrix} \alpha_0 & 1 - \alpha_0 & & \\ \alpha_1 & 1 - \alpha_1 & \alpha_0 & 1 - \alpha_0 \\ & & \alpha_1 & (1 - \alpha_1) \end{bmatrix}$$

$\mathbf{x} = [\mathbf{p}_1, \mathbf{p}_2, \mathbf{r}_1, \mathbf{r}_2]^T$, and $\mathbf{b} = [\mathbf{rhs1}, \mathbf{rhs2}, \mathbf{rhs3}]^T$.

Since the rows of A are linearly independent, the underdetermined system of Eq. (19.8) always has solutions. One way of finding one is as follows: suppose we already have an estimate $\bar{\mathbf{x}} = [\bar{\mathbf{p}}_1, \bar{\mathbf{p}}_2, \bar{\mathbf{r}}_1, \bar{\mathbf{r}}_2]$ for the unknowns. Then one solves the system

$$AA^T\mathbf{y} = \mathbf{b} - A\bar{\mathbf{x}}$$

for $\mathbf{y}$, and the solution is given by

$$\mathbf{x} = \bar{\mathbf{x}} + A^T\mathbf{y}.$$

This solution stays as close as possible to the initial guess $\bar{\mathbf{x}}$; see [75].

The $\bar{\mathbf{r}}_i$ and $\bar{\mathbf{p}}_i$ could, for instance, be generated by Adini's twists (see Section 17.3). This idea was carried out by Sarraga [418], [419] in a slightly different context.

19.6 "Filling in" Triangular Patches

Let us reconsider Figure 19.1. Do our G^1 conditions allow us to complete the cubic curve network such that the resulting surface will be G^1? In our notation, the undetermined points are $\mathbf{p}_1$ and $\mathbf{r}_1$. There are two G^1 equations that involve these points. They are of the form

$$\begin{aligned} \alpha_1\mathbf{p}_1 + (1 - \alpha_1)\mathbf{r}_1 &= \mathbf{rhs1}, \\ \alpha_0\mathbf{p}_1 + (1 - \alpha_0)\mathbf{r}_1 &= \mathbf{rhs2}. \end{aligned}$$

The coefficient matrix of this system:

$$A = \begin{bmatrix} \alpha_1 & 1 - \alpha_1 \\ \alpha_0 & 1 - \alpha_0 \end{bmatrix}$$

is singular if $\alpha_0 = \alpha_1$. Therefore, in this case, we are not guaranteed to have a solution (see Piper [375] for an explicit counterexample).

We can, however, solve the problem if we resort to quartics. After we degree elevate all boundary curves, we now have to determine unknown control points $\mathbf{p}_1, \mathbf{p}_2$ and $\mathbf{r}_1, \mathbf{r}_2$. This is exactly the situation from the preceding section and is solved in exactly the same way. B. Piper [375] first used quartics to solve this kind of Hermite interpolation problem.

19.7   Theoretical Aspects

We have developed an approach to geometric continuity that is powerful enough to solve several applications-oriented problems. It is practical, but it is not *general*: there are G^1 surfaces that do not satisfy the condition of Eq. (19.2); see Problems. T. DeRose has developed conditions that are both necessary and sufficient for G^1 continuity of adjacent Bézier patches.

Several authors (consult the surveys by Boehm [69], Herron [257], and Gregory [235]) define geometric continuity of surfaces in the following way: two surfaces that share a common boundary curve are called G^r if, for every point of the boundary curve, a reparametrization exists such that both surfaces are C^r in a neighborhood of that point. For the case $r = 1$, this definition yields tangent plane continuity. Its advantage is that it also works for higher orders; the price to be paid is that it is rather abstract. For the case of G^2 continuity, another popular definition is to require a common Dupin's indicatrix along the boundary curve (see Section 22.12 and Kahmann [284] and Pegna and Wolter [358]).

19.8   Problems

1. In Section 19.1 we saw an example of triangular patch boundaries that could not be completed to an overall C^1 surface. Find similar examples for tensor product patches.

2. Show that the G^1 conditions of Eq. (19.2) include the case of strict C^1 continuity.

3. Construct surfaces that are G^1 yet do not satisfy (19.2).

4. Consider eight triangular patches, assembled so as to form eight octants of a sphere-like surface. Show that this closed surface cannot be C^1, i.e., one cannot find a region in the plane that is composed of eight triangles that have a C^1 map that maps them to the surface.

Chapter 20

Coons Patches

We have already encountered design tools that originated in car companies; Bézier curves and surfaces were developed by Citroën and Renault in Paris. Two other major concepts also emerged from the automotive field: Coons patches (S. Coons consulted for Ford, Detroit) and Gordon surfaces (W. Gordon worked for General Motors, Detroit).[1] These methods have a different flavor than Bézier or B-spline methods: instead of being described by control nets, they "fill in" curve networks in order to generate surfaces.

Let us review the part of the design process that gave rise to the methods of this chapter. Even in the days of CAD/CAM, the first step in the design of a new car is (and certainly was in the 1960s) the manual production of a clay or wooden model of that car. The CAD process begins with the task of communicating the form of this model to the CAD database.

This communication process starts with data acquisition; in this case, the model is *digitized*: a digitizing device records points (in a fixed coordinate system) where a sensor touches the model surface. This sensor is moved over the model along certain predefined lines, called *feature lines*. Each of these lines is thus broken down into a sequence of digitized points. Once these are stored, curves are fitted through them, for example, by using interpolating splines. The resulting network of curves now must be completed in order to generate a full surface description of the model. This process – generating a surface from a network of curves – is solved using Coons and Gordon surfaces.

Additional literature on Coons patches includes: Coons's "little red book" [102] (also available in a French translation [105]) and Barnhill [15], [16]. In

---

[1]Just for the record – in the late 1960s, Chrysler began to develop a curve and surface scheme that was based on Chebychev polynomials.

the area of numerical grid generation for computational fluid dynamics, Coons patches are also frequently employed; here, they are known as *transfinite interpolants*; see [455].

Before we start with their description, we need to discuss an important "building block."

20.1   Ruled Surfaces

Ruled surfaces, also called "lofted surfaces"[2] are both simple and fundamental to surface design. They are of considerable importance in their own right, in particular for the design of "functional" surfaces in mechanical engineering. Ruled surfaces solve the following problem: given two space curves $\mathbf{c}_1$ and $\mathbf{c}_2$, both defined over the same parameter interval $u \in [0, 1]$, find a surface $\mathbf{x}$ that contains both curves as "opposite" boundary curves. More precisely: find $\mathbf{x}$ such that

$$\mathbf{x}(u, 0) = \mathbf{c}_1(u), \quad \mathbf{x}(u, 1) = \mathbf{c}_2(u). \tag{20.1}$$

Clearly, the stated problem has infinitely many solutions, so we pick the "simplest" one:

$$\mathbf{x}(u, v) = (1 - v)\mathbf{c}_1(u) + v\mathbf{c}_2(u), \tag{20.2}$$

or, with (20.1):

$$\mathbf{x}(u, v) = (1 - v)\mathbf{x}(u, 0) + v\mathbf{x}(u, 1). \tag{20.3}$$

Ruled surfaces have the familiar flavor of *linear interpolation*: every isoparametric line $u = const$ is a straight line segment, as illustrated in Figure 20.1.

The difference from earlier occurrences of linear interpolation is that now we interpolate to whole *curves*, not just discrete points – thus this process is often referred to as *transfinite interpolation*. It is still quite manageable, however: note how the linear terms in v are kept separate from the data terms in u.

An important aspect of ruled surfaces of the form of Eq. (20.3) is the generality that is allowed for the input curves $\mathbf{x}(u, 0)$ and $\mathbf{x}(u, 1)$: there is

---

[2]The word "lofted" has an interesting history: In the days of completely manual ship design, full-scale drawings were difficult to handle in the design office. These drawings were stored and dealt with in large attics, called "lofts."

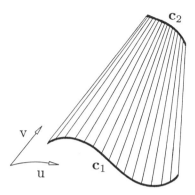

Figure 20.1: Ruled surfaces: two arbitrary curves c_1, c_2 are given. A surface is fitted between them by linear interpolation.

virtually no restriction on them other than having to be defined over the same parameter interval. [We chose the interval $[0, 1]$, but any other interval $[a, b]$ will do – we will then have to use formula (2.10) for general linear interpolation.] For instance, one of the input curves might be a cubic polynomial curve, the other a spline curve or even a polygon. More information on ruled surfaces can be found in Section 22.10.

20.2 Coons Patches: Bilinearly Blended

A ruled surface interpolates to *two* boundary curves – a rectangular surface, however, has *four* boundary curves, and that is precisely to what a Coons patch interpolates. This first instance of Coons patches was also developed first by Coons [101].

To be more precise: given are four arbitrary curves $\mathbf{c}_1(u), \mathbf{c}_2(u)$ and $\mathbf{d}_1(v), \mathbf{d}_2(v)$, defined over $u \in [0, 1]$ and $v \in [0, 1]$, respectively. Find a surface $\mathbf{x}$ that has these four curves as boundary curves:

$$\mathbf{x}(u, 0) = \mathbf{c}_1(u), \qquad \mathbf{x}(u, 1) = \mathbf{c}_2(u), \qquad (20.4)$$

$$\mathbf{x}(0, v) = \mathbf{d}_1(v), \qquad \mathbf{x}(1, v) = \mathbf{d}_2(v). \qquad (20.5)$$

We have just developed ruled surfaces, so let us utilize them for this new problem. The four boundary curves define two ruled surfaces:

$$\mathbf{r}_c(u, v) = (1 - v)\mathbf{x}(u, 0) + v\mathbf{x}(u, 1)$$

and

$$\mathbf{r}_d(u, v) = (1 - u)\mathbf{x}(0, v) + u\mathbf{x}(1, v).$$

Both interpolants are shown in Figure 20.2, and we see that $\mathbf{r}_c$ interpolates to the $\mathbf{c}$-curves, yet fails to reproduce the $\mathbf{d}$-curves. The situation for $\mathbf{r}_d$ is similar, and therefore equally unsatisfactory. Both $\mathbf{r}_c$ and $\mathbf{r}_d$ do well on two sides, yet fail on the other two, where they are *linear*. Our strategy is therefore as follows: let us try to retain what each ruled surface interpolates to and let us try to eliminate what it fails to interpolate to. A little thought reveals that the "interpolation failures" are captured by one surface: the bilinear interpolant $\mathbf{r}_{cd}$ to the four corners (see also Section 16.1):

$$\mathbf{r}_{cd}(u, v) = \begin{bmatrix} 1 - u & u \end{bmatrix} \begin{bmatrix} \mathbf{x}(0, 0) & \mathbf{x}(0, 1) \\ \mathbf{x}(1, 0) & \mathbf{x}(1, 1) \end{bmatrix} \begin{bmatrix} 1 - v \\ v \end{bmatrix}.$$

We are now ready to create a Coons patch $\mathbf{x}$. It is given by

$$\mathbf{x} = \mathbf{r}_c + \mathbf{r}_d - \mathbf{r}_{cd}, \qquad (20.6)$$

or, in the form of a recipe: "$\mathrm{loft}_u + \mathrm{loft}_v - \mathrm{bilinear}$." The involved surfaces and the solution are illustrated in Figure 20.2. Writing (20.6) in full detail gives

$$\mathbf{x}(u, v) = \begin{bmatrix} 1 - u & u \end{bmatrix} \begin{bmatrix} \mathbf{x}(0, v) \\ \mathbf{x}(1, v) \end{bmatrix}$$

$$+ \begin{bmatrix} \mathbf{x}(u, 0) & \mathbf{x}(u, 1) \end{bmatrix} \begin{bmatrix} 1 - v \\ v \end{bmatrix} \qquad (20.7)$$

$$- \begin{bmatrix} 1 - u & u \end{bmatrix} \begin{bmatrix} \mathbf{x}(0, 0) & \mathbf{x}(0, 1) \\ \mathbf{x}(1, 0) & \mathbf{x}(1, 1) \end{bmatrix} \begin{bmatrix} 1 - v \\ v \end{bmatrix}.$$

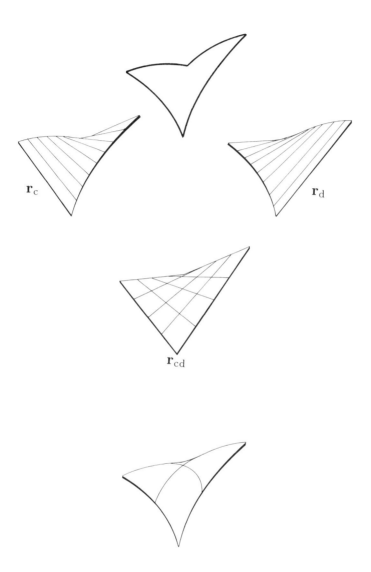

Figure 20.2: Coons patches: a bilinearly blended Coons patch is composed of two lofted surfaces and a bilinear surface.

It is left as an exercise for the reader to verify that (20.7) does indeed interpolate to all four boundary curves.

We can now justify the name "bilinearly blended" for the above Coons patch: a ruled surface "blends" together the two defining boundary curves; this blending takes place in both directions. However, the Coons patch is *not* generally itself a bilinear surface – the name refers purely to the method of construction.

The functions $1 - u, u$ and $1 - v, v$ are called *blending functions*. A close inspection of (20.7) reveals that many other pairs of blending functions, say, $f_1(u), f_2(u)$ and $g_1(v), g_2(v)$, could also be used to construct a generalized Coons patch. It would then be of the general form

$$
\begin{aligned}
\mathbf{x}(u, v) = \quad & \left[\begin{array}{cc} f_1(u) & f_2(u) \end{array} \right] \left[\begin{array}{c} \mathbf{x}(0, v) \\ \mathbf{x}(1, v) \end{array} \right] \\
+ & \left[\begin{array}{cc} \mathbf{x}(u, 0) & \mathbf{x}(u, 1) \end{array} \right] \left[\begin{array}{c} g_1(v) \\ g_2(v) \end{array} \right] \qquad (20.8) \\
- & \left[\begin{array}{cc} f_1(u) & f_2(u) \end{array} \right] \left[\begin{array}{cc} \mathbf{x}(0, 0) & \mathbf{x}(0, 1) \\ \mathbf{x}(1, 0) & \mathbf{x}(1, 1) \end{array} \right] \left[\begin{array}{c} g_1(v) \\ g_2(v) \end{array} \right].
\end{aligned}
$$

There are only two restrictions on the f_i and g_i: each pair must sum to one identically: otherwise we would generate nonbarycentric combinations of points (see Section 2.1). Also, we must have $f_1(0) = g_1(0) = 1, f_1(1) = g_1(1) = 0$ in order to actually interpolate. The shape of the blending functions has a predictable effect on the shape of the resulting Coons patch. Typically, one requires f_1 and g_1 to be monotonically decreasing; this produces surfaces of predictable shape, but is not necessary for theoretical reasons. Surface modelers that employ Coons patches typically allow designers to change the blending functions as a way to model the interior of the patch.

20.3   Coons Patches: Partially Bicubically Blended

The bilinearly blended Coons patch solves a problem of considerable importance with very little effort, but we pay for that by an annoying drawback. Consider Figure 20.3: it shows two bilinearly blended Coons patches, defined over $u \in [0, 2], v \in [0, 1]$. The boundary curves $v = 0$ and $v = 1$, both

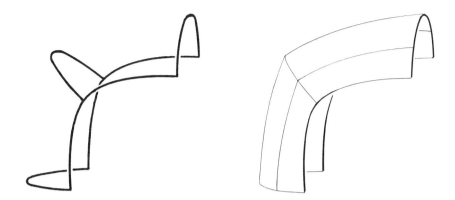

Figure 20.3: Coons patches: the input curves for two neighboring patches may have C^1 boundary curves (left), yet the two Coons patches determined by them do not form a smooth surface (right).

composite curves, are differentiable. However, the cross boundary derivative is clearly discontinuous along $u = 1$; also see Problems.[3]

Analyzing this problem, we see that it can be blamed on the fact that cross boundary tangents along one boundary depend on *data not pertaining to that boundary*. For example, for any given bilinearly blended Coons patch, a change in the boundary curve $\mathbf{x}(1, v)$ will affect the derivatives across the boundary $\mathbf{x}(0, v)$.

How can we separate the derivatives across one boundary from information along the opposite boundary? The answer: use different blending functions, namely, some that have zero slopes at the endpoints. Striving for simplicity, as usual, we find two obvious candidates for such blending functions: the cubic Hermite polynomials H_0^3 and H_3^3 from Section 6.5, as defined by (6.14).

Let us investigate the effect of this choice of blending functions: we have set $f_1 = g_1 = H_0^3$ and $f_2 = g_2 = H_3^3$ in (20.8). The cross boundary derivative

---

[3]We also see that bilinearly blended Coons patches suffer from a shape defect: each of the two patches is too "flat." This effect of Coons patches has been studied by Nachman [342].

along, say, $u = 0$, now becomes:

$$\mathbf{x}_u(0, v) = \left[\begin{array}{cc} \mathbf{x}_u(0,0) & \mathbf{x}_u(0,1) \end{array} \right] \left[\begin{array}{c} H_0^3(v) \\ H_3^3(v) \end{array} \right]; \qquad (20.9)$$

all other terms vanish since $d/du H_{3i}^3(0) = d/du H_{3i}^3(1) = 0$ for $i = 0$ and $i = 1$. Thus, the only data that influence $\mathbf{x}_u$ along $u = 0$ are the two tangents $\mathbf{x}_u(0,0)$ and $\mathbf{x}_u(0,1)$ – we have achieved our goal of making the cross boundary derivative along one boundary depend only on information pertaining to that boundary. With our new blending functions, the two patches from Figure 20.3 would now be C^1.

Unfortunately, we have also created a new problem. At the patch corners, these patches often seem to have "flat spots." The reason: partially bicubically blended Coons patches,[4] constructed as above, suffer from *zero corner twists*:

$$\mathbf{x}_{uv}(i,j) = \mathbf{0}; \quad i,j \in \{0,1\}.$$

This is easily verified by simply taking the u, v-partial of (20.8) and evaluating at the patch corners.

To understand this situation better, let us resort to the simpler case of curves – to functional curves for added simplicity. The two graphs $x(u)$ and $y(u)$ in Figure 12.3 illustrate what happens if we replace the linear functions $1 - t, t$ in piecewise linear interpolation by $H_0^3(t), H_3^3(t)$. We clearly see the introduction of "flat spots." The reason for this poor performance lies in the fact that we only use two functions, H_0^3 and H_3^3 from the full set of four Hermite polynomials. Both have zero derivatives at the interval endpoints, and pass that property on to the interpolant.

We will now modify the partially bicubically blended Coons patch in order to avoid the flat spots at the corners.

20.4   Coons Patches: Bicubically Blended

Cubic Hermite interpolation needs more input than positional data – first derivative information is needed. Since our positional input consists of whole curves, not just points, the obvious data to supply are derivatives along those input curves. Our given data now consist of

$$\mathbf{x}(u,0), \quad \mathbf{x}(u,1), \quad \mathbf{x}(0,v), \quad \mathbf{x}(1,v)$$

---

[4]We use the term *partially* bicubically blended since only a part of all cubic Hermite basis functions is utilized.

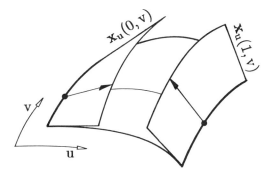

Figure 20.4: Coons patches: for the bicubically blended case, the concept of the lofted surface is generalized. In addition to the given boundary curves, cross boundary derivatives are supplied.

and

$$\mathbf{x}_v(u,0), \quad \mathbf{x}_v(u,1), \quad \mathbf{x}_u(0,v), \quad \mathbf{x}_u(1,v).$$

We can think of the now prescribed cross boundary derivatives as "tangent ribbons," illustrated in Figure 20.4 (only two of the four "ribbons" are shown there).

The derivation of the bicubically blended Coons patch is analogous to the one in Section 20.2: we must simply generalize the concept of a ruled surface appropriately. This is almost trivial; we obtain

$$\mathbf{h}_c(u,v) = H_0^3(u)\mathbf{x}(0,v) + H_1^3(u)\mathbf{x}_u(0,v) + H_2^3(u)\mathbf{x}_u(1,v) + H_3^3(u)\mathbf{x}(1,v)$$

for the u-direction (this surface is shown in Figure 20.4) and

$$\mathbf{h}_d(u,v) = H_0^3(v)\mathbf{x}(u,0) + H_1^3(v)\mathbf{x}_v(u,0) + H_2^3(v)\mathbf{x}_v(u,1) + H_3^3(v)\mathbf{x}(u,1).$$

Proceeding as in the bilinearly blended case, we define the interpolant to the corner data. This gives the tensor product bicubic Hermite interpolant $\mathbf{h}_{cd}$ from Section 17.6:

$$\mathbf{h}_{cd}(u,v) = \begin{bmatrix} H_0^3(u) & H_1^3(u) & H_2^3(u) & H_3^3(u) \end{bmatrix} \times \begin{bmatrix} \mathbf{x}(0,0) & \mathbf{x}_v(0,0) & \mathbf{x}_v(0,1) & \mathbf{x}(0,1) \\ \mathbf{x}_u(0,0) & \mathbf{x}_{uv}(0,0) & \mathbf{x}_{uv}(0,1) & \mathbf{x}_u(0,1) \\ \mathbf{x}_u(1,0) & \mathbf{x}_{uv}(1,0) & \mathbf{x}_{uv}(1,1) & \mathbf{x}_u(1,1) \\ \mathbf{x}(1,0) & \mathbf{x}_v(1,0) & \mathbf{x}_v(1,1) & \mathbf{x}(1,1) \end{bmatrix} \begin{bmatrix} H_0^3(v) \\ H_1^3(v) \\ H_2^3(v) \\ H_3^3(v) \end{bmatrix}.$$

$$(20.10)$$

The bicubically blended Coons patch now becomes

$$\mathbf{x} = \mathbf{h}_c + \mathbf{h}_d - \mathbf{h}_{cd}. \tag{20.11}$$

Before closing this section, we need to take a closer look at the $\mathbf{h}_{cd}$ part of (20.11). On closer inspection, we find that it wants data that we were not willing (or able) to provide in our initial problem description, namely, the central "twist partition" of the 4×4 matrix in (20.10). The bicubically blended Coons patch needs these quantities as input, and this has caused CAD software developers many headaches since Coons proposed his surface scheme in 1964. The most popular "solution" seems to be simply to define each of the four corner twists to be the zero vector. The drawbacks of that choice were already discussed in Section 16.8, but alternatives are pointed out in that section, too.

20.5   Piecewise Coons Surfaces

We will now apply the bicubically blended patch to the situation that it was intended to solve: we assume that we are given a network of curves as shown in Figure 21.5 and that we want to fill in this curve network with bicubically blended Coons patches. The resulting surface will be C^1.

To apply (20.11), we must create twist vectors and cross boundary derivatives (tangent ribbons) from the given curve network. As a preprocessing step, we estimate a twist vector $\mathbf{x}_{uv}(u_i, v_j)$ at each patch corner. This can be done by using any of the twist vector estimators discussed in Section 17.3. In that section, we assumed that the boundary curves of each patch were cubics; that assumption does not affect the computation of the twist vectors at all, however.

Having found a twist vector for each data point, we now need to create cross boundary derivatives for each boundary curve. Let us focus our attention on one patch of our network, and let us assume for simplicity (but without loss of generality!) that the parameters u and v vary between 0 and 1. We shall now construct the tangent ribbon $\mathbf{x}_v(u, 0)$. We have four pieces of information about $\mathbf{x}_v(u, 0)$: the value of $\mathbf{x}_v(u, 0)$ at $u = 0$ and at $u = 1$, and the derivatives (with respect to u) there, which are the twists that we made up above, namely, $\mathbf{x}_{uv}(0, 0)$ and $\mathbf{x}_{uv}(1, 0)$.

We therefore have the input data for a univariate cubic Hermite interpolant, and the desired tangent ribbon assumes the form

$$\mathbf{x}_v(u, 0) = \mathbf{x}_v(0, 0)H_0^3(u) + \mathbf{x}_{uv}(0, 0)H_1^3(u) + \mathbf{x}_{uv}(1, 0)H_2^3(u) + \mathbf{x}_v(1, 0)H_3^3(u). \tag{20.12}$$

The remaining three tangent ribbons are computed analogously.

We have thus found a (new) way to pass a C^1 surface through a C^1 network of curves. All we need is the ability to estimate the twists at the data points.

20.6 Problems

1. Verify the caption to Figure 20.3 algebraically.

2. Show that the bilinearly blended Coons patch is not in the convex hull of its boundary curves. Is this a good or a bad property?

3. Show that the bilinearly blended Coons patch, when applied to cubic boundary curves, yields a bicubic patch.

4. Show that Adini's twist from Section 17.3 is the twist of the bilinearly blended Coons patch for the four boundary cubics.

5. As we have seen, two adjacent bilinearly blended Coons patches are not C^1 in general. What are the conditions for the boundary curves of the two patches such that the Coons patches *are* C^1?

Chapter 21

Coons Patches: Additional Material

21.1 Compatibility

A situation may arise where one is not free to generate the tangent ribbon information as we did in the previous section. In some cases, the cross boundary derivatives may be prescribed and must then be adhered to. The bicubically blended Coons patch of Eq. (20.11) was designed to cope with exactly that situation, so we should be ready to address that problem.

Let us first return to the case of the bilinearly blended Coons patch. It is an obvious requirement that the four prescribed boundary curves meet in the corners; in other words, we must exclude data configurations as shown in Figure 21.1. This condition on the prescribed data is known as a *compatibility condition*. An incompatibility of that form can usually be overcome by adjusting boundary curves so that they meet at the patch corners.

The bicubically blended Coons patch suffers from a more difficult compatibility problem. It results from the appearance of the twist terms in the tensor product term $\mathbf{h}_{cd}$ in (20.10). The problem was not recognized by Coons, and only later did R. Barnhill and J. Gregory discover it; see Gregory [233].

From calculus, we know that we can usually interchange the order of differentiation when taking mixed partials: we can set $\mathbf{x}_{uv} = \mathbf{x}_{vu}$ if $\mathbf{x}(u, v)$ is twice continuously differentiable. Unfortunately, this simplification does not apply to our situation. Let us examine why: at $\mathbf{x}(0, 0)$, two given "tangent ribbons"

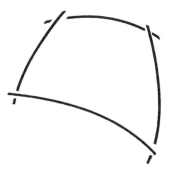

Figure 21.1: Compatibility problems: in the case of a bilinearly blended Coons patch, compatible boundary curves must be prescribed. Data as shown lead to ill-defined interpolants.

meet. We can obtain the twist at $\mathbf{x}(0,0)$ by differentiating the "ribbon" $\mathbf{x}_v(u,0)$ with respect to u:

$$\mathbf{x}_{uv}(0,0) = \lim_{u \to 0} \frac{\partial}{\partial u} \mathbf{x}_v(u,0),$$

or the other way around:

$$\mathbf{x}_{vu}(0,0) = \lim_{v \to 0} \frac{\partial}{\partial v} \mathbf{x}_u(0,v).$$

If the two twists $\mathbf{x}_{uv}(0,0)$ and $\mathbf{x}_{vu}(0,0)$ are equal, there are no problems: enter this twist term into the matrix in (20.10), and the bicubically blended Coons patch is well-defined.

However, as Figure 21.2 illustrates, these two terms need by no means be equal. Now we have a serious dilemma: entering either one of the two values yields a surface that only partially interpolates to the given data. Entering zero twist vectors only aggravates matters, since they will in general not agree with even one of the two twists above.

There are two ways out of this dilemma. One is to try to adjust the given data so that the incompatibilities disappear. Or, if the data cannot be changed, one can use a method known as *Gregory's square*. This method replaces the constant twist terms in the matrix in (20.10) by *variable twists*. The variable twists are computed from the tangent ribbons:

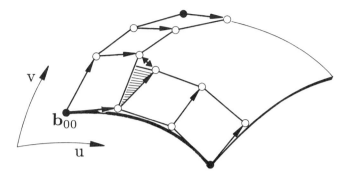

Figure 21.2: Compatibility problems: we show the example of tangent ribbons
that are represented in cubic Bézier form. Note how we obtain two different
interior Bézier points, and thus two different corner twists.

$$\mathbf{x}_{uv}(0,0) = \frac{u\frac{\partial}{\partial v}\mathbf{x}_u(0,0) + v\frac{\partial}{\partial u}\mathbf{x}_v(0,0)}{u+v},$$

$$\mathbf{x}_{uv}(0,1) = \frac{-u\frac{\partial}{\partial v}\mathbf{x}_u(0,1) + (v-1)\frac{\partial}{\partial u}\mathbf{x}_v(0,1)}{-u+v-1},$$

$$\mathbf{x}_{uv}(1,0) = \frac{(1-u)\frac{\partial}{\partial v}\mathbf{x}_u(1,0) + v\frac{\partial}{\partial u}\mathbf{x}_v(1,0)}{1-u+v},$$

$$\mathbf{x}_{uv}(1,1) = \frac{(u-1)\frac{\partial}{\partial v}\mathbf{x}_u(1,1) + (v-1)\frac{\partial}{\partial u}\mathbf{x}_v(1,1)}{u-1+v-1}.$$

The resulting surface does not have a continuous twist at the corners.
In fact, it is *designed* to be discontinuous: it assumes two different values,
depending on from where the corner is approached. If we approach $\mathbf{x}(0,0)$,
say, along the isoparametric line $u = 0$, we should get the u-partial of the
given tangent ribbon $\mathbf{x}_v(u,0)$ as the twist $\mathbf{x}_{uv}(0,0)$. If we approach the same
corner along $v = 0$, we should get the v-partial of the given ribbon $\mathbf{x}_u(0,v)$
to be $\mathbf{x}_{uv}(0,0)$.

An interesting application of Gregory's square was developed by Chiyokura and Kimura [93]: suppose we are given four boundary curves of a patch
in cubic Bézier form, and suppose that the cross boundary derivatives also
vary cubically. Let us consider the corner $\mathbf{x}_{00}$ and the two boundary curves
that meet there. These curves define the Bézier points $\mathbf{b}_{0j}$ and $\mathbf{b}_{i0}$. The cross
boundary derivatives determine $\mathbf{b}_{1j}$ and $\mathbf{b}_{i1}$. Note that $\mathbf{b}_{11}$ is defined twice!

This situation is illustrated in Figure 21.2. Chiyokura and Kimura made $\mathbf{b}_{11}$ a function of u and v:

$$\mathbf{b}_{11} = \mathbf{b}_{11}(u,v) = \frac{u\mathbf{b}_{11}(v) + v\mathbf{b}_{11}(u)}{u+v},$$

where $\mathbf{b}_{11}(u)$ denotes the point $\mathbf{b}_{11}$ that would be obtained from the cross boundary derivative $\mathbf{x}_u(0,v)$, etc. Similar expressions hold for the remaining three interior Bézier points, all following the pattern of Gregory's square.

Although a solution to the posed problem, one should note that Gregory's square (or the Chiyokura and Kimura application) is not free of problems. First, even with polynomial input data, it will produce a rational patch. This may not be acceptable in certain environments, since it requires its own kind of data structure.[1] Secondly, there are singularities at the corners. These are removable, but require special attention. In situations where one is not forced to use incompatible cross boundary derivatives, it is therefore advisable first to estimate corner twists and then to use (20.12) as a cross boundary derivative generator.

21.2   Control Nets from Coons Patches

Consider the following design situation: four boundary curves of a surface are given, all four in B-spline or Bézier form, i.e., by their control polygons. Let us assume that opposite boundary curves are defined over the same knot sequence and are of the same degrees. The problem: find the control net of a B-spline or Bézier surface that fits between the boundary curves. This situation is illustrated in Figure 21.3.

That control net may be obtained in a surprisingly simple way: interpret the boundary control polygons as piecewise linear curves and compute the bilinearly blended Coons patch that interpolates to them. This Coons patch is piecewise bilinear. Its vertices can be interpreted as vertices of a control net for a B-spline or Bézier surface. As it turns out, that surface is precisely the bilinearly blended Coons patch to the original boundary curves! The proof is straightforward and relies on the fact that both the B-spline and Bézier methods have linear precision (Farin [163]) .

The same principle is also applicable to interpolating spline curves and surfaces. Suppose we are given points on all four boundary curves of a surface

---

[1]It is not the data structure that describes a rational patch from Chapter 15, but a more specialized one that is closer to the standard bicubic Hermite form.

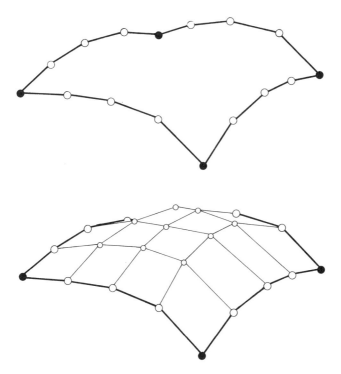

Figure 21.3: Coons patches: the bilinearly blended Coons patch may be applied to boundary control polygons.

(same number of points and same parametrization for opposite curves, of course!). We can construct the cubic spline interpolant to all four point sets. For the sake of concreteness, suppose that the spline curves are represented as piecewise Bézier curves. These four boundary spline curves define a bilinearly blended Coons patch. We may obtain its piecewise bicubic representation in two ways: first, we could compute an array of points, obtained from the given boundary points by applying the bilinearly blended Coons method to them. We could then apply bicubic spline interpolation to them.

That same surface could be obtained more easily by applying the bilinearly blended Coons method directly to the boundary curves (i.e., to their B-spline, Bézier, or Hermite representations).

Using the Coons technique in conjunction with other surface forms is one of the most significant applications of the Coons method, implemented in `netcoons` below. It can reduce computational costs considerably and is a quick way of fitting surfaces between boundary curves.[2] If the boundaries happened to be rational, then Coons blending should be applied to the homogeneous representation in four-space.

21.3   Translational Surfaces

There is an alternative way to derive the bilinearly blended Coons patch, based on the two concepts of *translational surfaces* and *convex combinations*.

A translational surface has the simple structure of being generated by two curves: let $c_1(u)$ and $c_2(v)$ be two such curves, intersecting at a common point $a = c_1(0) = c_2(0)$. A translational surface $t(u,v)$ is now defined by

$$t(u,v) = c_1(u) + c_2(v) - a. \tag{21.1}$$

Why the name "translational?" It is justified by considering an arbitrary isoparametric line of the surface, say, $u = \hat{u}$. We obtain $t(\hat{u},v) = c_2(v) + [-a + c_1(\hat{u})]$, that is, all isoparametric lines are *translates* of one of the input curves; see also Figure 21.4.

An interesting property of translational surfaces is that their twist is identically zero everywhere:

$$\frac{\partial^2}{\partial u \partial v} t(u,v) \equiv 0.$$

This property follows directly from the definition of Eq. (21.1). Since both input curves may be arbitrarily shaped, the resulting surface may well have high curvatures. This dispels the myth that zero twists are identical to flat spots. In fact, twists are not related to the shape of a surface – rather, they are a result of a particular parametrization. See also Section 17.3 on twist generation.

How are translational surfaces related to Coons patches? A translational surface can be viewed as the solution to an interpolation problem: given two intersecting curves, find a surface that contains them as boundary curves. If four boundary curves are given, as in the problem definition for the bilinearly blended Coons patch, we can form four translational surfaces, one for each corner. Let us denote by $t_{i,j}$ the translational surface that interpolates to the boundary curves meeting at the corner (i,j); $i,j \in \{0,1\}$.

---

[2]A warning: it does not always produce "optimal" shapes; see Nachman [342].

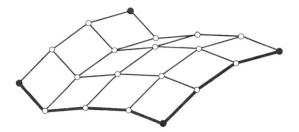

Figure 21.4: Translational surfaces: the Bézier net of a translational tensor product surface. The control polygons in each direction are translates of each other.

Now the bilinearly blended Coons patch $\mathbf{x}(u,v)$ can be written as

$$\mathbf{x}(u,v) = \left[\begin{array}{cc} 1-u & u \end{array}\right] \left[\begin{array}{cc} \mathbf{t}_{00}(u,v) & \mathbf{t}_{01}(u,v) \\ \mathbf{t}_{10}(u,v) & \mathbf{t}_{11}(u,v) \end{array}\right] \left[\begin{array}{c} 1-v \\ v \end{array}\right]. \qquad (21.2)$$

The form of Eq. (21.2) of the bilinearly blended Coons patch is called a *convex combination*. It blends together four surfaces, weighting each with a *weight function*. The weight functions sum to one (a necessity: nonbarycentric combinations are disallowed) and are nonnegative for $u, v \in \{0, 1\}$. Note that the weight functions are zero where the corresponding $\mathbf{t}_{i,j}$ is "wrong."

The weight functions in (21.2) are linear in both u and v, another justification for the term *bilinearly* blended Coons patch.

21.4 Gordon Surfaces

Gordon surfaces are a generalization of Coons patches. They were developed in the late 1960s by W. Gordon [226], [228], [225], [227], who was then working for the General Motors Research labs. He coined the term "transfinite interpolation" for this kind of surfaces.

It is often not sufficient to model a surface from only four boundary curves. A more complicated (and realistic) situation arises when a *network* of curves is prescribed, as shown in Figure 21.5. We will construct a surface $\mathbf{g}$ that interpolates to all these curves – they will then be isoparametric curves $\mathbf{g}(u_i, v); i = 0, \ldots, m$ and $\mathbf{g}(u, v_j); j = 0, \ldots, n$. We shall therefore refer to these input curves in terms of the final surface $\mathbf{g}$. The idea behind the construction of this *Gordon surface* $\mathbf{g}$ is the same as for the Coons patch: find

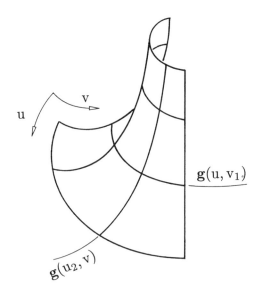

Figure 21.5: Gordon surfaces: a rectilinear network of curves is given and an interpolating surface is sought.

a surface $\mathbf{g}_1$ that interpolates to one family of isoparametric curves, for instance to the $\mathbf{g}(u_i, v)$. Next, find a surface $\mathbf{g}_2$ that interpolates to the $\mathbf{g}(u, v_j)$. Finally, add both together and subtract a surface $\mathbf{g}_{12}$.

Let us start with the task of finding the surface $\mathbf{g}_1$. If there are only two curves $\mathbf{g}(u_0, v)$ and $\mathbf{g}(u_1, v)$, the surface $\mathbf{g}_1$ reduces to the lofted surface $\mathbf{g}_1(u, v) = L_0^1(u)\mathbf{g}(u_0, v) + L_1^1(u)\mathbf{g}(u_1, v)$, where the L_i^1 are the linear Lagrange polynomials from Section 6.2. If we have more than two input curves, we might want to try higher degree Lagrange polynomials:

$$\mathbf{g}_1(u, v) = \sum_{i=0}^{m} \mathbf{g}(u_i, v) L_i^m(u). \tag{21.3}$$

Simple algebra verifies that we have successfully generalized the concept of a lofted surface.

Let us return to the construction of the Gordon surface, for which $\mathbf{g}_1$ will only be a building block. The second building block, $\mathbf{g}_2$, is obtained by analogy:

$$\mathbf{g}_2(u, v) = \sum_{j=0}^{n} \mathbf{g}(u, v_j) L_j^n(v).$$

The third building block, $\mathbf{g}_{12}$, is simply the interpolating tensor product surface

$$\mathbf{g}_{12}(u, v) = \sum_{i=0}^{m} \sum_{j=0}^{n} \mathbf{g}(u_i, v_j) L_i^m(u) L_j^n(v).$$

The Gordon surface $\mathbf{g}$ now becomes

$$\mathbf{g} = \mathbf{g}_1 + \mathbf{g}_2 - \mathbf{g}_{12}. \tag{21.4}$$

It is left as an exercise for the reader to verify that (21.4) in fact interpolates to all given curves. Note that for the actual computation of $\mathbf{g}$, we do not have to use the Lagrange polynomials. We only have to be able to solve the univariate polynomial interpolation problem, for example, by using the Vandermonde approach.

We have derived Gordon surfaces as based on polynomial interpolation. Much more generality is available. Equation (21.4) is also true if we use interpolation methods other than polynomial interpolation. The essence of (21.4) may be stated as follows: take a univariate interpolation scheme, apply it to all curves $\mathbf{g}(u, v_j)$ and to all curves $\mathbf{g}(u_i, v)$, add the resulting two surfaces, and subtract the tensor product interpolant that is defined by the univariate scheme. We may replace polynomial interpolation by spline interpolation, in which case we speak of *spline-blended* Gordon surfaces. The basis functions of the univariate interpolation scheme are called *blending functions*.

21.5 Boolean Sums

Our development of Coons patches was quite straightforward, yet it is slightly flawed from a geometric viewpoint. When we derived the basic equation (20.6), we *added* the two surfaces $\mathbf{r}_c$ and $\mathbf{r}_d$ as an intermediate step. This is illegal – the sum of two surfaces would depend on the choice of a coordinate system (see the discussion in Section 2.1). Although the situation is straightened out by subtracting the bilinear surface $\mathbf{r}_{cd}$, one might ask for a cleaner development. It is provided by the use of *Boolean sums*.

Let us define an operator $\mathcal{P}_1$ that, when applied to a rectangular surface $\mathbf{x}$, returns the ruled surface through $\mathbf{x}(u, 0)$ and $\mathbf{x}(u, 1)$:

$$[\mathcal{P}_1 \mathbf{x}](u, v) = (1 - v)\mathbf{x}(u, 0) + v\mathbf{x}(u, 1).$$

Similarly, we define $\mathcal{P}_2$ to return the ruled surface through $\mathbf{x}(0, v)$ and $\mathbf{x}(1, v)$:

$$[\mathcal{P}_2\mathbf{x}](u, v) = (1 - u)\mathbf{x}(0, v) + u\mathbf{x}(1, v).$$

In terms of Section 20.2, $\mathcal{P}_1$ and $\mathcal{P}_2$ yield the surfaces $\mathbf{r}_c$ and $\mathbf{r}_d$.

We would like to formulate the bilinearly blended Coons patch – which we now call $\mathcal{P}\mathbf{x}$ – in terms of $\mathcal{P}_1$ and $\mathcal{P}_2$.

Let us take $\mathcal{P}_1\mathbf{x}$ as a first building block for the Coons patch. Since $\mathcal{P}_1\mathbf{x}$ only interpolates on two boundaries, we will try to add another surface to it, such that the final result will interpolate to all four boundaries. Such a *correction surface* must interpolate to all four boundaries of the error surface $\mathbf{x} - \mathcal{P}_1\mathbf{x}$.[3] It may be obtained by applying $\mathcal{P}_2$ to the error surface. We then obtain

$$\mathcal{P}\mathbf{x} = \mathcal{P}_1\mathbf{x} + \mathcal{P}_2(\mathbf{x} - \mathcal{P}_1\mathbf{x}).$$

This expression for the bilinearly blended Coons patch may be shortened by showing only the involved operators:

$$\mathcal{P} = \mathcal{P}_1 + \mathcal{P}_2(\mathcal{I} - \mathcal{P}_1), \tag{21.5}$$

where $\mathcal{I}$ is the identity operator. This means of obtaining one operator $\mathcal{P}$ as a combination of two operators $\mathcal{P}_1, \mathcal{P}_2$ is called a Boolean sum and is often written

$$\mathcal{P} = \mathcal{P}_1 \oplus \mathcal{P}_2. \tag{21.6}$$

Of course, one may also multiply out the terms in (21.5). We then obtain

$$\mathcal{P} = \mathcal{P}_1 \oplus \mathcal{P}_2 = \mathcal{P}_1 + \mathcal{P}_2 - \mathcal{P}_1\mathcal{P}_2.$$

We now see that, even with the use of an operator calculus, we still have the same old Coons patch as defined by (20.6): the term $\mathcal{P}_1\mathcal{P}_2$ is simply the bilinear interpolant to the patch corners.

Let us summarize the essence of the Boolean sum approach: An interpolant to the given data is built by applying $\mathcal{P}_1$. A second operator $\mathcal{P}_2$ is then applied to the "failures" of $\mathcal{P}_1$, and the result is added back to the output from $\mathcal{P}_1$. The interpolant $\mathcal{P}_2$ may actually be of a simpler nature than $\mathcal{P}_1$, since it only has to act on zero data where $\mathcal{P}_1$ was "successful." We can illustrate this for the example of univariate cubic Hermite interpolation: we define $\mathcal{P}_1$ to be the (point-valued) linear interpolant between two points $\mathbf{x}_0$

---

[3]Note that this error surface is *vector*-valued, since both $\mathbf{x}$ and $\mathcal{P}_1\mathbf{x}$ are *point*-valued.

and $\mathbf{x}_1$ and $\mathcal{P}_2$ to be the (vector-valued) cubic Hermite interpolant to a data set $\mathbf{0}, \mathbf{m}_0, \mathbf{m}_1, \mathbf{0}$. Then $\mathcal{P}_1 \oplus \mathcal{P}_2$ is the standard cubic Hermite interpolant.

A note on the notation used in this section: the letter $\mathcal{P}$ that we used to denote our building block interpolants is due to the term *projector*. A projector is an operator, which, if applied to its own output, will not change the result.[4] For example, $\mathcal{P}_1\mathbf{x}$ is a ruled surface, and $\mathcal{P}_1\mathcal{P}_1\mathbf{x}$ is the same ruled surface. Operators with the property of being projectors are also called *idempotent*.

It was W. Gordon who realized the underlying algebraic structure of Coons patches. That discovery then led him to the generalization that now bears his name (Section 21.4). Boolean sums may be utilized in the development of many surface interpolation schemes – for an excellent survey, see Barnhill [15].

21.6 Triangular Coons Patches

Just as triangular Bézier patches provide an alternative to the rectangular variety, one may devise a triangular version of Coons patches. Several solutions have been proposed through the years; we will briefly explain the ones by Barnhill, Birkhoff, and Gordon [20] and by Nielson [347].

The C^0 Barnhill, Birkhoff, and Gordon (short: BBG) approach can be explained as follows. Suppose we are given three boundary curves, as shown in Figure 21.6. We seek a surface that interpolates to all three of them, i.e., a *transfinite triangular interpolant*. The construction follows the standard Coons patch development in that it consists of several building blocks, which are then combined in a clever way.

Let us denote[5] the three boundary curves by $\mathbf{x}(0, v, w)$, $\mathbf{x}(u, 0, w)$, $\mathbf{x}(u, v, 0)$. We define three building blocks, each being a ruled surface that interpolates to two boundary curves:

$$
\begin{aligned}
\mathcal{P}_1\mathbf{x}(\mathbf{u}) &= (1 - r)\mathbf{x}(u, 0, w) + r\mathbf{x}(u, v, 0); & r &= \tfrac{v}{v+w}, \\
\mathcal{P}_2\mathbf{x}(\mathbf{u}) &= (1 - s)\mathbf{x}(u, v, 0) + s\mathbf{x}(0, v, w); & s &= \tfrac{w}{w+u}, \\
\mathcal{P}_3\mathbf{x}(\mathbf{u}) &= (1 - t)\mathbf{x}(u, 0, w) + t\mathbf{x}(0, v, w); & t &= \tfrac{v}{v+u}.
\end{aligned}
\tag{21.7}
$$

---

[4]The term comes from geometry: if a 3D object is projected into a plane, we may then repeat that projection, yet it will not change the image.

[5]We use the the concept of barycentric coordinates as outlined in Section 18.1.

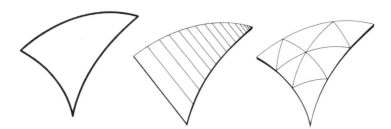

Figure 21.6: BBG interpolation: three boundary curves are given, left. Three ruled surfaces are constructed from them (only one shown, middle). They are combined to yield the final surface, right.

There are several combinations of these surfaces that yield an interpolant $\mathcal{P}\mathbf{x}$ to all three boundaries: the Boolean sum of any two – e.g., $\mathcal{P} = \mathcal{P}_1 \oplus \mathcal{P}_3$ – will have that property.

Another possibility is to define $\mathcal{P}$ as a convex combination of the three $\mathcal{P}_i$:

$$\mathcal{P}\mathbf{x} = u\mathcal{P}_1\mathbf{x} + v\mathcal{P}_2\mathbf{x} + w\mathcal{P}_3\mathbf{x}. \tag{21.8}$$

The building blocks that are used in (21.7) are rational in u, v, w, but they are linear in r, s, t. If we were to incorporate cross boundary derivative data, i.e., to build a C^1 BBG interpolant, we would define $\mathcal{P}_1, \mathcal{P}_2, \mathcal{P}_3$ to be cubic Hermite interpolants in r, s, t:

$$
\begin{aligned}
\mathcal{P}_1\mathbf{x}(\mathbf{u}) &= H_0^3(r)\mathbf{x}(u,0,w) + H_1^3(r)\mathbf{x}_1(u,0,w) \\
&\quad + H_2^3(r)\mathbf{x}_1(u,v,0) + H_3^3(r)\mathbf{x}(u,v,0), \\
\mathcal{P}_2\mathbf{x}(\mathbf{u}) &= H_0^3(s)\mathbf{x}(u,v,0) + H_1^3(s)\mathbf{x}_2(u,v,0) \\
&\quad + H_2^3(s)\mathbf{x}_2(0,v,w) + H_3^3(s)\mathbf{x}(0,v,w), \\
\mathcal{P}_3\mathbf{x}(\mathbf{u}) &= H_0^3(t)\mathbf{x}(u,0,w) + H_1^3(t)\mathbf{x}_3(u,0,w) \\
&\quad + H_2^3(t)\mathbf{x}_3(u,0,w) + H_3^3(t)\mathbf{x}(0,v,w).
\end{aligned}
\tag{21.9}
$$

The terms $\mathbf{x}_i$ are shorthand for directional derivatives of $\mathbf{x}$ taken in a direction parallel to edge i, more precisely:

$$
\begin{aligned}
\mathbf{x}_1(\mathbf{u}) &= (v+w)D_{\mathbf{e2}-\mathbf{e3}}\mathbf{x}(\mathbf{u}), \\
\mathbf{x}_2(\mathbf{u}) &= (u+w)D_{\mathbf{e3}-\mathbf{e1}}\mathbf{x}(\mathbf{u}), \\
\mathbf{x}_3(\mathbf{u}) &= (u+v)D_{\mathbf{e2}-\mathbf{e1}}\mathbf{x}(\mathbf{u}).
\end{aligned}
$$

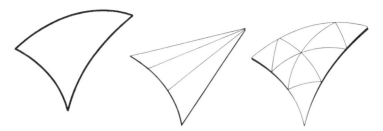

Figure 21.7: Nielson's side-vertex method: three boundary curves are given, left. Three ruled surfaces are constructed from them (only one shown, middle). They are combined to yield the final surface, right.

The factors $(v + w)$, etc., appear because cubic Hermite interpolation is sensitive to interval lengths. A reminder: the terms $\mathbf{e1}, \mathbf{e2}, \mathbf{e3}$ refer to points, *not* to edges!

Again, a Boolean sum of any two of the above operators will provide a solution – provided the cross boundary derivatives are compatible (which typically they won't be!).

A different approach is due to G. Nielson [347]. He considers – for a C^0 interpolant – *radial* curves, connecting a patch vertex with a point on the opposite edge, as shown in Figure 21.7. We have

$$\begin{aligned}
\mathcal{P}_1\mathbf{x}(\mathbf{u}) &= u\mathbf{x}(1,0,0) + (1-u)\mathbf{x}(0,r,1-r); & r &= \tfrac{v}{v+w}, \\
\mathcal{P}_2\mathbf{x}(\mathbf{u}) &= v\mathbf{x}(0,1,0) + (1-v)\mathbf{x}(1-s,0,s); & s &= \tfrac{u}{u+w}, \quad (21.10) \\
\mathcal{P}_3\mathbf{x}(\mathbf{u}) &= w\mathbf{x}(0,0,1) + (1-w)\mathbf{x}(1-t,t,0); & t &= \tfrac{u}{u+v}.
\end{aligned}$$

The final interpolant may then be written as a *triple Boolean sum:*

$$\begin{aligned}
\mathcal{P} &= \mathcal{P}_1 \oplus \mathcal{P}_2 \oplus \mathcal{P}_3 \\
&= \mathcal{P}_1 + \mathcal{P}_2 + \mathcal{P}_3 \\
&\quad -\mathcal{P}_1\mathcal{P}_2 - \mathcal{P}_1\mathcal{P}_3 - \mathcal{P}_2\mathcal{P}_3 \\
&\quad +\mathcal{P}_1\mathcal{P}_2\mathcal{P}_3.
\end{aligned}$$

To make this scheme C^1, one again replaces the linear interpolants in (21.10) by cubic Hermite interpolants, now with directional derivatives supplied in the radial directions.

For more literature on triangular Coons-type interpolants, consult the following: Barnhill [14], Barnhill and Gregory [26], [25], Gregory and Charrot [236], Marshall and Mitchell [328], Lacombe and Bédard [296], and Nielson [348].

21.7   Implementation

The following is a routine that fits a bilinearly blended Coons patch in between four boundary control polygons, as described in Section 21.2. The routine works on one coordinate only, and will have to be called separately for the x-, y-, and z-components of a control net.

```
void netcoons(net,rows,columns)
/* Uses bilinear Coons blending to complete a control
net of which only the four boundary polygons are used as input.
Works for one coordinate only.
Input:      net:            control net.
            rows, columns: dimensions of net.
Output:     net:            the completed net, with the old boundaries.
*/
```

21.8   Problems

1. What exactly *does* the bilinearly blended Coons patch interpolate to when applied to data as in Figure 21.1?

2. Equation (20.12) generates tangent ribbons from the given boundary curve network. Verify that the resulting surface does not suffer from twist incompatibilities.

3. Translational surfaces have zero twists. Show that the inverse statement is also true: every surface with identically vanishing twists is a translational surface.

4. Find a form analogous to (21.2) for the partially bicubically blended Coons patch and for the bicubically blended Coons patch.

5. Show that bilinearly blended Coons patches have *translational* precision: if the four boundary curves are boundaries of a translational surface, then the bilinearly blended Coons patch reproduces that translational surface.

6. Equation (21.8) shows how one can combine the three surfaces from (21.7) in order to obtain a complete interpolant. How would one have to blend together the three surfaces from (21.9)?

Chapter 22

W. Boehm: Differential Geometry II

22.1 Parametric Surfaces and Arc Element

A surface may be given by an *implicit* form $f(x, y, z) = 0$ or, more useful for CAGD, by its parametric form

$$\mathbf{x} = \mathbf{x}(u, v) = \begin{bmatrix} x(u, v) \\ y(u, v) \\ z(u, v) \end{bmatrix}; \quad \mathbf{u} = \begin{bmatrix} u \\ v \end{bmatrix} \in [\mathbf{a}, \mathbf{b}] \subset \mathbb{R}^2, \qquad (22.1)$$

where the cartesian coordinates x, y, z of a surface point are differentiable functions of the parameters u and v and $[\mathbf{a}, \mathbf{b}]$ denotes a rectangle in the u, v-plane; see Figure 22.1 (sometimes other domains are used, for example, triangles). To avoid potential problems with undefined normal vectors, we will assume

$$\mathbf{x}_u \wedge \mathbf{x}_v \neq \mathbf{0} \quad \text{for} \quad \mathbf{u} \in [\mathbf{a}, \mathbf{b}],$$

i.e., that both families of isoparametric lines are regular (see Section 11.1) and are nowhere tangent to each other. Such a parametrization is called *regular*.[1]

Any change $\mathbf{r} = \mathbf{r}(\mathbf{u})$ of the parameters will not change the shape of the surface; the new parametrization is regular if $\det[\mathbf{r}_u, \mathbf{r}_v] \neq 0$ for $\mathbf{u} \in [\mathbf{a}, \mathbf{b}]$, i.e., if one can find the inverse $\mathbf{u} = \mathbf{u}(\mathbf{r})$ of $\mathbf{r}$.

A regular curve $\mathbf{u} = \mathbf{u}(t)$ in the u, v-plane defines a regular curve $\mathbf{x}[\mathbf{u}(t)]$ on the surface. One can easily compute the (squared) arc element (see Section

---

[1]Examples of irregular parametrizations are shown in Figures 16.10, 16.11, and 16.12.

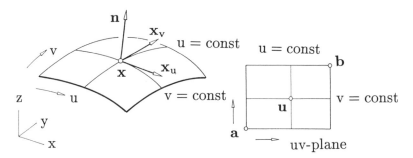

Figure 22.1: A parametric surface.

11.1) of this curve: from $\dot{\mathbf{x}} = \mathbf{x}_u \dot{u} + \mathbf{x}_v \dot{v}$ one immediately obtains

$$\mathrm{d}s^2 = ||\dot{\mathbf{x}}||^2 \mathrm{d}t^2 = (\mathbf{x}_u^2 \dot{u}^2 + 2\mathbf{x}_u \mathbf{x}_v \dot{u}\dot{v} + \mathbf{x}_v^2 \dot{v}^2)\mathrm{d}t^2,$$

which will be written as

$$\mathrm{d}s^2 = E\mathrm{d}u^2 + 2F\mathrm{d}u\mathrm{d}v + G\mathrm{d}v^2, \tag{22.2}$$

where

$$\begin{aligned}
E &= E(u, v) = \mathbf{x}_u \mathbf{x}_u, \\
F &= F(u, v) = \mathbf{x}_u \mathbf{x}_v, \\
G &= G(u, v) = \mathbf{x}_v \mathbf{x}_v.
\end{aligned}$$

The squared arc element of Eq. (22.2) is called the *first fundamental form* in classical differential geometry. It is of great importance for the further development of our material. Note that the arc element ds, being a geometric invariant of the curve through the point $\mathbf{x}$, does not depend on the particular parametrization chosen for the representation (22.1) of the surface.

For the arc length of the surface curve defined by $\mathbf{u} = \mathbf{u}(t)$, one obtains

$$\int_{t_0}^t ||\dot{\mathbf{x}}||\mathrm{d}t = \int_{t_0}^t \sqrt{E\dot{u}^2 + 2F\dot{u}\dot{v} + G\dot{v}^2}\mathrm{d}t.$$

Remark 1: The *area element* corresponding to the element dudv of the u, v-plane is given by

$$\mathrm{d}A = ||\mathbf{x}_u \mathrm{d}u \wedge \mathbf{x}_v \mathrm{d}v|| = ||\mathbf{x}_u \wedge \mathbf{x}_v||\mathrm{d}u\mathrm{d}v;$$

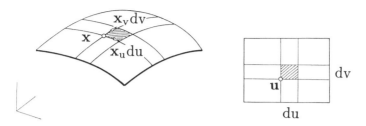

Figure 22.2: Area element.

see Figure 22.2. From $||\mathbf{a} \wedge \mathbf{b}||^2 = \mathbf{a}^2\mathbf{b}^2 - (\mathbf{ab})^2$, one obtains

$$D = ||\mathbf{x}_u \wedge \mathbf{x}_v|| = \sqrt{EG - F^2}. \tag{22.3}$$

The quantity D is called *discriminant* of (22.2). Thus the surface area A corresponding to a region U of the u, v-plane is given by

$$A = \int\int_U \sqrt{EG - F^2}\,du\,dv.$$

Remark 2: If $F = 0$ at a point of the surface, the two isoparametric lines that meet there are orthogonal to each other. Moreover, if $F \equiv 0$ at every point of the surface, the net of isoparametric lines is orthogonal everywhere.

Remark 3: Note that for any real[2] du, dv, the first fundamental form ds^2 is strictly positive. However, if $ds^2 = 0$, we have two *imaginary* directions. These are called *isotropic directions* at $\mathbf{x}$.

Remark 4: Let $\mathbf{u}_1 = \mathbf{u}_1(t_1)$ and $\mathbf{u}_2 = \mathbf{u}_2(t_2)$ define two surface curves, intersecting at $\mathbf{x}$. Both curves are intersecting orthogonally if the *polar form* of $\dot{\mathbf{x}}^2$, given by

$$\dot{\mathbf{x}}_1\dot{\mathbf{x}}_2 = E\dot{u}_1\dot{u}_2 + F(\dot{u}_1\dot{v}_2 + \dot{u}_2\dot{v}_1) + G\dot{v}_1\dot{v}_2,$$

vanishes at $\mathbf{x}$.

22.2 The Local Frame

The partials $\mathbf{x}_u$ and $\mathbf{x}_v$ at a point $\mathbf{x}$ span the tangent plane to the surface at $\mathbf{x}$. Let $\mathbf{y}$ be any point on this plane. Then

$$\det[\mathbf{y} - \mathbf{x}, \mathbf{x}_u, \mathbf{x}_v] = 0$$

---

[2]Note that the vector $[du, dv]$ defines a direction at a point $\mathbf{x}$.

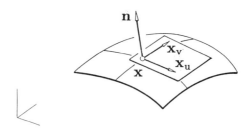

Figure 22.3: The local frame and the tangent plane.

is the implicit equation of the tangent plane. The parametric equation is

$$\mathbf{y}(u, v) = \mathbf{x} + \Delta u \mathbf{x}_u + \Delta v \mathbf{x}_v.$$

The normal $\mathbf{x}_u \wedge \mathbf{x}_v$ of the tangent plane coincides with the normal to the surface at $\mathbf{x}$. The normalized normal

$$\mathbf{n} = \frac{\mathbf{x}_u \wedge \mathbf{x}_v}{\|\mathbf{x}_u \wedge \mathbf{x}_v\|} = \frac{1}{D}[\mathbf{x}_u \wedge \mathbf{x}_v]$$

together with the unnormalized vectors $\mathbf{x}_u$, $\mathbf{x}_v$ form a local coordinate system, a *frame*, at $\mathbf{x}$ (see Figure 22.3). This frame plays the same important role for surfaces as does the Frenet frame (see Section 11.2) for curves. The normal is of unit length and is perpendicular to $\mathbf{x}_u$ and $\mathbf{x}_v$, i.e., $\mathbf{n}^2 = 1$ and $\mathbf{n}\mathbf{x}_u = \mathbf{n}\mathbf{x}_v = 0$. In general, the local coordinate system with origin $\mathbf{x}$ and axes $\mathbf{x}_u, \mathbf{x}_v$ forms only an affine system; it is also (unlike the Frenet frame) dependent on the parametrization (22.1).

22.3   The Curvature of a Surface Curve

Let $\mathbf{u}(t)$ define a curve on the surface $\mathbf{x}(\mathbf{u})$. From curve theory we know that its curvature $\kappa = \frac{1}{\rho}$ is defined by $\mathbf{t}' = \kappa \mathbf{m}$; the prime denotes differentiation with respect to the arc length of the curve. We will now reformulate this expression in surface terms. Since $\mathbf{t} = \mathbf{x}'$ and $u' = du/ds$, $v' = dv/ds$, we have

$$\mathbf{t}' = \mathbf{x}'' = \mathbf{x}_{uu}u'^2 + 2\mathbf{x}_{uv}u'v' + \mathbf{x}_{vv}v'^2 + \mathbf{x}_u u'' + \mathbf{x}_v v''.$$

Let ϕ be the angle between the main normal $\mathbf{m}$ of the curve and the surface normal $\mathbf{n}$ at the point $\mathbf{x}$ under consideration, as illustrated in Figure 22.4. Then

$$\mathbf{t}'\mathbf{n} = \kappa \mathbf{m}\mathbf{n} = \kappa \cos \phi.$$

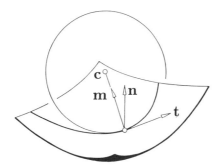

Figure 22.4: Osculating circle.

Inserting $\mathbf{t}'$ from above and keeping in mind that $\mathbf{n}\mathbf{x}_u = \mathbf{n}\mathbf{x}_v = 0$, we have

$$\kappa \cos \phi = \mathbf{n}\mathbf{x}_{uu}u'^2 + 2\mathbf{n}\mathbf{x}_{uv}u'v' + \mathbf{n}\mathbf{x}_{vv}v'^2. \tag{22.4}$$

Furthermore, $\mathbf{n}\mathbf{x}_u = 0$ implies $\mathbf{n}_u\mathbf{x}_u + \mathbf{n}\mathbf{x}_{uu} = 0$ etc. Thus, using the abbreviations

$$
\begin{aligned}
L &= L(u,v) &&= -\mathbf{x}_u\mathbf{n}_u &&= \mathbf{n}\mathbf{x}_{uu}, \\
M &= M(u,v) &&= -\tfrac{1}{2}(\mathbf{x}_u\mathbf{n}_v + \mathbf{x}_v\mathbf{n}_u) &&= \mathbf{n}\mathbf{x}_{uv}, \\
N &= N(u,v) &&= -\mathbf{x}_v\mathbf{n}_v &&= \mathbf{n}\mathbf{x}_{vv},
\end{aligned}
\tag{22.5}
$$

Eq. (22.4) can be written as

$$\kappa \cos \phi \, ds^2 = L du^2 + 2M du dv + N dv^2. \tag{22.6}$$

This expression is called the *second fundamental form* in classical differential geometry. For any given direction du/dv in the u, v-plane and any given angle ϕ, the second fundamental form, together with the first fundamental form (22.2), allows us to compute the curvature κ of a surface curve having that tangent direction.

Remark 5: Note that the arc length in the above development was only used in a theoretical context; for applications, it does not have to be actually computed.

Remark 6: Note that κ depends only on the tangent direction and the angle ϕ. It will change its sign, however, if there is a change in the orientation of $\mathbf{n}$.

22.4   Meusnier's Theorem

The right-hand side of (22.4) does not contain terms involving ϕ. For $\phi = 0$, i.e., $\cos \phi = 1$, we have that $\mathbf{m} = \mathbf{n}$: the osculating plane of the curve is perpendicular to the surface tangent plane at $\mathbf{x}$. The curvature κ_0 of such a curve is called the *normal curvature* of the surface at $\mathbf{x}$ in the direction of $\mathbf{t}$ (defined by du/dv). The normal curvature is given by

$$\kappa_0 = \kappa_0(\mathbf{x}; \mathbf{t}) = \frac{1}{\rho_0} = \frac{2^{\text{nd}} \text{fundamental form}}{1^{\text{st}} \text{fundamental form}}. \tag{22.7}$$

Now (22.6) takes the very short form

$$\rho = \rho_0 \cos \phi. \tag{22.8}$$

This simple formula has an interesting and important interpretation, known as Meusnier's theorem. It is illustrated in Figure 22.5: the osculating circles of all surface curves through $\mathbf{x}$ having the same tangent $\mathbf{t}$ there form a sphere. This sphere and the surface have a common tangent plane at $\mathbf{x}$; the radius of the sphere is ρ_0.

As a consequence of Meusnier's theorem, it is sufficient to study curves at $\mathbf{x}$ with $\mathbf{m} = \mathbf{n}$; moreover, these curves may be planar. Such curves, called *normal sections*, can be thought of as the intersection of the surface with a plane through $\mathbf{x}$ and containing $\mathbf{n}$, as illustrated in Figure 22.6.

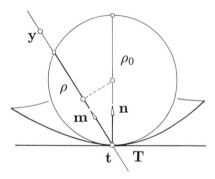

Figure 22.5: Meusnier's sphere viewed in the direction of $\mathbf{t}$.

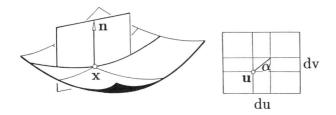

Figure 22.6: Normal section of a surface.

Remark 7: If the direction of the normal is chosen as in Figure 22.5, we have $0 \le \rho \le \rho_0$, and $\rho = 0$ only if $\phi = \pi/2$, i.e., if the osculating plane **O** coincides with the tangent plane.

22.5   Lines of Curvature

For Meusnier's theorem, we considered (osculating) planes that contained a fixed tangent at a point on a surface; we will now look at (osculating) planes containing the normal vector at a fixed point **x**. We will drop the subscript of κ_0 to simplify the notation. Setting $\lambda = dv/du = \tan \alpha$ (see Figure 22.6), we can rewrite (22.7) as

$$\kappa(\mathbf{x}, \mathbf{t}) = \kappa(\lambda) = \frac{L + 2M\lambda + N\lambda^2}{E + 2F\lambda + G\lambda^2}.$$

In the special case where $L : M : N = E : F : G$, the normal curvature κ is independent of λ. Points **x** with that property are called *umbilical points*.

In the general case, where κ changes as λ changes, $\kappa = \kappa(\lambda)$ is a rational quadratic function, as illustrated in Figure 22.7. The extreme values κ_1 and κ_2 of $\kappa(\lambda)$ occur at the roots λ_1 and λ_2 of

$$\det \begin{bmatrix} \lambda^2 & -\lambda & 1 \\ E & F & G \\ L & M & N \end{bmatrix} = 0. \tag{22.9}$$

It can be shown that λ_1 and λ_2 are always real. The extreme values κ_1 and κ_2 are the roots of

$$\det \begin{bmatrix} \kappa E - L & \kappa F - M \\ \kappa F - M & \kappa G - N \end{bmatrix} = 0. \tag{22.10}$$

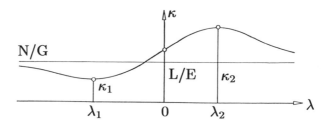

Figure 22.7: The function $\kappa = \kappa(\lambda)$.

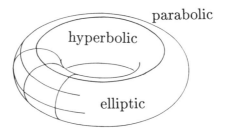

Figure 22.8: Lines of curvature on a torus. Also shown are the regions of elliptic, parabolic, and hyperbolic points.

The quantities λ_1 and λ_2 define directions in the (u, v)-plane; the corresponding directions in the tangent plane are called *principal directions*. The net of lines that have these directions at all of their points is called the net of *lines of curvature*. If necessary, it may be constructed by integrating (22.9).

Therefore, this net of lines of curvature can be used as a parametrization of the surface; then (22.9) must be satisfied by $du = 0$ and by $dv = 0$. This implies, excluding umbilical points, that

$$F \equiv 0 \quad \text{and} \quad M \equiv 0.$$

The first equation, $F \equiv 0$, states that lines of curvature are orthogonal to each other; the second equation states that they are conjugate to each other as defined in Section 22.9.

At an umbilical point, the principal directions are undefined; see also Remark 9.

Remark 8: For a surface of revolution, the net of lines of curvature is defined by the meridians and the parallels; an example is shown in Figure 22.8.

22.6 Gaussian and Mean Curvature

The extreme values κ_1 and κ_2 of $\kappa = \kappa(\lambda)$ are called *principal curvatures* of the surface at $\mathbf{x}$. A comparison of (22.10) with $\kappa^2 - (\kappa_1 + \kappa_2)\kappa + \kappa_1\kappa_2 = 0$ yields

$$\kappa_1\kappa_2 = \frac{LN - M^2}{EG - F^2} \qquad (22.11)$$

and

$$\kappa_1 + \kappa_2 = \frac{NE - 2MF + LG}{EG - F^2}. \qquad (22.12)$$

The term $K = \kappa_1\kappa_2$ is called *Gaussian curvature*, while $H = \frac{1}{2}(\kappa_1 + \kappa_2)$ is called *mean curvature*. Note that both κ_1 and κ_2 change sign if the normal $\mathbf{n}$ is reversed, but K is not affected by such a reversal.

If κ_1 and κ_2 are of the same sign, i.e., if $K > 0$, the point $\mathbf{x}$ under consideration is called *elliptic*. For example, all points of an ellipsoid are elliptic points. If κ_1 and κ_2 have different signs, i.e., $K < 0$, the point $\mathbf{x}$ under consideration is called *hyperbolic*. For example, all points of a hyperboloid are hyperbolic points. Finally, if either $\kappa_1 = 0$ or $\kappa_2 = 0$, K vanishes, the point $\mathbf{x}$ under consideration is called *parabolic*. For example, all points of a cylinder are parabolic points. In the special case where both K and H vanish, one has a *flat* point.

The Gaussian curvature K depends on the coefficients of the first and second fundamental forms. It is a very important result, due to Gauss, that K can also be expressed only in terms of E, F, and G and their derivatives. This is known as the *Theorema Egregium* and states that K depends only on the *intrinsic geometry* of the surface. This means it does not change if the surface is deformed in a way that does not change length measurement within it.

Remark 9: All points of a sphere are umbilic. The Gaussian and mean curvatures of a sphere are constant.

Remark 10: Any *developable*, i.e., a surface that can be deformed to planar shape without changing length measurements in it, must have

$K \equiv 0$. Conversely, every surface with $K \equiv 0$ can be developed into a plane (if necessary, by applying cuts). See also Section 22.10.

22.7   Euler's Theorem

The normal curvatures in different directions **t** at a point **x** are not independent of each other. For simplicity, let the isoparametric curves of a surface be lines of curvature; then we have $F \equiv M \equiv 0$ (see Section 22.5). As a consequence, we have

$$\kappa_1 = \frac{L}{E} \quad \text{and} \quad \kappa_2 = \frac{N}{G},$$

and $\kappa(\lambda)$ may be written as

$$\kappa(\lambda) = \frac{L + N\lambda^2}{E + G\lambda^2} = \kappa_1 \frac{E}{E + G\lambda^2} + \kappa_2 \frac{G\lambda^2}{E + G\lambda^2}. \qquad (22.13)$$

The coefficients of κ_1 and κ_2 have a nice geometric meaning: let Ψ denote the angle between $\mathbf{x}_u$ and the tangent vector $\dot{\mathbf{x}} = \mathbf{x}_u \dot{u} + \mathbf{x}_v \dot{v}$ of the curve under consideration, as illustrated in Figure 22.9. We obtain

$$\cos \Psi = \frac{\dot{\mathbf{x}} \mathbf{x}_u}{\|\dot{\mathbf{x}}\| \, \|\mathbf{x}_u\|} = \frac{\sqrt{E}}{\sqrt{E + G\lambda^2}}$$

and

$$\sin \Psi = \frac{\dot{\mathbf{x}} \mathbf{x}_v}{\|\dot{\mathbf{x}}\| \, \|\mathbf{x}_v\|} = \frac{\sqrt{G}\lambda}{\sqrt{E + G\lambda^2}},$$

where $\lambda = \dot{v}/\dot{u}$ as before. Hence $\kappa(\lambda)$ may be written as

$$\kappa(\lambda) = \kappa_1 \cos^2 \Psi + \kappa_2 \sin^2 \Psi.$$

This important result was found by L. Euler.

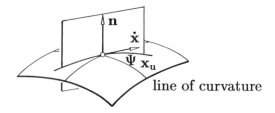

Figure 22.9: Configuration for Euler's theorem.

22.8   Dupin's Indicatrix

Euler's theorem has the following geometric implication. If we introduce polar coordinates $r = \sqrt{\rho}$ and Ψ for a point $\mathbf{y}$ of the tangent plane at $\mathbf{x}$ by setting $y_1 = \sqrt{\rho}\cos\Psi$, $y_2 = \sqrt{\rho}\sin\Psi$, then setting $\kappa_1 = \frac{1}{\rho_1}$ and $\kappa_2 = \frac{1}{\rho_2}$, Euler's theorem can be written

$$\frac{y_1^2}{\rho_1} + \frac{y_2^2}{\rho_2} = 1.$$

This is the equation of a conic section, the *Dupin's indicatrix* (see Figure 22.10). Its points $\mathbf{y}$ in the direction given by Ψ have distance $\sqrt{\rho}$ from $\mathbf{x}$. Taking into account that a reversal of the direction of $\mathbf{n}$ will effect a change in the sign of ρ, this conic section is an ellipse if $K > 0$, a pair of hyperbolas if $K < 0$ (corresponding to $\sqrt{\rho}$ and $\sqrt{-\rho}$), and a pair of parallel lines if $K = 0$ (but $H \neq 0$).

Dupin's indicatrix has a nice geometric interpretation: we may approximate our surface at $\mathbf{x}$ by a paraboloid, that is, a Taylor expansion with terms up to second order. Then Remark 7 of Chapter 11 leads to a very simple interpretation of Dupin's indicatrix: the indicatrix, scaled down by a factor of $1 : m$, can be viewed as the intersection of the surface with a plane parallel to the tangent plane at $\mathbf{x}$ in the distance $\epsilon = \frac{1}{2m^2}$. This is illustrated in Figure 22.11.

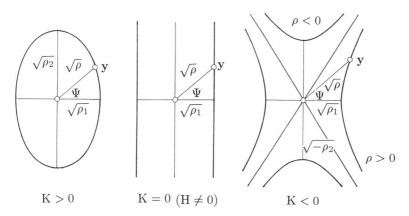

Figure 22.10: Dupin's indicatrix for an elliptic, a parabolic, and a hyperbolic point.

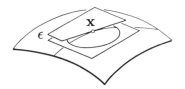

Figure 22.11: Dupin's indicatrix, scale $1 : m$.

Remark 11: This illustrates the appearance of a pair of hyperbolas in Figure 22.10; they appear when intersecting the surface in distances $\epsilon = \pm\frac{1}{2m^2}$. We can thus assign a *sign* to the Dupin indicatrix, depending on its being "above" or "below" the tangent plane.

22.9 Asymptotic Lines and Conjugate Directions

The asymptotic directions of Dupin's indicatrix have a simple geometric meaning: surface curves passing through $\mathbf{x}$ and having a tangent in an asymptotic direction there have zero curvature at $\mathbf{x}$; in other words, these directions are defined by

$$L\mathrm{d}u^2 + 2M\,\mathrm{d}u\mathrm{d}v + N\mathrm{d}v^2 = 0. \tag{22.14}$$

They are real and different if $K < 0$, real but coalescing if $K = 0$, and complex if $K > 0$.

The net of lines having these directions in all their points is called the *net of asymptotic lines*. If necessary, it may be calculated by integrating (22.14). In a hyperbolic region of the surface, it is real and regular and can be used for a real parametrization.

For this parametrization, one has

$$L \equiv 0 \text{ and } N \equiv 0,$$

and vice versa.

As earlier, let $\mathbf{y}$ be a point on Dupin's indicatrix at a point $\mathbf{x}$. Let $\dot{\mathbf{y}}$ denote its tangent direction at $\mathbf{y}$. The direction $\dot{\mathbf{y}}$ is called *conjugate* to the direction $\dot{\mathbf{x}}$ from $\mathbf{x}$ to $\mathbf{y}$. Consider two surface curves $\mathbf{u}_1(t_1)$ and $\mathbf{u}_2(t_2)$ that have tangent directions $\dot{\mathbf{x}}_1$ and $\dot{\mathbf{x}}_2$ at $\mathbf{x}$. Some elementary calculations yield that $\dot{\mathbf{x}}_1$ is conjugate to $\dot{\mathbf{x}}_2$ if

$$L\dot{u}_1\dot{u}_2 + M(\dot{u}_1\dot{v}_2 + \dot{u}_2\dot{v}_1) + N\dot{v}_1\dot{v}_2 = 0.$$

Note that this expression is symmetric in $\dot{\mathbf{u}}_1, \dot{\mathbf{u}}_2$. By definition asymptotic directions are *self-conjugate*.

Remark 12: Isoparametric curves of a surface are conjugate if $M \equiv 0$ and vice versa.

Remark 13: The principal directions, defined by (22.9), are orthogonal and conjugate; they bisect the angles between the asymptotic directions; i.e., they are the axis directions of Dupin's indicatrix (see Figure 22.10).

Remark 14: The tangent planes of two "consecutive" points on a surface curve intersect in a straight line $\mathbf{s}$. Let the curve have direction $\mathbf{t}$ at a point $\mathbf{x}$ on the surface. Then $\mathbf{s}$ and $\mathbf{t}$ are conjugate to each other. In particular, if $\mathbf{t}$ is an asymptotic direction, $\mathbf{s}$ coincides with $\mathbf{t}$. If $\mathbf{t}$ is one of the principal directions at $\mathbf{x}$, then $\mathbf{s}$ is orthogonal to $\mathbf{t}$ and vice versa. These properties characterize lines of curvature and asymptotic lines and may be used to define them geometrically.

22.10 Ruled Surfaces and Developables

If a surface contains a family of straight lines it is called a *ruled surface*. It is convenient to use these straight lines as one family of isoparametric lines. Then the ruled surface may be written

$$\mathbf{x} = \mathbf{x}(t, v) = \mathbf{p}(t) + v\mathbf{q}(t), \qquad (22.15)$$

where $\mathbf{p}$ is a point and $\mathbf{q}$ is a vector, both depending on t. The isoparametric lines $t = const$ are called the *generatrices* of the surface; see Figure 22.12.

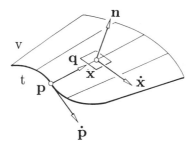

Figure 22.12: Ruled surface.

The partials of a ruled surface are given by $\mathbf{x}_t = \dot{\mathbf{p}} + v\dot{\mathbf{q}}$ and $\mathbf{x}_v = \mathbf{q}$. The normal $\mathbf{n}$ at $\mathbf{x}$ is given by

$$\mathbf{n} = \frac{(\dot{\mathbf{p}} + v\dot{\mathbf{q}}) \wedge \mathbf{q}}{\|(\dot{\mathbf{p}} + v\dot{\mathbf{q}}) \wedge \mathbf{q}\|}.$$

A point $\mathbf{y}$ on the tangent plane at $\mathbf{x}$ satisfies

$$\det[\mathbf{y} - \mathbf{p}, \dot{\mathbf{p}}, \mathbf{q}] + v\det[\mathbf{y} - \mathbf{p}, \dot{\mathbf{q}}, \mathbf{q}] = 0;$$

in other words, the tangent planes along a generatrix form a pencil of planes. However, if $\dot{\mathbf{p}}, \dot{\mathbf{q}}$, and $\mathbf{q}$ are linearly dependent, i.e., if

$$\det[\dot{\mathbf{p}}, \dot{\mathbf{q}}, \mathbf{q}] = 0, \tag{22.16}$$

the tangent plane does not vary with v.

If (22.16) holds for all t, the tangent planes are fixed along each generatrix; hence all tangent planes of the surface form a one-parameter family of planes. Conversely, any one-parameter family of planes envelopes a developable surface that may be written as a ruled surface (22.15), satisfying condition (22.16); see also Remark 10.

Remark 15: The generatrices of any ruled surface coalesce with one family of its asymptotic lines. As a consequence of asymptotic lines being real, one has $K \leq 0$.

Remark 16: In particular, the generatrices of a developable surface agree with its coalescing asymptotic lines, also forming one family of its lines of curvature. The second family of lines of curvature is formed by the orthogonal trajectories of the generatrices.

Remark 17: It can be shown that any developable surface is a cone $\mathbf{p} = \mathbf{const}$, a cylinder $\mathbf{q} = \mathbf{const}$, or a surface formed by all tangents of a space curve, that is, $\mathbf{q} = \dot{\mathbf{p}}$; see Figure 22.13.

Remark 18: The normals along a line of curvature of any surface form a developable surface. This property characterizes and defines lines of curvature.

Remark 19: The tangent planes along a curve $\mathbf{x} = \mathbf{x}[\mathbf{u}(t)]$ on any surface form a developable surface. It may be developed into a plane; if by this process the curve $\mathbf{x}[\mathbf{u}(t)]$ happens to be developed into a straight line, the curve $\mathbf{x}[\mathbf{u}(t)]$ is called a *geodesic*. At any point $\mathbf{x}$ of a geodesic, one finds that

$$\det[\dot{\mathbf{x}}, \ddot{\mathbf{x}}, \mathbf{n}] = 0. \tag{22.17}$$

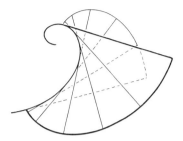

Figure 22.13: General developable surface.

Equation (22.17) is the differential equation of a geodesic; it is of second order. Geodesics may also be characterized as providing the shortest path between two points on the surface.

22.11 Nonparametric Surfaces

Let $z = f(x, y)$ be a function of two variables as shown in Figure 22.14. The surface

$$\mathbf{x} = \mathbf{x}(u, v) = \begin{bmatrix} u \\ v \\ z(u, v) \end{bmatrix}$$

is then called a nonparametric surface. From the above, one immediately finds

$$E = 1 + z_u^2, \quad F = z_u z_v, \quad G = 1 + z_v^2,$$

$$D^2 = EG - F^2 = 1 + z_u^2 + z_v^2,$$

$$\mathbf{n} = \frac{1}{D} \begin{bmatrix} -z_u \\ -z_v \\ 1 \end{bmatrix},$$

$$L = \frac{1}{D} z_{uu}, \quad M = \frac{1}{D} z_{uv}, N = \frac{1}{D} z_{vv},$$

$$K = \frac{1}{D^2} (z_{uu} z_{vv} - z_{uv}^2).$$

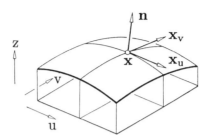

Figure 22.14: Nonparametric surface.

22.12   Composite Surfaces

A surface $\mathbf{x} = \mathbf{x}(u, t)$ with global parameters u and t may be composed of patches or segments of different surfaces. Let $\mathbf{x}_- = \mathbf{x}_-(t)$ denote the right boundary curve and $\mathbf{x}_+ = \mathbf{x}_+(t)$ the left boundary curve of two such patches, connected along $\mathbf{x} = \mathbf{x}(t)$, $t \in [a, b]$, as illustrated in Figure 22.15. Both patches are tangent plane continuous if $\mathbf{n}_- = \pm\mathbf{n}_+$ at all $\mathbf{x}(t)$. This may also be written

$$\alpha\mathbf{x}_{-u} = \beta\mathbf{x}_{+v} + \gamma\dot{\mathbf{x}}, \qquad (22.18)$$

where α, β, γ are functions of t and the product $\alpha\beta$ is nonvanishing.

The two patches are curvature continuous if they are tangent plane continuous and both Dupin's indicatrices agree along the common boundary, in the sense of Remark 11 and as illustrated in Figure 22.16.

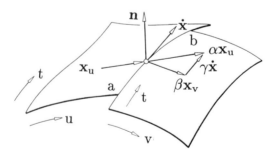

Figure 22.15: Composite surface.

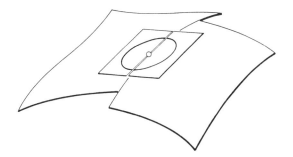

Figure 22.16: Common Dupin's indicatrix.

If the common boundary is C^1, both indicatrices have a pair of points in common which are opposed to each other in the direction of the boundary tangent. Although a conic section is defined by five points (see Section 14.5), or by its midpoint and three nondiametrical points, it can be shown that both Dupin's indicatrices coincide if there exists a family of curvature continuous surface curves crossing the common boundary. In particular, this family may be one of asymptotic lines (Pegna and Wolter [359]) or any family of isolines (Boehm [66]) or even any family of unordered directions (Pegna and Wolter [358]). Moreover, if the boundary is C^0 only at a point, there are two directions of curvature continuity there, and so no further conditions have to be met.

Remark 20: A surface is called tangent or curvature continuous if any plane section is tangent or curvature continuous, respectively.

Remark 21: Note that the asymptotic directions of both patches may coincide even if they are imaginary.

Remark 22: A consequence of (22.18) is the following:

$$\alpha \mathbf{x}_{-ut} = \beta \mathbf{x}_{+ut} + \gamma \ddot{\mathbf{x}} - \dot{\alpha} \mathbf{x}_{-u} + \dot{\beta} \mathbf{x}_{+v} + \dot{\gamma} \dot{\mathbf{x}}.$$

Remark 23: Note that, although Dupin's indicatrix is a euclidean invariant only, curvature continuity of surfaces is an affinely and projectively invariant property.

Chapter 23

Interrogation and Smoothing

We discussed many methods for curve and surface generation. In this chapter, we shall discuss some ways to inspect the geometric quality of those curves and surfaces and develop a few ideas on how to remove shape imperfections.

23.1   Use of Curvature Plots

A spline curve is typically obtained in one of two ways: as a curve that interpolates to given data points, or as the result of interactive manipulation of a B-spline polygon. In both cases, it is hard to tell from the display on the screen if the shape of the curve is acceptable or not: two curves may look identical on the screen, yet reveal significant shape differences when plotted to full scale on a large flatbed plotter. Such plots are both expensive and time-consuming – one needs a tool to analyze curve shape at the CAD terminal.

Such a tool is provided by the *curvature plot* of the curve. For a given curve, we can plot curvature versus arc length. The resulting graph is the curvature plot. We have already used curvature plots in Chapter 9. All three curves from Figures 9.2, 9.4, and 9.6 look very similar, yet their curvature plots reveal substantial differences. The same is true for Figures 9.10 and 9.12. What actually constitutes a "substantial" difference depends on the application at hand, of course.

The curvature of a space curve is nonnegative by definition (11.7).[1] Very often, one is interested in the detection of inflection points of the current

---

[1]See also the problems section at the end of Chapter 15.

planar projection, i.e., the points of inflection of the curve as it appears on
the screen. If we introduce signed curvature by

$$\kappa(u) = \frac{\ddot{x}(u)\dot{y}(u) - \ddot{y}(u)\dot{x}(u)}{[(\dot{x}(u))^2 + (\dot{y}(u))^2]^{3/2}}, \tag{23.1}$$

where x, y are the two components of the curve, it is easy to point to changes
in the sign of curvature, which indicate inflection points. (Those sign changes
can be marked by special symbols on the plot.) Signed curvature is used in
all examples in this book.

We now go one step further and use curvature plots for the definition of
fair curves: *A curve is fair if its curvature plot is continuous and consists of
only a few monotone pieces.*[2] Regions of monotone curvature are separated by
points of extreme curvature. The number of curvature extrema of a fair curve
should thus be small – curvature extrema should only occur where explicitly
desired by the designer, and nowhere else!

This definition of fairness (also suggested by Dill [139], Birkhoff [58], and
Su and Liu [453] in similar form) is certainly subjective; however, it has proven
to be a practical concept. Once a designer has experienced that "flat spots"
on the curve correspond to "almost zero" curvature values and that points
of inflection correspond to crossings of the zero curvature line, he or she will
use curvature plots as an everyday tool.

23.2   Curve and Surface Smoothing

A typical problem in the design process of many objects is that of *digitizing
errors:* data points have been obtained from some digitizing device (a tablet
being the simplest), and a fair curve is sought through them. In many cases,
however, the digitized data are inaccurate, and this presence of digitizing
error manifests itself in a "rough" curvature plot of an interpolating spline
curve.[3]

For a given curvature plot of a C^2 cubic spline, we may now search for the
largest slope discontinuity of $\kappa(s)$ (s being arc length) and try to "fair" the

---

[2]M. Sabin has suggested that "a frequency analysis of the radius of curvature plotted
against arc length might give some measure of fairness – the lower the dominant frequency,
the fairer the curve." Quoted from Forrest [190].

[3]Typically, splines that are obtained from interactive adjustment of control polygons
exhibit rough curvature plots as well.

curve there. Let this largest slope discontinuity occur at $u = u_j$. The slope κ' is given by

$$\frac{d\kappa}{ds} = \frac{\det[\dot{\mathbf{x}}, \dddot{\mathbf{x}}]}{||\dot{\mathbf{x}}||^4} - 3\dot{\mathbf{x}}\ddot{\mathbf{x}}\frac{\det[\dot{\mathbf{x}}, \ddot{\mathbf{x}}]}{||\dot{\mathbf{x}}||^6}, \qquad (23.2)$$

where, as usual, dots denote derivatives with respect to the given parameter u (see Pottmann [378]). Note that this formula applies for 2D curves only.

The data point $\mathbf{x}(u_j)$ is potentially in error; so why not move $\mathbf{x}(u_j)$ to a more favorable position? It seems that a more favorable position should be such that the spline curve through the new data no longer exhibits a slope discontinuity.

We now make the following observation: if a spline curve is three (instead of just two) times differentiable at a point $\mathbf{x}(u_j)$, then certainly its curvature is differentiable at u_j; i.e., it does not have a slope discontinuity there (assuming that the tangent vector does not vanish at u_j). Also, the two cubic segments corresponding to the intervals (u_{j-1}, u_j) and (u_j, u_{j+1}) are now part of *one* cubic segment: the knot u_j is only a pseudo-knot, which could be removed from the knot sequence without changing the curve.

We will thus try to *remove* the "offending knot" u_j from the knot sequence, thereby fairing the curve, and then reinsert the knot in order to keep a spline curve with the same number of intervals as the initial one. We discussed knot insertion in Chapter 10. The inverse process, *knot removal*, has no unique solution. Several possibilities are discussed in Farin *et al.* [169], Sapidis [414] and Farin and Sapidis [170], [415]. We present here a simple yet effective solution to the knot removal problem. Let the offending knot u_j be associated with the vertex $\mathbf{d}_j$.[4] We now formulate our knot removal problem: to what position $\hat{\mathbf{d}}_j$ must we move $\mathbf{d}_j$ such that the resulting new curve becomes three times differentiable at u_j? After some calculation (equating the left and the right third derivative of the new spline curve), one verifies that the new vertex $\hat{\mathbf{d}}_j$ is given by

$$\hat{\mathbf{d}}_j = \frac{(u_{j+2} - u_j)\mathbf{l}_j + (u_j - u_{j-2})\mathbf{r}_j}{u_{j+2} - u_{j-2}},$$

where the auxiliary points $\mathbf{l}_j, \mathbf{r}_j$ are given by

$$\mathbf{l}_j = \frac{(u_{j+1} - u_{j-3})\mathbf{d}_{j-1} - (u_{j+1} - u_j)\mathbf{d}_{j-2}}{u_j - u_{j-3}}$$

---

[4]This uses the notation from Chapter 7.

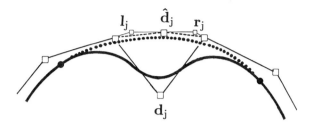

Figure 23.1: Knot removal: if $\mathbf{d}_j$ is moved to $\hat{\mathbf{d}}_j$, the new curve is three times differentiable at u_j.

and

$$\mathbf{r}_j = \frac{(u_{j+3} - u_{j-1})\mathbf{d}_{j+1} - (u_j - u_{j-1})\mathbf{d}_{j+2}}{u_{j+3} - u_j}.$$

The geometry underlying these equations is illustrated in Figure 23.1.

The faired curve now differs from the old curve between $\mathbf{x}(u_{j-2})$ and $\mathbf{x}(u_{j+2})$ – thus this fairing procedure is *local*.

Figures 23.2 and 23.3 illustrate an application of this algorithm, although it is not used locally, but for all knots. Note that the initial and the smoothed curves look almost identical, and only their curvature plots reveal significant shape differences. The initial curve has an inflection point, which is not visible without the use of curvature plots. The faired curve does not have this shape defect any more.

In practice, the improved vertex $\hat{\mathbf{d}}_j$ may be further away from the original vertex $\mathbf{d}_j$ than a prescribed tolerance allows. In that case, one restricts a realistic $\hat{\mathbf{d}}_j$ to be in the direction toward the optimal $\hat{\mathbf{d}}_j$, but within tolerance to the old $\mathbf{d}_j$.

Other methods for curve fairing exist. We mention Kjellander's method [288], which moves a data point to a more favorable location and then interpolates the changed data set with a C^2 cubic spline. This method is global. A method that fairs only data points, not spline curves, is presented by Renz [393]. This method computes second divided differences, smoothes them, and "integrates" back up. Methods that aim at the smoothing of single Bézier curves are discussed by Hoschek [267], [269]. Variations on the described method are given in [170]. A method that tries to reduce the degrees of each cubic segment to quadratic is given in [164].

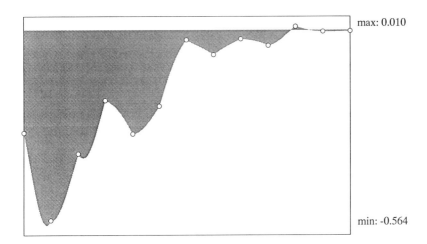

max: 0.010

min: -0.564

Figure 23.2: Curve smoothing: an initial curve and B-spline polygon with its curvature plot.

23.3 Surface Interrogation

Curvature plots are useful for curves; it is reasonable, therefore, to investigate the analogous concepts for surfaces. Several authors have done this, including Beck *et al.* [43], Farouki [174], Dill [139], Munchmeyer [341], [340], and Forrest [193]. An interesting early example is on page 197 of Hilbert and Cohn-Vossen [258]. Surfaces have two major kinds of curvature: Gaussian and mean; see Section 22.6. Both kinds can be used for the detection of surface imperfections. Another type of curvature can be useful, too: *absolute curvature* κ_{abs}. It is defined by

$$\kappa_{\mathrm{abs}} = |\kappa_1| + |\kappa_2|,$$

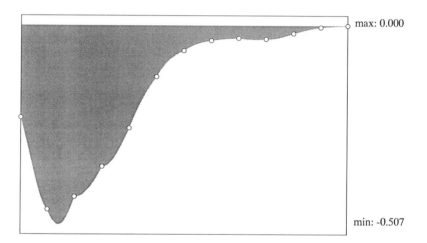

max: 0.000

min: -0.507

Figure 23.3: Curve smoothing: the smoothed curve and its curvature plot.

where κ_1 and κ_2 are the maximal and minimal normal curvatures at the point under consideration.

Gaussian curvature does not offer much information about generalized cylinders of the form

$$\mathbf{c}(u,v) = (1-u)\mathbf{x}(v) + u[\mathbf{x}(v) + \mathbf{v}].$$

Even if the generating curve $\mathbf{x}(v)$ is highly curved, we still have $K \equiv 0$ for these surfaces. A similar statement can be made about the mean curvature H, which is always zero for minimal surfaces, no matter how complicated.

All three kinds of curvatures are shown in Plate V. Note how the surface (a wire frame rendering) looks perfectly flat, yet the curvatures detect many concave and convex regions.[5]

---

[5]Plates V, VI, VII, and VIII are taken from L. Fayard's master's thesis [181].

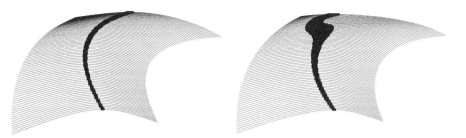

Figure 23.4: Reflection lines: left, a surface with a "perfect" reflection line; right, after a perturbation was applied to the surface.

Another example is shown in Plate VI: both Gaussian and mean curvature detect regions of uneven curvature where the wire frame rendering (and also absolute curvature) indicate nothing.

Three points of the surface from Plate VI were then perturbed by less than 0.3 mm (the surface itself having a side length of ≈ 800 mm). Again, Gaussian and mean curvatures detect these irregularities easily, while absolute curvature does not.

Plate VII shows the ability of curvature interrogation to detect the effect of a bad choice of twist vectors in surface generation: Gaussian curvature is displayed for the choice of C^2 twists (upper left), Adini's twist (upper right), FMILL twist (lower left), and zero twists (lower right). While the first three methods yield similar plots, the zero twist clearly gives rise to an imperfect surface.

Another method for surface interrogation is the use of *reflection lines*, first described in Klass [290] and later by Poeschl [377]. Reflection lines are a standard surface interrogation tool in the styling shop of a car manufacturer. They are the pattern that is formed on the polished car surface by the mirror images of a number of parallel fluorescent strip lights. If the mirror images are "nice," then the corresponding surface is deemed acceptable. Reflection lights can easily be simulated on a raster graphics device (mark points whose normal points to one of the light sources). With some more effort, they can also be computed on a line drawing device (see Klass [290]).

Figure 23.4 shows a surface with one reflection line and the effect that a small perturbation can have on that reflection line.

Reflection lines and curvature "paintings" have different usages: reflection lines are not as fine-tuned as curvatures; they are prone to miss local shape defects of a surface.[6] On the other hand, curvatures of a surface may look perfect, yet it might not have a "pleasant" overall shape – reflection lines have a better chance of flagging global imperfections.

Once imperfections are detected in a tensor product B-spline surface, one would want methods to remove them without time-consuming interactive adjustment of control polygons. One such method is to apply the curve smoothing method from Section 23.2 in a tensor product way: smooth all control net rows, then all control net columns. The resulting surface is usually smoother than the original surface. Figure 23.4 is an example of this method: the right figure is a B-spline surface; four iterations of the program `fair_surf` produced the figure on the left.

More involved methods for surface fairing exist; they aim for the enforcement of convexity constraints in tensor product spline surfaces. We mention Andersson *et al.* [7], Jones [282], and Kaufmann and Klass [287].

23.4   Implementation

The routine `curvatures` may be used to generate curvature values of a rational Bézier curve. It writes the values into a file that might be read by another program that generates a curvature plot.

To compute the curvature at the parameter value t, the curve is subdivided using the (rational) de Casteljau algorithm. Of the two subpolygons that are generated, the larger one is selected, and its beginning curvature is computed. Since the subdivision routine `rat_subdiv` orders both subpolygons beginning at the subdivision point, only one curvature routine `curvature_0` is needed.

```
void curvatures(coeffx,coeffy,weight,degree,dense)
/*     writes  signed  curvatures of  a rational  Bezier curve into
       a file.
input:
    coeffx, coeffy:   2D Bezier polygon
    weight:           the weights
    degree:           the degree
    dense:            how many curvature values to compute
```

---

[6] This is because reflection lines may be viewed as a first-order interrogation tool (involving only first derivatives), while curvature plots are second-order interrogation tools.

```
output:
                        written into file outfile
*/
```

The routine `curvature_0` is a simple application of Eq. (11.10):

```
float curvature_0(bez_x,bez_y,weight,degree)
/* computes curvature of rational  Bezier curve at t=0
   Input: bez_x, bez_y, weight: control polygon and weights
          degree:              degree of curve
*/
```

The following **area** routine is included for completeness:

```
float area(p1,p2,p3)
/* find area of triangle p1,p2,p3 */
```

Note, however, that **area** returns a negative value if the input points are ordered clockwise!

The following routine generates the "raw data" that are needed to create the curvature plot of a rational B-spline curve. Of course, one may simply set all weights to unity for the polynomial case.

```
void bspl_kappas(bspl_x,bspl_y,bspl_w,knot,l,dense)
/*      writes curvatures of cubic rational  B-spline curve into
        a file.
  input:
        bspl_x,bspl_y:  2D rat. B-spline polygon
        bspl_w:         the  B-spline weights
        knot:           the knot sequence
        dense:          how many curvature values to compute per interval
        l:              no. of intervals
  output:
                        written into file outfile
*/
```

The preceding programs are utilized by the main program `plot_b_kappa.c` in order to produce postscript output for a curvature plot.

Now the programs to fair curves and surfaces: first, the curve case:

```
void fair_bspline(bspl,knot,l,from,to)
/* Fairs a cubic rational B-spline curve by knot removal/reinsertion.

Input: bspl:      cubic B-spline control polygon (one coord.)
knot:       knot sequence
l:          no. of intervals
from, to:   from where to where to fair
Output: same as input, but hopefully fairer.
*/
```

Second, the surface case:

```
void fair_surf(bspl,lu,lv,knot_u, knot_v)
/*  Fairs B-spline control net.
Input: bspl:   B-spline control net (one coordinate only)
lu,lv:  no. of intervals in u- and v-direction
knots_u, knots_v:   knot vectors in u- and v-direction
Output: as input

Note:   Has to be called once for each x-,y-,z- coordinate.
```

23.5   Problems

1. Write a program to compute the torsion of a Bézier or a spline curve. Then produce torsion plots as an additional interrogation tool for space curves.

2. The routine curvatures produces a file that contains pairs t_i, κ_i, i.e., it can be used to plot curvature versus parameter. Modify the program so it can be used to produce plots of arc length versus parameter.

3. Show that a planar cubic curve may have two points of inflection, i.e., points where curvature changes sign.

Chapter 24

Evaluation of Some Methods

In this chapter, we will examine some of the many methods that have been presented. We will try to point out the relative strengths and weaknesses of each, a task that is necessarily influenced by personal experience and opinion.

24.1 Bézier Curves or B-spline Curves?

Taken at face value, this is a meaningless question: Bézier curves are a special case of B-spline curves. Any system that contains B-splines in their full generality, allowing for multiple knots, is capable of representing a Bézier curve or, for that matter, a piecewise Bézier curve.

In fact, several systems use both concepts simultaneously. A curve may be stored as an array holding B-spline control vertices, knots, and knot multiplicities. For evaluation purposes, the curve may then be converted to the piecewise Bézier form.

24.2 Spline Curves or B-spline Curves?

This question is often asked, yet it does not make much sense. B-splines form a basis for all splines, so any spline curve can be written as a B-spline curve.

What is often meant is the following: if we want to design a curve, should we pass an interpolating spline curve through data points, or should we design a curve by interactively manipulating a B-spline polygon? Now the question has become one concerning curve *generation* methods, rather than curve *representation* methods.

A flexible system should have both: interpolation and interactive manipulation. The interpolation process may of course be formulated in terms of B-splines. Since many designers do not favor interactive manipulation of control polygons, one should allow them to generate curves using interpolation. Subsequent curve modification may also take place without display of a control polygon: for instance, the designer might move one (interpolation) point to a new position. The system could then compute the B-spline polygon modification that would produce exactly that effect. So a user might actually work with a B-spline package, but a system that is adapted to his or her needs might hide that fact. See "inverse design" in Section 7.9 for details.

We finally note that every C^2 B-spline curve may be generated as an interpolating spline curve: read off junction points, end tangent vectors, and knot sequence from the B-spline curve. Feed these data into a C^2 cubic spline interpolator, and the original curve will be regenerated.

24.3   The Monomial or the Bézier Form?

We have made the point in this book that the monomial form[1] is *less geometric* than the Bézier form for a polynomial curve. A software developer, however, might not care much about the beauty of geometric ideas – in the workplace, the main priority is performance. Since the fundamental work by Farouki and Rajan [177], [178], [175], one important performance issue has been resolved: the Bézier form is *numerically more stable* than the monomial form. Farouki and Rajan observed that numerical inaccuracies, unavoidable with the use of finite precision computers, affect curves in the monomial form significantly more than those in Bézier form. More precisely, they show that the condition number of simple roots of a polynomial[2] is smaller in the Bernstein basis than in the monomial basis. If one decides to use the Bézier form for stability reasons, then it is essential that no conversions be made to other representations; these will destroy the accuracy gained by the use of the Bézier form. For example, it is not advisable from a stability viewpoint to store data in the monomial form and to convert to Bézier form to perform certain operations. More details are given in Daniel and Daubisse [109].

---

[1]This form is also called the *power basis form*.

[2]This number indicates by how much the location of a root is perturbed as a result of a perturbation of the coefficients of the given polynomial.

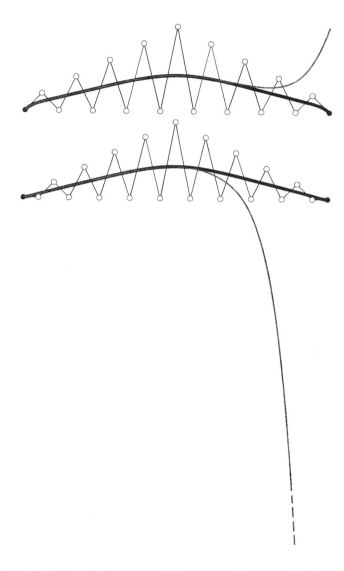

Figure 24.1: Stability of the monomial form: slight perturbations in the co-efficients affect the monomial form (gray) much more than the Bézier form (black). Top: a degree 18 curve; bottom: a degree 20 curve.

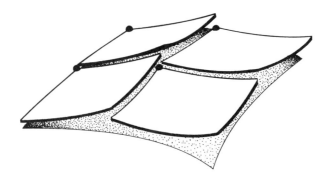

Figure 24.2: A piecewise monomial surface: the patches miss the points (1,1) due to roundoff.

Figure 24.1 shows a numerical example, carried out using the routine bez_to_pow with single precision arithmetic. A degree 18 Bézier curve (top) was converted to the monomial form. Then, the coefficients of the Bézier and of the monomial form were perturbed by a random error, less than 0.001 in each coordinate. (The x-values of the control polygon extend from 0 to 20.) The Bézier curve shows no visual sign of perturbation, whereas the monomial form is not very reliable near $t = 1$. The experiment was then repeated for a degree 20 curve (bottom) with even more disastrous results. While degrees like 18 or 20 look high, one should not forget that such degrees already appear in harmless-looking tensor product surfaces: for example, the diagonal $u = v$ of a patch with $n = m = 9$ is of degree 18!

As a consequence of its numerical instability, the monomial form is not very reliable for the representation of curves or surfaces. For the case of surfaces, Figure 24.2 gives an illustration (in somewhat exaggerated form). Since the monomial form is essentially a Taylor expansion around the local coordinate (0,0) of each patch, it is quite close to the intended surface there. Further away from (0,0), however, roundoff takes its toll. The point (1,1) is *computed* and therefore missed. For an adjacent patch, the actual point is *stored* as a patch corner, thus giving rise to the discontinuities shown in Figure 24.2. The significance of this phenomenon increases dramatically when curves or surfaces of high degrees are used.

One should not forget to mention the main attraction of the monomial form: speed. Horner's method is faster than the de Casteljau algorithm; it is also faster than the routine **hornbez**. There is a trade-off therefore between stability and speed. (Given modern hardware, things are not quite that clear-cut, however: T. DeRose and T. Holman [138] have developed a multiprocessor architecture that hardwires the de Casteljau algorithm into a network of processors and now outperforms Horner's method.)

24.4 The B-spline or the Hermite Form?

Cubic B-spline curves are numerically more stable than curves in the piecewise cubic Hermite form. This comes as no surprise, since some of the Hermite basis functions are negative, giving rise to numerically unstable nonconvex combinations. However, there is an argument in favor of the piecewise Hermite form: it stores interpolation points explicitly. In the B-spline form, they must be computed. Even if this computation is stable, it may produce roundoff.

As another argument in favor of the Hermite form, one might add that end conditions for C^2 spline interpolation are more easily formulated in the Hermite form than in the B-spline form.

A significant argument against the use of the Hermite form points to its lack of invariance under affine parameter transformations. Everyone who has programmed the Hermite form has probably experienced the trauma resulting from miscalculated tangent lengths. A programmer should not be burdened with the subtleties of the interplay between tangent lengths and parameter interval lengths.

An important advantage of the B-spline form is *storage*. For B-spline curves, one needs a control point array of length $L + 2$ plus a knot vector array of length $L + 1$ for a curve with L data points, resulting in an overall storage requirement of $4L + 7$ reals.[3] For the piecewise Hermite form, one needs a data array of length $2L$ (position and tangent at each data point) plus a knot vector of length $L + 1$, resulting in a storage requirement of $7L + 1$ reals. For surfaces (with same degrees in u and v for simplicity), the discrepancy becomes even larger: $3(L + 2)^2 + 2(L + 1)$ *vs* $12L^2 + 2(L + 1)$ reals. (For the Hermite form, we have to store position, u- and v-tangents, and twist for each data point.)

---

[3]We are not storing knot multiplicities. We would then be able to represent curves that are only C^0, which the cubic Hermite form is not capable of.

When both forms are used for tensor product interpolation, the Hermite form must solve three sets of linear equations (see Section 17.6) while the B-spline form must solve only two sets (see Section 17.4).

24.5   Triangular or Rectangular Patches?

Most of the early CAD efforts were developed in the car industry, and this is perhaps the main reason for the predominance of rectangular patches in most CAD systems; the first applications of CAD methods to car body design were to the outer panels such as roof, doors, hood, etc. These parts basically have a rectangular geometry, hence it is natural to break them down into smaller rectangles. These smaller rectangles were then represented by rectangular patches.

Once a CAD system had been successfully applied to a design problem, it seemed natural to extend its use to other tasks: the design of the interior car body panels, for instance. Such structures do not possess a notably rectangular structure, and rectangular patches are therefore not a natural choice for modeling these complicated geometries. However, rectangle-based schemes already existed, and the obvious approach was to make them work in "unnatural" situations also. They do the job, although with some difficulties, which arise mainly in the case of degenerate rectangular patches.

Triangular patches do not suffer from such degeneracies and are thus better suited to describe complex geometries than are rectangular patches. It seems obvious, therefore, to advise any CAD system developer to add triangular patches to the system.

There is a catch: the very addition of a new patch type to an existing system is a formidable task in itself. This new patch type must be interfaced with every existing function that the system offers: a routine must be written for triangular patch/plane intersection, for the offset surface of a triangular patch, and so on. Adding the nice features of triangular patches to an existing system has its price, both in the development of new code and in its subsequent maintenance.

A different situation arises if a completely new system is to be developed. The extra cost for the inclusion of triangular patches into an emerging system is not nearly as high as that for the addition to an existing system. Such a new system would most likely profit enough from the additional flexibility offered by triangular patches to justify a higher implementation cost. Similar arguments hold for other non-four-sided patch types, as described in Sabin [404], Sabin and Kimura [405], Hosaka and Kimura [266], Gregory [234], and Varady [461]. A method to convert (exactly) a triangular patch into three rational rectangular patches is described in [276].

We should add that a development of rectangular patches has blurred the clear distinction between patch types: This is the trimmed surface as discussed in Section 17.10. It can mimic multisided patches and seems to be a reasonable way to upgrade a rectangular surface system toward more versatility.

Chapter 25

Quick Reference of Curve and Surface Terms

[1] **ab initio design**   Latin: from the beginning. Used to describe design processes in which the designer inputs his or her ideas directly into the computer, without constraints such as →interpolatory constraints.

Affine combination   Same as a →barycentric combination.

Affine invariance   A property of a curve or surface generation scheme: it does not matter if computation of a point on a curve or surface occurs before or after an affine map is applied to the input data.

Affine map   Any map that is composed of translations, rotations, scalings, and shears. Maps parallels to parallels. Leaves ratios of →collinear points unchanged.

Approximation   Fitting a curve or surface to given data. As opposed to →interpolation, the curve or surface approximation only has to be close to data.

Barycentric combination   A weighted average where the sum of the weights equals one.

Barycentric coordinates   A point in $I\!E^2$ may be written as a unique →barycentric combination of three points. The coefficients in this combination are its barycentric coordinates.

Basis function   Functions form linear spaces, which have bases. The elements of these bases are the basis functions.

Bernstein polynomial   The basis functions for →Bézier curves.

---

[1] Arrows refer to terms that are defined elsewhere in this glossary.

Beta-spline curve   A $\rightarrow G^2$ piecewise cubic curve that is defined over a uniform knot sequence.

Bézier curve   A polynomial curve that is expressed in terms of $\rightarrow$Bernstein polynomials.

Bézier polygon   The coefficients in the expansion of a $\rightarrow$Bézier curve in terms of $\rightarrow$Bernstein polynomials are $\rightarrow$ points. Connected according to their natural numbering, they form the Bézier polygon.

Bilinear patch   A patch that is $\rightarrow$ruled in two directions. Or: a hyperbolic paraboloid.

Blossom   A multivariate polynomial that is associated with a given polynomial through the process of $\rightarrow$blossoming.

Blossoming   The procedure of applying n (the polynomial degree) $\rightarrow$de Casteljau algorithm steps or several de Boor algorithm steps to a polynomial (or to a segment of a spline curve), but each one for a different parameter value.

B-spline   A piecewise polynomial function. It is defined over a $\rightarrow$knot partition, has $\rightarrow$local support, and is nonnegative. If a $\rightarrow$spline curve is expressed in terms of B-splines, it is called a B-spline curve.

B-spline polygon   The coefficients in the expansion of a $\rightarrow$B-spline curve in terms of $\rightarrow$B-splines are $\rightarrow$points. Connected according to their natural numbering, they form the B-spline polygon.

Breakpoint   Same as a $\rightarrow$knot.

CAGD   Computer-aided geometric design.

Collinear   Being on a straight line.

Compatibility   For some interpolation problems, the input data may not be arbitrary but must satisfy some consistency constraints, called compatibility conditions.

Conic section   The intersection curve between a cone and a plane. Or: the projective image of a parabola. A nondegenerate conic is either an ellipse, a parabola, or a hyperbola.

Control polygon   See $\rightarrow$Bézier polygon or $\rightarrow$B-spline polygon.

Convex curve   A planar curve that is a subset of the boundary of its convex hull.

Convex hull   The smallest convex set that contains a given set.

Convex set   A set such that the straight line connecting any two points of the set is completely contained within the set.

Coons patch   A →patch that is fitted between four arbitrary boundary curves.

Cross plot   Breaking down the plot of a parametric curve into the plots of each coordinate function.

Curve   The path of a point moving through space. Or: the image of the real line under a continuous map.

Curvature   At a point on a curve, curvature is the inverse of the radius of the →osculating circle. Also: curvature measures by how much a curve deviates from a straight line at a given point.

de Boor algorithm   The algorithm that recursively computes a point on a →B-spline curve.

de Casteljau algorithm   The algorithm that recursively computes a point on a →Bézier curve.

Direct G^2 splines G^2 piecewise cubics that are generated by specifying a control polygon and some Bézier points.

Domain   The preimage of a curve or surface.

End condition   In cubic →spline curve interpolation, one can supply an extra condition at each of the two endpoints. Examples of such end conditions: prescribed tangents or zero curvature.

Frenet frame   At each point of a (nondegenerate) curve, the first, second, and third derivative vectors are linearly independent. Applying Gram-Schmidt orthonormalization to them yields the Frenet frame of the curve at the given point.

Functional curve or surface: A curve of the form $y = f(x)$ or a surface of the form $z = f(x, y)$.

G^2 spline curve   A C^1 piecewise cubic curve that is twice differentiable with respect to arc length.

γ-spline   A $\rightarrow G^2$ spline that is C^1 over a given knot sequence.

Geometric continuity   Smoothness properties of a curve or a surface that are more general than its order of differentiability.

Gordon surface   A generalization of → Coons patches. Interpolates to a rectilinear network of curves.

Hermite interpolation   Generating a curve or surface from data that consist of points and first and/or second derivatives.

Hodograph   The first derivative curve of a parametric curve.

Horner's scheme   An efficient method to evaluate a polynomial in → monomial form by nested multiplication.

Inflection point   A point on a curve where the tangent intersects the curve. Often corresponds to points with zero curvature.

Interior Bézier points   For curves, those Bézier points that are not →junction points. For surfaces, those Bézier points that are not boundary points.

Interpolation   Finding a curve or surface that satisfies some imposed constraints exactly. The most common constraint is the requirement of passing through a set of given points.

Junction point   A →spline curve is composed of →segments. The common point shared by two segments is called the junction point. See also →knot.

Knot   A →spline curve is defined over a partition of an interval of the real line. The points that define the partition are called knots. If evaluated at a knot, the spline curve passes through a →junction point.

Knot insertion   Adding a new knot to the knot sequence of a →B-spline curve without changing the graph of the curve.

Lagrange interpolation   Finding a polynomial curve through a given set of data points.

Linear precision   A property of many curve schemes: if the curve generation scheme is applied to equidistant points on a straight line, that straight line is reproduced.

Local control   A curve or surface scheme has the local control property if a change in the input data only changes the curve or surface in a region near the changed data.

Lofting   Creating a →ruled surface between two given curves.

Minmax box   Smallest 2D or 3D box with edges parallel to the coordinate axes that completely contains a given object.

Monomial form   A polynomial is in monomial form if it is expressed in terms of the monomials $1, t, t^2, \ldots$.

Node   A term that is used inconsistently in the literature: it sometimes refers to a knot, sometimes to a control vertex.

NURB   Nonuniform rational B-spline curve.

ν-spline   An →interpolating → G^2 spline curve that is C^1 over a given knot sequence.

Osculating circle   At a given point, the osculating circle approximates the curve "better" than any other circle.

Osculating plane   The plane that contains the →osculating circle of a curve at a given point.

Oslo algorithm   The process of simultaneously inserting several →knots into a →B-spline curve.

Parametrization   Assigning parameter values to →junction points in →spline curves. Also used with a different meaning: the function that describes the speed of a point traversing a curve.

Patch   Complicated →surfaces are usually broken down into smaller units, called patches. For example, a bicubic spline surface consists of a collection of bicubic patches.

Point   A location in →space. If one uses coordinate systems to describe space, a point is represented as an n-tuple of real numbers.

Precision   A curve or surface generation scheme has n^{th} order precision if it reproduces polynomials of degree n.

Projective map   A map that is composed of →affine maps and central projections. Leaves cross ratios of →collinear points unchanged. Does (in general) not map parallels to parallels.

Quadric   A surface with the implicit representation $f(x,y,z) = 0$, where f is a quadratic polynomial. Or: the projective image of an elliptic paraboloid, a hyperbolic paraboloid, or a parabolic cylinder.

Rational curves and surfaces   Projections of nonrational (integral) curves or surfaces from four-space into three-space.

Ruled surface   A surface containing a family of straight lines. Obtained as linear interpolation between two given curves.

Segment   An individual polynomial (or rational polynomial) curve in an assembly of such curves to form a →spline curve. The bivariate analog of a segment is a →patch.

Shape parameter   A degree of freedom (usually one or several real numbers) in a curve or surface representation that can be used to fine-tune the shape of that curve or surface.

Solid Modeling   The description of objects that are bounded by surfaces.

Space   The collection of all →points.

Spline curve   A continuous curve that is composed of several polynomial →segments. Spline curves are often represented in terms of →B-spline functions. They may be the result of an →interpolation process or of an →ab initio design process. If the segments are rational polynomials, we have a rational spline curve.

Subdivision   Breaking a curve or surface down into smaller pieces, usually recursively.

Support   The region over which a function is nonzero.

Surface   The locus of all points of a moving and deforming →curve. Or: the image of a region in two-space under a continuous map. A surface is often broken down into →patches.

Tangent   The straight line that best approximates a smooth curve at a point on it. This straight line is parallel to the →tangent vector.

Tangent vector   The first derivative of a differentiable curve at a point on it. The length of the tangent vector depends on the →parametrization of the curve.

Tensor product   A method to generate rectangular surfaces using curve methods.

Torsion   A measure of how much a curve "curves away" from the →osculating plane at a given point.

Transfinite interpolation   Interpolating to curves, with infinitely, i.e., transfinitely many points on it, as opposed to discrete interpolation, which only interpolates to finitely many points.

Translational surface   A surface that is obtained by sweeping one curve along another one.

Triangular patch   A →patch whose →domain is a triangle.

Trimmed surface   If the domain of a parametric surface is divided into "valid" and "invalid" regions, the image of the valid regions is called a trimmed surface.

Twist vector   The mixed second partial of a surface at a point. Note: not a geometric property of the surface, but parametrization dependent.

Variation diminution   Intuitively: a curve or surface scheme has this property if its output "wiggles less" than the data from which it is constructed.

Vector   A direction. Usually the difference of two →points.

Volume deformation   A surface or a collection of surfaces may be embedded in a cube. That cube may then be deformed using some trivariate Bézier or B-spline method – this is the volume distortion – in order to change the shape of the initial surface(s).

Appendix 1

List of Programs

The following list contains all programs that are contained in the text. Some more are included in the diskette; they are auxiliary routines.

Some of the programs generate postscript output. They want scaling factors `scale_x, scale_y`. The use of these factors is demonstrated in the main program `decasmain.c`. Other routines want them as input; they have then to be computed from min-max boxes in the calling program.

The diskette also contains several `*.dat` files. These are sample input files. The `*.ps` files contain postscript code and were used to generate some of the figures in the text.

aitken:	Section 6.8.
area:	Section 23.4.
bessel_ends:	Section 9.6.
bez_to_points:	Section 3.5.
bezier_to_power:	Section 4.9.
bspline_to_bezier:	Section 7.10.
bspl_to_points:	Section 10.11.
c2_spline:	Section 9.6.
check_flat:	Section 4.9.
curvature_0:	Section 23.4.
curvatures:	Section 23.4.
deboor:	Section 10.11.
deboormain:	Section 10.11.
decas:	Section 3.5.
decasmain:	Section 3.5.

Appendix 2

Notation

The following notation is used in this book:

$\wedge$	cross product
$\dot{}$, $\ddot{}$	curve derivatives with respect to the current parameter
$'$, $''$	curve derivatives with respect to arc length
a, b, α, β	real numbers or real-valued functions
$\mathbf{0}$	short for $(0,0,0)$
$\mathbf{a}, \mathbf{b}$	points or vectors (possibly in terms of barycentric coordinates)
$\mathbf{b}[x_1, \ldots, x_n]$	blossom
$t^{<i>}, \mathbf{u}^{<i>}$	i-fold repetition of t or $\mathbf{u}$
A, B	matrices
$\mathbf{A}, \mathbf{B}$	matrices whose elements are points ("hypermatrices")
$\mathbf{b}_i^r, \mathbf{b}_\mathbf{i}^r$	intermediate points in the de Casteljau algorithm
$B_i^n, B_\mathbf{i}^n$	Bernstein polynomials of degree n
$\mathbf{e1}, \mathbf{e2}, \mathbf{e3}$	short for $(1,0,0)$, $(0,1,0)$, and $(0,0,1)$
$\mathbb{E}^d$	d-dimensional euclidean space
$D_\mathbf{d} f$	directional derivative of f in the direction $\mathbf{d}$
Δ_i	difference in parameter intervals (i.e., $\Delta_i = u_{i+1} - u_i$)
Δ^r	iterated forward difference
H_i^3	cubic Hermite polynomials
$\mathbf{P}$	control polygon
Φ	an affine map
$\mathcal{P}_i$	operators
$||\mathbf{v}||$	(euclidean) length of the vector $\mathbf{v}$
$\mathbf{x}_u$	u-partial of $\mathbf{x}(u,v)$

Bibliography

[1] S. Abi-Ezzi. *The Graphical Processing of B-splines in a Highly Dynamic Environment.* PhD thesis, RPI, 1989. Rensselaer Design Research Center.

[2] T. Ackland. On osculatory interpolation, where the given values of the function are at unequal intervals. *J Inst. Actuar.*, 49:369–375, 1915.

[3] J. Ahlberg, E. Nilson, and J. Walsh. *The Theory of Splines and their Applications.* Academic Press, 1967.

[4] H. Akima. A new method of interpolation and smooth curve fitting based on local procedures. *J ACM*, 17(4):589–602, 1970.

[5] P. Alfeld. A bivariate C^2 Clough-Tocher scheme. *Computer Aided Geometric Design*, 1(3):257–267, 1984.

[6] P. Alfeld and L. Schumaker. The dimension of bivariate spline spaces of smoothness r and degree $d \geq 4r + 1$. *Constructive Approximation*, 3:189–197, 1987.

[7] R. Andersson, E. Andersson, M. Boman, B. Dahlberg, T. Elmroth, and B. Johansson. The automatic generation of convex surfaces. In R. Martin, editor, *The Mathematics of Surfaces II*, pages 427–446, Oxford University Press, 1987.

[8] G. Aumann. Interpolation with developable Bezier patches. *Computer Aided Geometric Design*, 8(5):409–420, 1991.

[9] A. Ball. Consurf I: introduction of the conic lofting tile. *Computer Aided Design*, 6(4):243–249, 1974.

[10] A. Ball. Consurf II: description of the algorithms. *Computer Aided Design*, 7(4):237–242, 1975.

[11] A. Ball. Consurf III: how the program is used. *Computer Aided Design*, 9(1):9–12, 1977.

[12] A. Ball. Reparametrization and its application in computer-aided geometric design. *Int. J for Numer. Methods in Eng.*, 20:197–216, 1984.

[13] A. Ball. The parametric representation of curves and surfaces using rational polynomial functions. In R. Martin, editor, *The Mathematics of Surfaces II*, pages 39–62, Oxford University Press, 1987.

[14] R. Barnhill. Smooth interpolation over triangles. In R. Barnhill and R. Riesenfeld, editors, *Computer Aided Geometric Design*, pages 45–70, Academic Press, 1974.

[15] R. Barnhill. Representation and approximation of surfaces. In J. R. Rice, editor, *Mathematical Software III*, pages 69–120, Academic Press, 1977.

[16] R. Barnhill. Coons' patches. *Computers in Industry*, 3:37–43, 1982.

[17] R. Barnhill. A survey of the representation and design of surfaces. *IEEE Computer Graphics and Applications*, 3:9–16, 1983.

[18] R. Barnhill. Surfaces in computer aided geometric design: a survey with new results. *Computer Aided Geometric Design*, 2(1-3):1–17, 1985.

[19] R. Barnhill, editor. *Geometry Processing*. SIAM, Philadelphia, 1992.

[20] R. Barnhill, G. Birkhoff, and W. Gordon. Smooth interpolation in triangles. *J Approx. Theory*, 8(2):114–128, 1973.

[21] R. Barnhill, W. Boehm, and J. Hoschek, editors. *Surfaces in CAGD '89*. North Holland, Amsterdam, 1990.

[22] R. Barnhill, J. Brown, and I. Klucewicz. A new twist in CAGD. *Computer Graphics and Image Processing*, 8(1):78–91, 1978.

[23] R. Barnhill and G. Farin. C^1 quintic interpolation over triangles: two explicit representations. *Int. J Numer. Methods in Eng.*, 17:1763–1778, 1981.

[24] R. Barnhill, G. Farin, M. Jordan, and B. Piper. Surface/surface intersection. *Computer Aided Geometric Design*, 4(1-2):3–16, 1987.

[25] R. Barnhill and J. Gregory. Compatible smooth interpolation in triangles. *J of Approx. Theory*, 15(3):214–225, 1975.

[26] R. Barnhill and J. Gregory. Polynomial interpolation to boundary data on triangles. *Math. of Computation*, 29(131):726–735, 1975.

[27] R. Barnhill and R. F. Riesenfeld, editors. *Computer Aided Geometric Design*. Academic Press, 1974.

[28] M. Barnsley. *Fractals Everywhere*. Academic Press, 1988.

[29] P. Barry. de Boor-Fix functionals and polar forms. *Computer Aided Geometric Design*, 7(5):425–430, 1990.

[30] P. Barry and R. Goldman. De Casteljau-type subdivision is peculiar to Bézier curves. *Computer Aided Design*, 20(3):114–116, 1988.

[31] P. Barry and R. Goldman. A recursive proof of a B-spline identity for degree elevation. *Computer Aided Geometric Design*, 5(2):173–175, 1988.

[32] B. Barsky. *The Beta-spline: A Local Representation Based on Shape Parameters and Fundamental Geometric Measures*. PhD thesis, Dept. of Computer Science, U. of Utah, 1981.

[33] B. Barsky. Exponential and polynomial methods for applying tension to an interpolating spline curve. *Computer Vision, Graphics and Image Processing*, 27:1–18, 1984.

[34] B. Barsky. *Computer Graphics and Geometric Modeling Using Beta-splines*. Springer Verlag, 1988.

[35] B. Barsky. Introducing the rational beta-spline. In *Proceedings of the Third Intnl. Conf. on Engineering Graphics and Descriptive Geometry*, 1988. Vienna, Austria.

[36] B. Barsky and J. Beatty. *Varying the Betas in Beta-splines*. Technical Report CS-82-49, U. of Waterloo, Waterloo, Ontario, Canada N31 3G1, December 1982.

[37] B. Barsky and T. DeRose. *Geometric continuity of parametric curves*. Technical Report UCB/CSB 84/205, Computer Science Division, U. of California, Berkley, 1984.

[38] B. Barsky and T. DeRose. The beta2-spline: a special case of the beta-spline curve and surface representation. *IEEE Computer Graphics and Applications*, 5(9):46–58, 1985.

[39] B. Barsky and D. Greenberg. Determining a set of B-spline control vertices to generate an interpolating surface. *Computer Graphics and Image Processing*, 14(3):203–2226, 1980.

[40] B. Barsky and S. Thomas. TRANSPLINE – a system for representing curves using transformations among four spline formulations. *The Computer J*, 24(3):271–277, 1981.

[41] R. Bartels and J. Beatty. *Beta-splines with a difference*. Technical Report CS-83-40, Computer Science Department, U. of Waterloo, Ontario, Canada, 1984.

[42] R. Bartels, J. Beatty, and B. Barsky. *An Introduction to Splines for Use in Computer Graphics and Geometric Modeling*. Morgan Kaufmann, 1987.

[43] J. Beck, R. Farouki, and J. Hinds. Surface analysis methods. *IEEE Computer Graphics and Applications*, 6(12):18–36, 1986.

[44] E. Beeker. Smoothing of shapes designed with free-form surfaces. *Computer Aided Design*, 18(4):224–232, 1986.

[45] G. Behforooz and N. Papamichael. End conditions for interpolatory cubic splines with unequally spaced knots. *J of Comp. Applied Math.*, 6(1), 1980.

[46] M. Berger. *Geometry I.* Springer-Verlag, 1987.

[47] S. Bernstein. Démonstration du théorème de Weierstrass fondeé sur le calcul des probabilités. *Harkov Soobs. Matem ob-va*, 13:1–2, 1912.

[48] H. Bez. On invariant curve forms. *Computer Aided Geometric Design*, 3(3):193–204, 1986.

[49] H. Bez and J. Edwards. Distributed algorithm for the planar convex hull algorithm. *Computer Aided Design*, 22(2):81–86, 1990.

[50] P. Bézier. Définition numérique des courbes et surfaces I. *Automatisme*, XI:625–632, 1966.

[51] P. Bézier. Définition numérique des courbes et surfaces II. *Automatisme*, XII:17–21, 1967.

[52] P. Bézier. Procédé de définition numérique des courbes et surfaces non mathématiques. *Automatisme*, XIII(5):189–196, 1968.

[53] P. Bézier. *Numerical Control: Mathematics and Applications.* Wiley, 1972. translated by R. Forrest.

[54] P. Bézier. Mathematical and practical possibilities of UNISURF. In R. Barnhill and R. Riesenfeld, editors, *Computer Aided Geometric Design*, pages 127–152, Academic Press, 1974.

[55] P. Bézier. *Essay de Définition Numérique des Courbes et des Surfaces Expérimentales.* PhD thesis, University of Paris VI, 1977.

[56] P. Bézier. General distortion of an ensemble of biparametric patches. *Computer Aided Design*, 10(2):116–120, 1978.

[57] P. Bézier. *The Mathematical Basis of the UNISURF CAD System.* Butterworths, London, 1986.

[58] G. Birkhoff. *Aesthetic Measure.* Harvard University Press, 1933.

[59] W. Boehm. Parameterdarstellung kubischer und bikubischer Splines. *Computing*, 17:87–92, 1976.

[60] W. Boehm. Cubic B-spline curves and surfaces in computer aided geometric design. *Computing*, 19(1):29–34, 1977.

[61] W. Boehm. Inserting new knots into B-spline curves. *Computer Aided Design*, 12(4):199–201, 1980.

[62] W. Boehm. Generating the Bézier points of B-spline curves and surfaces. *Computer Aided Design*, 13(6):365–366, 1981.

[63] W. Boehm. On cubics: a survey. *Computer Graphics and Image Processing*, 19:201–226, 1982.

[64] W. Boehm. Curvature continuous curves and surfaces. *Computer Aided Geometric Design*, 2(4):313–323, 1985.

[65] W. Boehm. Rational geometric splines. *Computer Aided Geometric Design*, 4(1-2):67–77, 1987.

[66] W. Boehm. Smooth curves and surfaces. In G. Farin, editor, *Geometric Modeling: Algorithms and New Trends*, pages 175–184, SIAM, Philadelphia, 1987.

[67] W. Boehm. On de Boor-like algorithms and blossoming. *Computer Aided Geometric Design*, 5(1):71–80, 1988.

[68] W. Boehm. On the definition of geometric continuity. *Computer Aided Design*, 20(7):370–372, 1988. Letter to the Editor.

[69] W. Boehm. Visual continuity. *Computer Aided Design*, 20(6):307–311, 1988.

[70] W. Boehm. On cyclides in geometric modeling. *Computer Aided Geometric Design*, 7(1-4):243–256, 1990.

[71] W. Boehm. Smooth rational curves. *Computer Aided Design*, 22(1):70, 1990. Letter to the Editor.

[72] W. Boehm and G. Farin. Concerning subdivison of bézier triangles. *Computer Aided Design*, 15(5):260–261, 1983. Letter to the Editor.

[73] W. Boehm, G. Farin, and J. Kahmann. A survey of curve and surface methods in CAGD. *Computer Aided Geometric Design*, 1(1):1–60, 1984.

[74] W. Boehm and D. Hansford. Bézier patches on quadrics. In G. Farin, editor, *NURBS for Curve and Surface Design*, pages 1–14, SIAM, 1991.

[75] W. Boehm and H. Prautzsch. *Numerical Methods*. Vieweg, 1992.

[76] G. Bol. *Projective Differential Geometry, Vol. 1*. Vandenhoeck and Ruprecht, Goettingen, 1950. Vol. 2 in 1954, Vol. 3 in 1967. In German.

[77] J. Brewer and D. Anderson. Visual interaction with Overhauser curves and surfaces. *Computer Graphics*, 11(2):132–137, 1977.

[78] I. Brueckner. Construction of Bézier points of quadrilaterals from those of triangles. *Computer Aided Design*, 12(1):21–24, 1980.

[79] P. Brunet. Increasing the smoothness of bicubic spline surfaces. *Computer Aided Geometric Design*, 2(1-3):157–164, 1985.

[80] B. Buchberger. Applications of Groebner bases in non-linear computational geometry. In J. Rice, editor, *Mathematical Aspects of Scientific Software*, Springer-Verlag, 1988.

[81] C. Calladine. Gaussian curvature and shell structures. In J. Gregory, editor, *The Mathematics of Surfaces*, pages 179–196, Clarendon Press, 1986.

[82] Y. Cao and X. Hua. The convexity of quadratic parametric triangular Bernstein-Bézier surfaces. *Computer Aided Geometric Design*, 8(1):1–6, 1991.

[83] M. Casale and J. Bobrow. A set operation algorithm for sculptured solids modeled with trimmed patches. *Computer Aided Geometric Design*, 6(3):235–248, 1989.

[84] E. Catmull and J. Clark. Recursively generated B-spline surfaces on arbitrary topological meshes. *Computer Aided Design*, 10(6):350–355, 1978.

[85] E. Catmull and R. Rom. A class of local interpolating splines. In R. Barnhill and R. Riesenfeld, editors, *Computer Aided Geometric Design*, pages 317–326, Academic Press, 1974.

[86] G. Chaikin. An algorithm for high speed curve generation. *Computer Graphics and Image Processing*, 3:346–349, 1974.

[87] G. Chang. Matrix formulation of Bézier technique. *Computer Aided Design*, 14(6):345–350, 1982.

[88] G. Chang and P. Davis. The convexity of Bernstein polynomials over triangles. *J Approx Theory*, 40:11–28, 1984.

[89] G. Chang and Y. Feng. An improved condition for the convexity of Bernstein-Bézier surfaces over triangles. *Computer Aided Geometric Design*, 1(3):279–283, 1985.

[90] S. Chang, M. Shantz, and R. Rochetti. Rendering cubic curves and surfaces with integer adaptive forward differencing. *Computer Graphics*, 23(3):157–166, 1989. SIGGRAPH '89 Proceedings.

[91] P. Charrot and J. Gregory. A pentagonal surface patch for computer aided geometric design. *Computer Aided Geometric Design*, 1(1):87–94, 1984.

[92] F. Cheng and B. Barsky. Interproximation: interpolation and approximation using cubic spline curves. *Computer Aided Design*, 23(10):700–706, 1991.

[93] H. Chiyokura and F. Kimura. Design of solids with free-form surfaces. *Computer Graphics*, 17(3):289–298, 1983.

[94] H. Chiyokura, T. Takamura, K. Konno, and T. Harada. g^1 surface interpolation over irregular meshes with rational curves. In G. Farin, editor, *NURBS for Curve and Surface Design*, pages 15–34, SIAM, 1991.

[95] R. Clough and J. Tocher. Finite element stiffness matrices for analysis of plates in blending. In *Proceedings of Conference on Matrix Methods in Structural Analysis*, 1965.

[96] J. Cobb. *A rational bicubic representation of the sphere.* Technical Report, Computer Science, U. of Utah, 1988.

[97] J. Cobb. Concerning Piegl's sphere approximation. *Computer Aided Geometric Design*, 6(1):85, 1989. Letter to the Editor.

[98] E. Cohen. A new local basis for designing with tensioned splines. *ACM Transactions on Graphics*, 6(2):81–122, 1987.

[99] E. Cohen, T. Lyche, and R. Riesenfeld. Discrete B-splines and subdivision techniques in computer aided geometric design and computer graphics. *Comp. Graphics and Image Process.*, 14(2):87–111, 1980.

[100] E. Cohen and L. Schumaker. Rates of convergence of control polygons. *Computer Aided Geometric Design*, 2(1-3):229–235, 1985.

[101] S. Coons. *Surfaces for computer aided design.* Technical Report, MIT, 1964. Available as AD 663 504 from the National Technical Information Service, Springfield, VA 22161.

[102] S. Coons. *Surfaces for computer aided design of space forms.* Technical Report, MIT, 1967. Project MAC-TR 41.

[103] S. Coons. *Rational bicubic surface patches.* Technical Report, MIT, 1968. Project MAC.

[104] S. Coons. Surface patches and B-spline curves. In R. Barnhill and R. Riesenfeld, editors, *Computer Aided Geometric Design*, pages 1–16, Academic Press, 1974.

[105] S. Coons. *Méthode Matricielle.* Hermes, Paris, 1987. Translation from English by P. Bézier and M. Moronval.

[106] M. Cox. The numerical evaluation of B-splines. *J Inst. Maths. Applics.*, 10:134–149, 1972.

[107] H. Coxeter. *Introduction to Geometry.* Wiley, 1961.

[108] W. Dahmen. Subdivision algorithms converge quadratically. *J. of Computtional and Applied Mathematics*, 16:145–158, 1986.

[109] M. Daniel and J. Daubisse. The numerical problem of using Bézier curves and surfaces in the power basis. *Computer Aided Geometric Design*, 6(2):121–128, 1989.

[110] P. Davis. *Interpolation and Approximation.* Dover, New York, 1975. First edition 1963.

[111] P. Davis. Lecture notes on CAGD. 1976. Given at the U. of Utah.

[112] C. de Boor. Bicubic spline interpolation. *J. Math. Phys.*, 41:212–218, 1962.

[113] C. de Boor. On calculating with B-splines. *J Approx. Theory*, 6(1):50–62, 1972.

[114] C. de Boor. *A Practical Guide to Splines*. Springer, 1978.

[115] C. de Boor. B-form basics. In G. Farin, editor, *Geometric Modeling: Algorithms and New Trends*, pages 131–148, SIAM, Philadelphia, 1987.

[116] C. de Boor. Cutting corners always works. *Computer Aided Geometric Design*, 4(1-2):125–131, 1987.

[117] C. de Boor. Local corner cutting and the smoothness of the limiting curve. *Computer Aided Geometric Design*, 7(5):389–398, 1990.

[118] C. de Boor and R. de Vore. A geometric proof of total positivity for spline interpolation. *Math. of Computation*, 45(172):497–504, 1985.

[119] C. de Boor and K. Hollig. B-splines without divided differences. In G. Farin, editor, *Geometric Modeling – Algorithms and New Trends*, pages 21–27, SIAM, Philadelphia, 1987.

[120] C. de Boor, K. Hollig, and M. Sabin. High accuracy geometric Hermite interpolation. *Computer Aided Geometric Design*, 4(4):269–278, 1987.

[121] P. de Casteljau. *Outillages méthodes calcul*. Technical Report, A. Citroen, Paris, 1959.

[122] P. de Casteljau. *Courbes et surfaces à poles*. Technical Report, A. Citroen, Paris, 1963.

[123] P. de Casteljau. *Shape Mathematics and CAD*. Kogan Page, London, 1986.

[124] P. de Casteljau. *Le Lissage*. Hermes, Paris, 1990.

[125] G. de Rham. Un peu de mathématique à propos d'une courbe plane. *Elem. Math.*, 2:73–76; 89–97, 1947. Also in Collected Works, 678-689.

[126] G. de Rham. Sur une courbe plane. *J Math. Pures Appl.*, 35:25–42, 1956. Also in Collected Works, 696-713.

[127] W. Degen. Some remarks on Bézier curves. *Computer Aided Geometric Design*, 5(3):259–268, 1988.

[128] W. Degen. Explicit continuity conditions for adjacent Bézier surface patches. *Computer Aided Geometric Design*, 7(1–4):181–190, 1990.

[129] Y. DeMontaudouin. Resolution of $p(x, y) = 0$. *Computer Aided Design*, 23(9):653–654, 1991.

[130] T. DeRose. *Geometric Continuity: A Parametrization Independent Measure of Continuity for Computer Aided Geometric Design*. PhD thesis, Dept. of Computer Science, U. Calif. at Berkeley, 1985. Also Technical Report UCB/CSD 86/255.

[131] T. DeRose. Composing Bézier simplices. *ACM Transactions on Graphics*, 7(3):198–221, 1988.

[132] T. DeRose. Geometric programming. In *SIGGRAPH '88 course notes*, 1988.

[133] T. DeRose. A coordinate-free approach to geometric programming. In W. Strasser and H. Seidel, editors, *Theory and Practice of Geometric Modeling*, pages 291–306, Springer-Verlag, Berlin, 1989.

[134] T. DeRose. Necessary and sufficient conditions for tangent plane continuity of Bézier surfaces. *Computer Aided Geometric Design*, 7(1-4):165–180, 1990.

[135] T. DeRose. Rational Bezier curves and surfaces on projective domains. In G. Farin, editor, *NURBS for Curve and Surface Design*, pages 35–46, SIAM, 1991.

[136] T. DeRose and B. Barsky. Geometric continuity, shape parameters, and geometric constructions for Catmull-Rom splines. *ACM Transactions on Graphics*, 7(1):1–41, 1988.

[137] T. DeRose and R. Goldman. A tutorial introduction to blossoming. In H. Hagen and D. Roller, editors, *Geometric Modeling*, Springer, 1991.

[138] T. DeRose and T. Holman. *The triangle: a multiprocessor architecture for fast curve and surface generation*. Technical Report 87-08-07, Computer Science Department, U. of Washington, 1987.

[139] J. Dill. An application of color graphics to the display of surface curvature. *Computer Graphics*, 15:153–161, 1981.

[140] M. do Carmo. *Differential Geometry of Curves and Surfaces*. Prentice Hall, Englewood Cliffs, 1976.

[141] T. Dokken. Finding intersections of B-spline represented geometries using recursive subdivision techniques. *Computer Aided Geometric Design*, 2(1-3):189–195, 1985.

[142] T. Dokken, M. Daehlen, T. Lyche, and K. Morken. Good approximation of circles by curvature-continuous Bézier curves. *Computer Aided Geometric Design*, 7(1-4):33–42, 1990.

[143] T. Dokken and A. Ytrehus. Recursive subdivision and iteration in intersections and related problems. In T. Lyche and L. Schumaker, editors, *Mathematical Methods in Computer Aided Geometric Design*, pages 207–214, Academic Press, 1989.

[144] D. Doo and M. Sabin. Behaviour of recursive division surfaces near extraordinary points. *Computer Aided Design*, 10(6):356–360, 1978.

[145] W-H. Du and F. Schmitt. On the G^1 continuity of piecewise Bézier surfaces: a review with new results. *Computer Aided Design*, 22(9):556–573, 1990.

[146] N. Dyn, J. Gregory, and D. Levin. Analysis of uniform binary subdivision schemes for curve design. *Constructive Approximation*, 7(2):127–148, 1991.

[147] N. Dyn, D. Levin, and C. Micchelli. Using parameters to increase smoothness of curves and surfaces generated by subdivision. *Computer Aided Geometric Design*, 7(1-4):129–140, 1990.

[148] N. Dyn and C. Micchelli. *Piecewise polynomial spaces and geometric continuity of curves*. Technical Report, IBM Report RCC11390, Yorktown Heights, 1985.

[149] M. Epstein. On the influence of parametrization in parametric interpolation. *SIAM J Numer. Analysis*, 13(2):261–268, 1976.

[150] G. Farin. *Konstruktion und Eigenschaften von Bézier-Kurven und -Flächen*. Master's thesis, Technical University Braunschweig, FRG, 1977.

[151] G. Farin. *Subsplines ueber Dreiecken*. PhD thesis, Technical University Braunschweig, FRG, 1979.

[152] G. Farin. *Bézier polynomials over triangles and the construction of piecewise C^r polynomials*. Technical Report TR/91, Brunel U., Uxbridge, England, 1980.

[153] G. Farin. A construction for the visual C^1 continuity of polynomial surface patches. *Computer Graphics and Image Processing*, 20:272–282, 1982.

[154] G. Farin. Designing C^1 surfaces consisting of triangular cubic patches. *Computer Aided Design*, 14(5):253–256, 1982.

[155] G. Farin. Visually C^2 cubic splines. *Computer Aided Design*, 14(3):137–139, 1982.

[156] G. Farin. Algorithms for rational Bézier curves. *Computer Aided Design*, 15(2):73–77, 1983.

[157] G. Farin. A modified Clough-Tocher interpolant. *Computer Aided Geometric Design*, 2(1-3):19–27, 1985.

[158] G. Farin. Some remarks on V^2 - splines. *Computer Aided Geometric Design*, 2(2):325–328, 1985.

[159] G. Farin. Piecewise triangular C^1 surface strips. *Computer Aided Design*, 18(1):45–47, 1986.

[160] G. Farin. Triangular Bernstein-Bézier patches. *Computer Aided Geometric Design*, 3(2):83–128, 1986.

[161] G. Farin. Curvature continuity and offsets for piecewise conics. *ACM Transactions on Graphics*, 8(2):89–99, 1989.

[162] G. Farin, editor. *NURBS for Curve and Surface Design*. SIAM, Philadelphia, 1991.

[163] G. Farin. Commutativity between tensor product and Coons-type operators. *Rocky Mtn. J of Math.*, 1992 (to appear).

[164] G. Farin. Degree reduction fairing of cubic B-spline curves. In R. Barnhill, editor, *Geometry Processing for Design and Manufacturing*, pages 87–99, SIAM, Philadelphia, 1992.

[165] G. Farin and P. Barry. A link between Lagrange and Bézier curve and surface schemes. *Computer Aided Design*, 18:525–528, 1986.

[166] G. Farin and H. Hagen. Optimal twist estimation. In H. Hagen, editor, *Surface Design*, SIAM, Philadelphia, 1992.

[167] G. Farin, D. Hansford, and A. Worsey. The singular cases for γ-spline interpolation. *Computer Aided Geometric Design*, 7(6):533–546, 1990.

[168] G. Farin, B. Piper, and A. Worsey. The octant of a sphere as a non-degenerate triangular Bézier patch. *Computer Aided Geometric Design*, 4(4):329–332, 1988.

[169] G. Farin, G. Rein, N. Sapidis, and A. Worsey. Fairing cubic B-spline curves. *Computer Aided Geometric Design*, 4(1-2):91–104, 1987.

[170] G. Farin and N. Sapidis. Curvature and the fairness of curves and surfaces. *IEEE Computer Graphics and Applications*, 9(2):52–57, 1989.

[171] G. Farin and A. Worsey. Reparametrization and degree elevation of rational Bézier curves. In G. Farin, editor, *NURBS for Curve and Surface Design*, pages 47–58, SIAM, 1991.

[172] R. Farouki. Exact offset procedures for simple solids. *Computer Aided Geometric Design*, 2(4):257–279, 1985.

[173] R. Farouki. The approximation of non-degenerate offset surfaces. *Computer Aided Geometric Design*, 3(1):15–43, 1986.

[174] R. Farouki. Direct surface section evaluation. In G. Farin, editor, *Geometric Modeling: Algorithms and New Trends*, pages 319–334, SIAM, Philadelphia, 1987.

[175] R. Farouki. On the stability of transformations between power and Bernstein polynomial forms. *Computer Aided Geometric Design*, 8(1):29–36, 1991.

[176] R. Farouki and J. Hinds. A hierarchy of geometric forms. *IEEE Computer Graphics and Applications*, 5(5):51–78, 1985.

[177] R. Farouki and V. Rajan. On the numerical condition of polynomials in Bernstein form. *Computer Aided Geometric Design*, 4(3):191–216, 1987.

[178] R. Farouki and V. Rajan. Algorithms for polynomials in Bernstein form. *Computer Aided Geometric Design*, 5(1):1–26, 1988.

[179] R. Farouki and T. Sakkalis. Real rational curves are not 'unit speed'. *Computer Aided Geometric Design*, 8(2):151–158, 1991.

[180] I. Faux and M. Pratt. *Computational Geometry for Design and Manufacture.* Ellis Horwood, 1979.

[181] L. Fayard. *Surface Interrogation Using Curvature Plots.* Master's thesis, Dept. of Computer Science, Arizona State U., 1988.

[182] D. Ferguson. Construction of curves and surfaces using numerical optimization techniques. *Computer Aided Design*, 18(1):15–21, 1986.

[183] J. Ferguson. Multivariable curve interpolation. *J ACM*, 11(2):221–228, 1964.

[184] D. Field. Mathematical problems in solid modeling: a brief survey. In G. Farin, editor, *Geometric Modeling: Algorithms and New Trends*, pages 91–107, SIAM, Philadelphia, 1987.

[185] D. Filip. Adaptive subdivision algorithms for a set of Bézier triangles. *Computer Aided Design*, 18(2):74–78, 1986.

[186] N. Fog. Creative definition and fairing of ship hulls using a B-spline surface. *Computer Aided Design*, 16(4):225–230, 1984.

[187] J. Foley and A. Van Dam. *Fundamentals of Interactive Computer Graphics.* Addison-Wesley, 1982.

[188] T. Foley. Local control of interval tension using weighted splines. *Computer Aided Geometric Design*, 3(4):281–294, 1986.

[189] T. Foley. Interpolation with interval and point tension controls using cubic weighted ν-splines. *ACM Trans. on Math. Software*, 13(1):68–96, 1987.

[190] A. Forrest. *Curves and Surfaces for Computer-aided Design.* PhD thesis, Cambridge, 1968.

[191] A. Forrest. Interactive interpolation and approximation by Bézier polynomials. *The Computer J*, 15(1):71–79, 1972. Reprinted in CAD 22(9):527-537,1990.

[192] A. Forrest. On Coons' and other methods for the representation of curved surfaces. *Computer Graphics and Image Processing*, 1(4):341–359, 1972.

[193] A. Forrest. On the rendering of surfaces. *Computer Graphics*, 13(2):253–259, 1979.

[194] A. Forrest. The twisted cubic curve: a computer-aided geometric design approach. *Computer Aided Design*, 12(4):165–172, 1980.

[195] L. Frederickson. *Triangular spline interpolation/generalized triangular splines.* Technical Reports 6/70 and 7/71, Dept. of Math., Lakehead U., Canada, 1971.

[196] F. Fritsch. Energy comparison of Wilson-Fowler splines with other interpolating splines. In G. Farin, editor, *Geometric Modeling: Algorithms and New Trends*, pages 185–201, SIAM, Philadelphia, 1987.

[197] F. Fritsch and R. Carlson. Monotone piecewise cubic interpolation. *SIAM J Numer. Analysis*, 17(2):238–246, 1980.

[198] Q. Fu. The intersection of a bicubic patch and a plane. *Computer Aided Geometric Design*, 7(6):475–488, 1990.

[199] D. Gans. *Transformations and Geometries.* Appleton-Century-Crofts, 1969.

[200] T. Garrity and J. Warren. Geometric continuity. *Computer Aided Geometric Design*, 8(1):51–66, 1991.

[201] G. Geise. Über berührende Kegelschnitte ebener Kurven. *ZAMM*, 42:297–304, 1962.

[202] G. Geise and U. Langbecker. Finite quadratic segments with four conic boundary curves. *Computer Aided Geometric Design*, 7(1-4):141–150, 1990.

[203] A. Geisow. *Surface interrogations.* PhD thesis, U. of East Anglia, 1983.

[204] A. Glassner, editor. *An Introduction to Ray Tracing.* Academic Press, 1989.

[205] M. Goldapp. Approximation of circular arcs by cubic polynomials. *Computer Aided Geometric Design*, 8(3):227–238, 1991.

[206] R. Goldman. Using degenerate Bézier triangles and tetrahedra to subdivide Bézier curves. *Computer Aided Design*, 14(6):307–311, 1982.

[207] R. Goldman. Illicit expressions in vector algebra. *ACM Transactions on Graphics*, 4(3):223–243, 1985.

[208] R. Goldman. The method of resolvents: a technique for the implicitization, inversion, and intersection of non-planar, parametric, rational cubic curves. *Computer Aided Geometric Design*, 2(4):237–255, 1985.

[209] R. Goldman. Blossoming and knot insertion algorithms for B-spline curves. *Computer Aided Geometric Design*, 7(1-4):69–82, 1990.

[210] R. Goldman and B. Barsky. On beta-continuous functions and their application to the construction of geometrically continuous curves and surfaces. In T. Lyche and L. Schumaker, editors, *Mathematical Methods in Computer Aided Geometric Design*, pages 299–312, Academic Press, 1989.

[211] R. Goldman and T. DeRose. Recursive subdivision without the convex hull property. *Computer Aided Geometric Design*, 3(4):247–265, 1986.

[212] R. Goldman and C. Micchelli. Algebraic aspects of geometric continuity. In T. Lyche and L. Schumaker, editors, *Mathematical Methods in Computer Aided Geometric Design*, pages 313–332, Academic Press, 1989.

[213] H. Gonska and J. Meier. A bibliography on approximation of functions by Bernstein type operators. In L. Schumaker and K. Chui, editors, *Approximation Theory IV*, Academic Press, 1983.

[214] T. Goodman. Properties of Beta-splines. *J Approx. Theory*, 44(2):132–153, 1985.

[215] T. Goodman. *Shape preserving interpolation by parametric rational cubic splines*. Technical Report, U. of Dundee, 1988. Department of Mathematics and Computer Science.

[216] T. Goodman. Constructing piecewise rational curves with Frenet frame continuity. *Computer Aided Geometric Design*, 7(1-4):15–32, 1990.

[217] T. Goodman. Closed surfaces defined from biquadratic splines. *Constructive Approximation*, 7(2):149–160, 1991.

[218] T. Goodman. Convexity of Bézier nets on triangulations. *Computer Aided Geometric Design*, 8(2):175–180, 1991.

[219] T. Goodman. Inflections on curves in two and three dimensions. *Computer Aided Geometric Design*, 8(1):37–51, 1991.

[220] T. Goodman, B. Ong, and K. Unsworth. Constrained interpolation using rational cubic splines. In G. Farin, editor, *NURBS for Curve and Surface Design*, SIAM, 1991.

[221] T. Goodman and H. Said. Properties of generalized ball curves and surfaces. *Computer Aided Design*, 23(8):554–560, 1991.

[222] T. Goodman and H. Said. Shape preserving properties of the generalised Ball basis. *Computer Aided Geometric Design*, 8(2):115–122, 1991.

[223] T. Goodman and K. Unsworth. Generation of β-spline curves using a recurrence relation. *NATO ASI series F*, 17:325–353, 1985.

[224] T. Goodman and K. Unsworth. Manipulating shape and producing geometric continuity in beta-spline surfaces. *IEEE Computer Graphics and Applications*, 6(2):50–56, 1986.

[225] W. Gordon. Blending-function methods of bivariate and multivariate interpolation and approximation. *SIAM J Numer. Analysis*, 8(1):158–177, 1969.

[226] W. Gordon. Distributive lattices and the approximation of multivariate functions. In I. Schoenberg, editor, *Approximation with Special Emphasis on Splines*, University of Wisconsin Press, Madison, 1969.

[227] W. Gordon. *Free-form Surface Interpolation through Curve Networks*. Technical Report GMR-921, General Motors Research Laboratories, 1969.

[228] W. Gordon. Spline-blended surface interpolation through curve networks. *J of Math. and Mechanics*, 18(10):931–952, 1969.

[229] W. Gordon and R. Riesenfeld. B-spline curves and surfaces. In R. E. Barnhill and R. F. Riesenfeld, editors, *Computer Aided Geometric Design*, pages 95–126, Academic Press, 1974.

[230] T. Gossing. Bulge, shear and squash: a representation for the general conic arc. *Computer Aided Design*, 13(2):81–84, 1981.

[231] J. Gourret, N. Magnenat-Thalmann, and D. Thalmann. Modeling of contact deformations between a synthetic human and its environment. *Computer Aided Design*, 23(7):514–520, 1991.

[232] T. Grandine. Computing zeros of spline functions. *Computer Aided Geometric Design*, 6(2):129–136, 1989.

[233] J. Gregory. Smooth interpolation without twist constraints. In R. E. Barnhill and R. F. Riesenfeld, editors, *Computer Aided Geometric Design*, pages 71–88, Academic Press, 1974.

[234] J. Gregory. N-sided surface patches. In J. Gregory, editor, *The Mathematics of Surfaces*, pages 217–232, Clarendon Press, 1986.

[235] J. Gregory. Geometric continuity. In T. Lyche and L. Schumaker, editors, *Mathematical Methods in Computer Aided Geometric Design*, pages 353–372, Academic Press, 1989.

[236] J. Gregory and P. Charrot. A C^1 triangular interpolation patch for computer-aided geometric design. *Computer Graphics and Image Processing*, 13(1):80–87, 1980.

[237] J. Gregory and J. Hahn. Geometric continuity and convex combination patches. *Computer Aided Geometric Design*, 4(1-2):79–90, 1987.

[238] J. Gregory and M. Safraz. A rational cubic spline with tension. *Computer Aided Geometric Design*, 7(1-4):1–14, 1990.

[239] J. Gregory and J. Zhou. Convexity of Bezier on sub-triangles. *Computer Aided Geometric Design*, 8(3):207–213, 1991.

[240] T. Greville. On the normalization of the B-splines and the location of the nodes for the case of unequally spaced knots. In O. Shisha, editor, *Inequalities*, Academic Press, 1967. Supplement to the paper 'On spline functions' by I. Schoenberg.

[241] T. Greville. Introduction to spline functions. In T. Greville, editor, *Theory and Applications of Spline Functions*, pages 1–36, Academic Press, 1969.

[242] E. Grosse. Tensor spline approximation. *Linear Algebra and its Applications*, 34:29–41, 1980.

[243] H. Hagen. Geometric spline curves. *Computer Aided Geometric Design*, 2(1-3):223–228, 1985.

[244] H. Hagen. Bézier-curves with curvature and torsion continuity. *Rocky Mtn. J of Math.*, 16(3):629–638, 1986.

[245] H. Hagen. Geometric surface patches without twist constraints. *Computer Aided Geometric Design*, 3(3):179–184, 1986.

[246] H. Hagen and G. Bonneau. Variational design of smooth rational Bezier curves. *Computer Aided Geometric Design*, 8(5):393–400, 1991.

[247] H. Hagen and H. Pottmann. Curvature continuous triangular interpolants. In T. Lyche and L. Schumaker, editors, *Mathematical Methods in Computer Aided Geometric Design*, pages 373–384, Academic Press, 1989.

[248] H. Hagen and G. Schulze. Automatic smoothing with geometric surface patches. *Computer Aided Geometric Design*, 4(3):231–236, 1987.

[249] J. Hahn. Geometric continuous patch complexes. *Computer Aided Geometric Design*, 6(1):55–67, 1989.

[250] B. Hamann, G. Farin, and G. Nielson. G^1 surface interpolation based on degree elevated conics. In G. Farin, editor, *NURBS for Curve and Surface Design*, pages 75–86, SIAM, 1991.

[251] J. Hands. Reparametrisation of rational surfaces. In R. Martin, editor, *The Mathematics of Surfaces II*, pages 87–100, Oxford University Press, 1987.

[252] S. Hanna, J. Abel, and D. Greenberg. Intersection of parametric surfaces by means of lookup tables. *IEEE Computer Graphics and Applications*, 3(7):39–48, 1983.

[253] P. Hartley and C. Judd. Parametrization of Bézier-type B-spline curves. *Computer Aided Design*, 10(2):130–134, 1978.

[254] P. Hartley and C. Judd. Parametrization and shape of B-spline curves. *Computer Aided Design*, 12(5):235–238, 1980.

[255] J. Hayes. *New shapes from bicubic splines*. Technical Report, National Physics Laboratory, 1974.

[256] L. Hering. Closed C^2 and C^3 continuous Bézier and B-spline curves with given tangents. *Computer Aided Design*, 15(1):3–6, 1983.

[257] G. Herron. Techniques for visual continuity. In G. Farin, editor, *Geometric Modeling*, pages 163–174, SIAM, Philadelphia, 1987.

[258] D. Hilbert and S. Cohn-Vossen. *Geometry and the Imagination*. Chelsea, New York, 1952.

[259] B. Hinds, J. McCartney, and G. Woods. Pattern development for 3d surfaces. *Computer Aided Design*, 23(8):583–592, 1991.

[260] H. Hochfeld and M. Ahlers. Role of Bézier curves and surfaces in the Volkswagen CAD approach from 1967 to today. *Computer Aided Design*, 22(9):598–608, 1990.

[261] G. Hoelzle. Knot placement for piecewise polynomial approximation of curves. *Computer Aided Design*, 15(5):295–296, 1983.

[262] C. Hoffmann. *Geometric & Solid Modeling*. Morgan Kaufmann, 1989.

[263] D. Hoitsma and M. Lee. Generalized rational B-spline surfaces. In G. Farin, editor, *NURBS for Curve and Surface Design*, pages 87–102, SIAM, 1991.

[264] J. Holladay. Smoothest curve approximation. *Math. Tables and Other Aids to Computation*, 11:233–243, 1957.

[265] K. Hollig and H. Mogerle. G-splines. *Computer Aided Geometric Design*, 7(1-4):197–208, 1990.

[266] M. Hosaka and F. Kimura. Non-four-sided patch expressions with control points. *Computer Aided Geometric Design*, 1(1):75–86, 1984.

[267] J. Hoschek. Detecting regions with undesirable curvature. *Computer Aided Geometric Design*, 1(2):183–192, 1984.

[268] J. Hoschek. Offset curves in the plane. *Computer Aided Design*, 17(2):77–82, 1985.

[269] J. Hoschek. Smoothing of curves and surfaces. *Computer Aided Geometric Design*, 2(1-3):97–105, 1985.

[270] J. Hoschek. Approximate conversion of spline curves. *Computer Aided Geometric Design*, 4(1-2):59–66, 1987.

[271] J. Hoschek. Spline approximation of offset curves. *Computer Aided Geometric Design*, 5(1):33–40, 1988.

[272] J. Hoschek and D. Lasser. *Grundlagen der Geometrischen Datenverarbeitung*. B.G. Teubner, Stuttgart, 1989. English translation: Fundamentals of Computer Aided Geometric Design, Jones and Bartlett, publishers.

[273] J. Hoschek and F. Schneider. Spline conversion for trimmed rational Bézier- and B-spline surfaces. *Computer Aided Design*, 22(9):580–590, 1990.

[274] J. Hoschek and N. Wissel. Optimal approximate conversion of spline curves and spline approximation of offset curves. *Computer Aided Design*, 20(8):475–483, 1988.

[275] E. Houghton, E. Emnett, R. Factor, and L. Sabharwal. Implementation of a divide-and-conquer-method for the intersection of parametric surfaces. *Computer Aided Geometric Design*, 2(1-3):173–184, 1985.

[276] K. Iino and D. Wilde. Subdivision of triangular Bézier patches into rectangular Bézier patches. *Transactions of the ASME*, 1992 (to appear).

[277] T. Jensen. Assembling triangular and rectangular patches and multivariate splines. In G. Farin, editor, *Geometric Modeling: Algorithms and New Trends*, pages 203–220, SIAM, Philadelphia, 1987.

[278] T. Jensen, C. Petersen, and M. Watkins. Practical curves and surfaces for a geometric modeler. *Computer Aided Geometric Design*, 8(5):357–370, 1991.

[279] B. Joe. Knot insertion for beta-spline curves and surfaces. *ACM Transactions on Graphics*, 9(1):41–66, 1990.

[280] S. Jolles. *Die Theorie der Oskulanten und das Sehnensystem der Raumkurve 4. Ordnung, 2. Spezies*. PhD thesis, Technical U. Aachen, 1886.

[281] A. Jones. *An algorithm for convex parametric splines*. Technical Report ETA-TR-29, Boeing Computer Services, 1985.

[282] A. Jones. Shape control of curves and surfaces through constrained optimization. In G. Farin, editor, *Geometric Modeling: Algorithms and New Trends*, pages 265–279, SIAM, Philadelphia, 1987.

[283] A. Jones. Nonrectangular surface patches with curvature continuity. *Computer Aided Design*, 20(6):325–335, 1988.

[284] J. Kahmann. Continuity of curvature between adjacent Bézier patches. In R. Barnhill and W. Boehm, editors, *Surfaces in Computer Aided Geometric Design*, pages 65–76, North-Holland, 1983.

[285] M. Kallay and B. Ravani. Optimal twist vectors as tools for interpolating a network of curves with a minimum surface energy. *Computer Aided Geometric Design*, 7(6):465–474, 1990.

[286] K. Kato. Generation of n-sided surface patches with holes. *Computer Aided Design*, 23(10):676–683, 1991.

[287] E. Kaufmann and R. Klass. Smoothing surfaces using reflection lines for families of splines. *Computer Aided Design*, 20(6):312–316, 1988.

[288] J. Kjellander. Smoothing of bicubic parametric surfaces. *Computer Aided Design*, 15(5):288–293, 1983.

[289] J. Kjellander. Smoothing of cubic parametric splines. *Computer Aided Design*, 15(3):175–179, 1983.

[290] R. Klass. Correction of local surface irregularities using reflection lines. *Computer Aided Design*, 12(2):73–77, 1980.

[291] R. Klass. An offset spline approximation for plane cubics. *Computer Aided Design*, 15(5):296–299, 1983.

[292] L. Kocić. Modification of Bézier curves and surfaces by degree elevation technique. *Computer Aided Design*, 23(10):692–699, 1991.

[293] P. Korovkin. *Linear Operators and Approximation Theory*. Hindustan Publishing Co., Delhi, 1960.

[294] M. Kosters. Curvature-dependent parametrization of curves and surfaces. *Computer Aided Design*, 23(8):569–578, 1991.

[295] M. Lachance and A. Schwartz. Four point parabolic interpolation. *Computer Aided Geometric Design*, 8(2):143–150, 1991.

[296] C. Lacombe and C. Bédard. Interpolation function over a general triangular mid-edge finite element. *Comp. Math. Appl.*, 12A(3):362–373, 1986.

[297] P. Lancaster and K. Salkauskas. *Curve and Surface Fitting*. Academic Press, 1986.

[298] J. Lane and R. Riesenfeld. A theoretical development for the computer generation and display of piecewise polynomial surfaces. *IEEE Trans. Pattern Analysis Machine Intell.*, 2(1):35–46, 1980.

[299] J. Lane and R. Riesenfeld. A geometric proof for the variation diminishing property of B-spline approximation. *J of Approx. Theory*, 37:1–4, 1983.

[300] D. Lasser. Bernstein-Bézier representation of volumes. *Computer Aided Geometric Design*, 2(1-3):145–150, 1985.

[301] D. Lasser. Intersection of parametric surfaces in the Bernstein-Bézier representation. *Computer Aided Design*, 18(4):186–192, 1986.

[302] D. Lasser and A. Purucker. B-spline-Bezier representations of rational geometric spline curves: quartics and quintics. In G. Farin, editor, *NURBS for Curve and Surface Design*, pages 115–130, SIAM, 1991.

[303] E. Lee. The rational Bézier representation for conics. In G. Farin, editor, *Geometric Modeling: Algorithms and New Trends*, pages 3–19, SIAM, Philadelphia, 1987.

[304] E. Lee. Choosing nodes in parametric curve interpolation. *Computer Aided Design*, 21(6), 1989. Presented at the SIAM Applied Geometry meeting, Albany, N.Y., 1987.

[305] E. Lee. A note on blossoming. *Computer Aided Geometric Design*, 6(4):359–362, 1989.

[306] E. Lee. Energy, fairness, and a counterexample. *Computer Aided Design*, 22(1):37–40, 1990.

[307] E. Lee and M. Lucian. Moebius reparametrizations of rational B-splines. *Computer Aided Geometric Design*, 8(3):213–216, 1991.

[308] J. Levin. Mathematical models for determining the intersection of quadric surfaces. *Computer Graphics and Image Processing*, 11:73–87, 1979.

[309] G. Levner, P. Tassinari, and D. Marini. A simple method for raytracing bicubic surfaces. In R. Earnshaw, editor, *Theoretical Foundations of Computer Graphics and CAD*, pages 805–820, Springer Verlag, 1988.

[310] J. Lewis. 'B-spline' bases for splines under tension, nu-splines, and fractional order splines. 1975. Presented at the SIAM-SIGNUM-meeting, San Francisco, Dec. 3-5.

[311] J. Li, J. Hoschek, and E. Hartmann. G^{n-1} functional splines for interpolation and approximation of curves, surfaces and solids. *Computer Aided Geometric Design*, 7(1-4):209–220, 1990.

[312] S. Lien, M. Shantz, and V. Pratt. Adaptive forward differencing for rendering curves and surfaces. *Computer Graphics*, 21, 1987. SIGGRAPH '87 proceedings.

[313] R. Liming. *Practical Analytical Geometry with Applications to Aircraft*. Macmillan, 1944.

[314] R. Liming. *Mathematics for Computer Graphics*. Aero Publishers, 1979.

[315] D. Liu. GC^1 continuity conditions between two adjacent rational Bézier surface patches. *Computer Aided Geometric Design*, 7(1-4):151–164, 1990.

[316] D. Liu and J. Hoschek. GC^1 continuity conditions between adjacent rectangular and triangular Bézier surface patches. *Computer Aided Design*, 21(4):194–200, 1989.

[317] S. Lodha and J. Warren. Bézier representation for quadric surface patches. *Computer Aided Design*, 22(9):574–579, 1990.

[318] C. Loop and T. DeRose. Generalized B-spline surfaces of arbitrary topology. *Computer Graphics*, 24(4):347–356, 1990.

[319] G. Lorentz. *Bernstein Polynomials*. Toronto Press, 1953. 2nd ed., Chelsea 1986.

[320] M. Lucian. Convergence of rational Bezier curve degree elevation. In G. Farin, editor, *NURBS for Curve and Surface Design*, pages 131–140, SIAM, Philadelphia, 1991.

[321] T. Lyche and V. Morken. Knot removal for parametric B-spline curves and surfaces. *Computer Aided Geometric Design*, 4(3):217–230, 1987.

[322] B. Mandelbrot. *The Fractal Geometry of Nature*. Freeman, San Francisco, 1983.

[323] J. Manning. Continuity conditions for spline curves. *The Computer J*, 17(2):181–186, 1974.

[324] D. Manocha and J. Canny. Rational curves with polynomial parametrization. *Computer Aided Design*, 23(9):653–653, 1991.

[325] M. Mantyla. *An Introduction to Solid Modeling*. Computer Science Press, Rockville, Md, 1988.

[326] R. Markot and R. Magedson. Solutions of tangential surface and curve intersections. *Computer Aided Design*, 21(7):421–429, 1989.

[327] R. Markot and R. Magedson. Procedural method for evaluating the intersection curves of two parametric surfaces. *Computer Aided Design*, 23(6):395–404, 1991.

[328] J. Marshall and A. Mitchell. Blending interpolants in the finite element method. *Int. J Numer. Meth. Eng.*, 12:77–83, 1978.

[329] D. McAllister and J. Roulier. Interpolation by convex quadratic splines. *Mathematics of Computation*, 32(144):1154–1162, 1978.

[330] D. McConalogue. A quasi-intrinsic scheme for passing a smooth curve through a discrete set of points. *The Computer J*, 13:392–396, 1970.

[331] D. McConalogue. Algorithm 66 – an automatic French-curve procedure for use with an incremental plotter. *The Computer J*, 14:207–209, 1971.

[332] H. McLaughlin. Shape preserving planar interpolation: an algorithm. *IEEE Computer Graphics and Applications*, 3(3):58–67, 1985.

[333] A. Meek and R. Thomas. A guided clothoid spline. *Computer Aided Geometric Design*, 8(2):163–174, 1991.

[334] C. Micchelli and H. Prautzsch. Computing surfaces invariant under subdivision. *Computer Aided Geometric Design*, 4(4):321–328, 1987.

[335] J. Miller. Sculptured surfaces in solid models: issues and alternative approaches. *IEEE Computer Graphics and Applications*, 6(12):37–48, 1986.

[336] C. Millham and A. Meyer. Modified Hermite quintic curves and applications. *Computer Aided Design*, 23(10):707–712, 1991.

[337] F. Moebius. *August Ferdinand Moebius, Gesammelte Werke*. Verlag von S. Hirzel, 1885. Also published by Dr. M. Saendig oHG, Wiesbaden, FRG, 1967.

[338] P. Montès. Kriging interpolation of a Bézier curve. *Computer Aided Design*, 23(10):713–716, 1991.

[339] M. Mortenson. *Geometric Modeling*. Wiley, 1985.

[340] F. Munchmeyer. On surface imperfections. In R. Martin, editor, *The Mathematics of Surfaces II*, pages 459–474, Oxford University Press, 1987.

[341] F. Munchmeyer. Shape interrogation: a case study. In G. Farin, editor, *Geometric Modeling: Algorithms and New Trends*, pages 291–301, SIAM, Philadelphia, 1987.

[342] L. Nachman. Blended tensor product B-spline surface. *Computer Aided Design*, 20(6):336–340, 1988.

[343] L. Nachman. A note on control polygons and derivatives. *Computer Aided Geometric Design*, 8(3):223–226, 1991.

[344] A. Nasri. Boundary-corner control in recursive-subdivision surfaces. *Computer Aided Design*, 23(6):405–410, 1991.

[345] A. Nasri. Surface interpolation on irregular networks with normal conditions. *Computer Aided Geometric Design*, 8(1):89–96, 1991.

[346] G. Nielson. Some piecewise polynomial alternatives to splines under tension. In R. E. Barnhill and R. F. Riesenfeld, editors, *Computer Aided Geometric Design*, pages 209–235, Academic Press, 1974.

[347] G. Nielson. The side-vertex method for interpolation in triangles. *J of Approx. Theory*, 25:318–336, 1979.

[348] G. Nielson. Minimum norm interpolation in triangles. *SIAM J Numer. Analysis*, 17(1):46–62, 1980.

[349] G. Nielson. A rectangular nu-spline for interactive surface design. *IEEE Computer Graphics and Applications*, 6(2):35–41, 1986.

[350] G. Nielson. Coordinate free scattered data interpolation. In L. Schumaker, editor, *Topics in Multivariate Approximation*, Academic Press, 1987.

[351] G. Nielson. A transfinite, visually continuous, triangular interpolant. In G. Farin, editor, *Geometric Modeling: Algorithms and New Trends*, pages 235–246, SIAM, Philadelphia, 1987.

[352] G. Nielson and T. Foley. A survey of applications of an affine invariant norm. In T. Lyche and L. Schumaker, editors, *Mathematical Methods in CAGD*, pages 445–467, Academic Press, 1989.

[353] H. Nowacki, D. Liu, and X. Lu. Fairing Bézier curves with constraints. *Computer Aided Geometric Design*, 7(1-4):43–56, 1990.

[354] A. Overhauser. *Analytic definition of curves and surfaces by parabolic blending*. Technical Report, Ford Motor Company, 1968.

[355] D. Parkinson and D. Moreton. Optimal biarc-curve fitting. *Computer Aided Design*, 23(6):411–419, 1991.

[356] R. Patterson. Projective transformations of the parameter of a rational Bernstein-Bézier curve. *ACM Transactions on Graphics*, 4:276–290, 1986.

[357] T. Pavlidis. Curve fitting with conic splines. *ACM Transactions on Graphics*, 2(1):1–31, 1983.

[358] J. Pegna and F. Wolter. A simple pratical criterion to guarantee second order smoothness of blend surfaces. 1988. Preprint MIT.

[359] J. Pegna and F. Wolter. Geometric criteria to guarantee curvature continuity of blend surfaces. *ASME Transactions, J. of Mech. Design*, 114, 1992.

[360] Q. Peng. An algorithm for finding the intersection lines between two B-spline surfaces. *Computer Aided Design*, 16(4):191–196, 1984.

[361] M. Penna and R. Patterson. *Projective Geometry and its Applications to Computer Graphics*. Prentice Hall, 1986.

[362] G. Peters. Interactive computer graphics application of the parametric bicubic surface to engineering design problems. In R. Barnhill and R. Riesenfeld, editors, *Computer Aided Geometric Design*, pages 259–302, Academic Press, 1974.

[363] J. Peters. Local smooth surface interpolation: a classification. *Computer Aided Geometric Design*, 7(1-4):191–196, 1990.

[364] J. Peters. Smooth mesh interpolation with cubic patches. *Computer Aided Design*, 22(2):109–120, 1990.

[365] J. Peters. Smooth interpolation of a mesh of curves. *Constructive Approximation*, 7(2):221–247, 1991.

[366] C. Petersen. Adaptive contouring of three-dimensional surfaces. *Computer Aided Geometric Design*, 1(1):61–74, 1984.

[367] J. Peterson. Degree reduction of Bézier curves. *Computer Aided Design*, 23(6):460–461, 1991. Letter to the Editor.

[368] L. Piegl. A geometric investigation of the rational Bézier scheme in computer aided geometric design. *Computers in Industry*, 7(5):401–410, 1986.

[369] L. Piegl. The sphere as a rational Bézier surface. *Computer Aided Geometric Design*, 3(1):45–52, 1986.

[370] L. Piegl. Interactive data interpolation by rational Bézier curves. *IEEE Computer Graphics and Applications*, 7(4):45–58, 1987.

[371] L. Piegl. On the use of infinite control points in CAGD. *Computer Aided Geometric Design*, 4(1-2):155–166, 1987.

[372] L. Piegl. Hermite- and Coons-like interpolants using rational Bézier approximation form with infinite control points. *Computer Aided Design*, 20(1):2–10, 1988.

[373] L. Piegl. On NURBS: a survey. *Computer Graphics and Applications*, 11(1):55–71, 1990.

[374] L. Piegl and W. Tiller. Curve and surface constructions using rational B-splines. *Computer Aided Design*, 19(9):485–498, 1987.

[375] B. Piper. Visually smooth interpolation with triangular Bézier patches. In G. Farin, editor, *Geometric Modeling: Algorithms and New Trends*, pages 221–233, SIAM, Philadelphia, 1987.

[376] A. Pobegailo. Local interpolation with weight functions for variable-smoothness curve design. *Computer Aided Design*, 23(8):579–582, 1991.

[377] T. Poeschl. Detecting surface irregularities using isophotes. *Computer Aided Geometric Design*, 1(2):163–168, 1984.

[378] H. Pottmann. Curves and tensor product surfaces with third order geometric continuity. In S. Slaby and H. Stachel, editors, *Proceedings of the Third International Conference on Engineering Graphics and Descriptive Geometry*, pages 107–116, 1988.

[379] H. Pottmann. Projectively invariant classes of geometric continuity for CAGD. *Computer Aided Geometric Design*, 6(4):307–322, 1989.

[380] H. Pottmann. A projectively invariant characterization of G^2 continuity for rational curves. In G. Farin, editor, *NURBS for Curve and Surface Design*, pages 141–148, SIAM, Philadelphia, 1991.

[381] M. Pratt. Cyclides in computer aided geometric design. *Computer Aided Geometric Design*, 7(1-4):221–242, 1990.

[382] M. Pratt and A. Geisow. Surface/surface intersection problems. In J. Gregory, editor, *The Mathematics of Surfaces*, Clarendon Press, 1986.

[383] H. Prautzsch. Degree elevation of B-spline curves. *Computer Aided Geometric Design*, 1(12):193–198, 1984.

[384] H. Prautzsch and C. Micchelli. Computing curves invariant under halving. *Computer Aided Geometric Design*, 4(1-2):133–140, 1987.

[385] H. Prautzsch and B. Piper. A fast algorithm to raise the degree of B-spline curves. *Computer Aided Geometric Design*, 8(4):253–266, 1991.

[386] L. Ramshaw. *Blossoming: a connect-the-dots approach to splines*. Technical Report, Digital Systems Research Center, Palo Alto, Ca, 1987.

[387] L. Ramshaw. Béziers and B-splines as multiaffine maps. In R. Earnshaw, editor, *Theoretical Foundations of Computer Graphics and CAD*, pages 757–776, Springer Verlag, 1988.

[388] L. Ramshaw. Blossoms are polar forms. *Computer Aided Geometric Design*, 6(4):323–359, 1989.

[389] T. Rando and J. Roulier. Designing faired parametric surfaces. *Computer Aided Design*, 23(7):492–497, 1991.

[390] D. Reese, M. Reidger, and R. Lang. *Flaechenhaftes glaetten und veraendern von Schiffsoberflaechen*. Technical Report MTK 0243, T. U. Berlin, 1983.

[391] G. Renner. Inter-patch continuity of surfaces. In R. Martin, editor, *The Mathematics of Surfaces II*, pages 237–254, Oxford University Press, 1987.

[392] A. Renyi. *Wahrscheinlichkeitsrechnung*. VEB Deutscher Verlag der Wissenschaften, 1962.

[393] W. Renz. Interactive smoothing of digitized point data. *Computer Aided Design*, 14(5):267–269, 1982.

[394] A. Requicha. Representations for rigid solids: theory, methods, and systems. *ACM Comp. Surveys*, 12:437–464, 1980.

[395] R. Riesenfeld. *Applications of B-spline Approximation to Geometric Problems of Computer-aided Design*. PhD thesis, Dept. of Computer Science, Syracuse U., 1973.

[396] R. Riesenfeld. On Chaikin's algorithm. *IEEE Computer Graphics and Applications*, 4(3):304–310, 1975.

[397] D. Rogers and L. Adlum. Dynamic rational B-spline surfaces. *Computer Aided Design*, 22(9):609–616, 1990.

[398] J. Rossignac and A. Requicha. Offsetting operations in solid modelling. *Computer Aided Geometric Design*, 3(2):129–148, 1986.

[399] J. Roulier and E. Passow. Monotone and convex spline interpolation. *SIAM J Numer. Analysis*, 14(5):904–909, 1977.

[400] C. Runge. Ueber empirische Funktionen und die Interpolation zwischen aequidistanten Ordinaten. *ZAMM*, 46:224–243, 1901.

[401] M. Sabin. *The Use of Piecewise Forms for the Numerical Representation of Shape*. PhD thesis, Hungarian Academy of Sciences, Budapest, Hungary, 1976.

[402] M. Sabin. A method for displaying the intersection curve of two quadric surfaces. *The Computer J*, 19:336–338, 1976.

[403] M. Sabin. Recursive subdivision. In J. Gregory, editor, *The Mathematics of Surfaces*, pages 269–281, Clarendon Press, 1986.

[404] M. Sabin. Some negative results in n-sided patches. *Computer Aided Design*, 18(1):38–44, 1986.

[405] M. Sabin and F. Kimura. Concerning n-sided patches. *Computer Aided Geometric Design*, 1(3):289–290, 1984. Letters to the Editor.

[406] P. Sablonnière. Spline and Bézier polygons associated with a polynomial spline curve. *Computer Aided Design*, 10(4):257–261, 1978.

[407] P. Sablonnière. *Bases de Bernstein et Approximants Splines*. PhD thesis, U. of Lille, 1982.

[408] P. Sablonnière. Interpolation by quadratic splines on triangles and squares. *Computers in Industry*, 3:45–52, 1982.

[409] P. Sablonnière. Bernstein-Bézier methods for the construction of bivariate spline approximants. *Computer Aided Geometric Design*, 2(1-3):29–36, 1985.

[410] P. Sablonnière. Composite finite elements of class C^k. *J of Computational and Appl. Math.*, 12,13:542–550, 1985.

[411] K. Salkauskas. C^1 splines for interpolation of rapidly varying data. *Rocky Mtn. J of Math.*, 14(1):239–250, 1984.

[412] J. Sánchez-Reyes. Single-valued curves in polar coordinates. *Computer Aided Design*, 22(1):19–26, 1990.

[413] J. Sánchez-Reyes. Single-valued surfaces in cylindrical coordinates. *Computer Aided Design*, 23(8):561–568, 1991.

[414] N. Sapidis. *Algorithms for Locally Fairing B-spline Curves*. Master's thesis, U. of Utah, 1987.

[415] N. Sapidis and G. Farin. Automatic fairing algorithm for B-spline curves. *Computer Aided Design*, 22(2):121–129, 1990.

[416] B. Sarkar and C-H. Meng. Smooth-surface approximation and reverse engineering. *Computer Aided Design*, 23(9):623–628, 1991.

[417] B. Sarkar and C. Menq. Parameter optimization in approximating curves and surfaces to measurement data. *Computer Aided Geometric Design*, 8(4):267–290, 1991.

[418] R. Sarraga. G^1 interpolation of generally unrestricted cubic Bézier curves. *Computer Aided Geometric Design*, 4(1-2):23–40, 1987.

[419] R. Sarraga. Errata: G^1 interpolation of generally unrestricted cubic Bézier curves. *Computer Aided Geometric Design*, 6(2):167–172, 1989.

[420] I. Schoenberg. Contributions to the problem of approximation of equidistant data by analytic functions. *Quart. Appl. Math.*, 4:45–99, 1946.

[421] I. Schoenberg. On variation diminishing approximation methods. In R. E. Langer, editor, *On Numerical Approx.*, pages 249–274, U. of Wisconsin Press, 1953.

[422] I. Schoenberg. On spline functions. In O. Shisha, editor, *Inequalities*, pages 255–291, Academic Press, 1967.

[423] L. Schumaker. *Spline Functions: Basic Theory.* Wiley, 1981.

[424] L. Schumaker. On shape preserving quadratic spline interpolation. *SIAM J Numer. Analysis*, 20(4):854–864, 1983.

[425] L. Schumaker and W. Volk. Efficient evaluation of multivariate polynomials. *Computer Aided Geometric Design*, 3(2):149–154, 1986.

[426] A. Schwartz. Subdividing Bézier curves and surfaces. In G. Farin, editor, *Geometric Modeling: Algorithms and New Trends*, pages 55–66, SIAM, Philadelphia, 1987.

[427] T. Sederberg. Improperly parametrized rational curves. *Computer Aided Geometric Design*, 3(1):67–75, 1986.

[428] T. Sederberg and D. Anderson. Steiner surface patches. *IEEE Computer Graphics and Applications*, 5(5):23–36, 1985.

[429] T. Sederberg, H. Christiansen, and S. Katz. Improved test for closed loops in surface intersections. *Computer Aided Design*, 21(8):505–508, 1989.

[430] T. Sederberg and M. Kakimoto. Approximating rational curves using polymial curves. In G. Farin, editor, *NURBS for Curve and Surface Design*, pages 149–158, SIAM, 1991.

[431] T. Sederberg and R. Meyers. Loop detection in surface patch intersections. *Computer Aided Geometric Design*, 5(2):161–171, 1988.

[432] T. Sederberg and T. Nishita. Geometric hermite approximation of surface patch intersection curves. *Computer Aided Geometric Design*, 8(2):97–114, 1991.

[433] T. Sederberg and S. Parry. A comparison of three curve intersection algorithms. *Computer Aided Design*, 18(1):58–63, 1986.

[434] T. Sederberg and S. Parry. Free-form deformation of solid geometric models. *Computer Graphics*, 20(4):151–160, 1986. SIGGRAPH proceedings.

[435] T. Sederberg and X. Wang. Rational hodographs. *Computer Aided Geometric Design*, 4(4):333–335, 1987.

[436] T. Sederberg, S. White, and A. Zundel. Fat arcs: a bounding region with cubic convergence. *Computer Aided Geometric Design*, 6(3):205–218, 1989.

[437] H. Seidel. Knot insertion from a blossoming point of view. *Computer Aided Geometric Design*, 5(1):81–86, 1988.

[438] H. Seidel. A general subdivision theorem for Bézier triangles. In T. Lyche and L. Schumaker, editors, *Mathematical Methods in Computer Aided Geometric Design*, pages 573–582, Academic Press, 1989.

[439] H. Seidel. Computing B-spline control points. In W. Strasser and H. Seidel, editors, *Theory and Practice of Geometric Modeling*, pages 17–32, Springer-Verlag, Berlin, 1989.

[440] H. Seidel. A new multiaffine approach to B-splines. *Computer Aided Geometric Design*, 6(1):23–32, 1989.

[441] H. Seidel. Symmetric triangular algorithms for curves. *Computer Aided Geometric Design*, 7(1-4):57–68, 1990.

[442] H. Seidel. Computing B-spline control points using polar forms. *Computer Aided Design*, 23(9):634–640, 1991.

[443] H. Seidel. Symmetric recursive algorithms for surfaces: B patches and the de Boor algorithm for polynomials over triangles. *Constructive Approximation*, 7(2):257–279, 1991.

[444] H-P. Seidel. Computing B-spline control points using polar forms. *Computer Aided Design*, 23(9):634–640, 1991.

[445] S. Selesnick. Local invariants and twist vectors in CAGD. *Computer Graphics and Image Processing*, 17(2):145–160, 1981.

[446] M. Shantz and S. Chang. Rendering trimmed NURBS with adaptive forward differencing. *Computer Graphics*, 22(4):189–198, 1988.

[447] S. Shetty and P. White. Curvature-continuous extensions for rational B-spline curves and surfaces. *Computer Aided Design*, 23(7):484–491, 1991.

[448] L. Shirman and C. Séquin. Local surface interpolation with Bézier patches. *Computer Aided Geometric Design*, 4(4):279–295, 1987.

[449] L. Shirman and C. Séquin. Local surface interpolation with shape parameters between adjoining Gregory patches. *Computer Aided Geometric Design*, 7(5):375–388, 1990.

[450] L. Shirman and C. Séquin. Local surface interpolation with Bezier patches: errata and improvements. *Computer Aided Geometric Design*, 8(3):217–222, 1991.

[451] E. Staerk. *Mehrfach differenzierbare Bézierkurven und Bézierflächen*. PhD thesis, T. U. Braunschweig, 1976.

[452] D. Stancu. Some Bernstein polynomials in two variables and their applications. *Soviet Mathematics*, 1:1025–1028, 1960.

[453] B.-Q. Su and D.-Y. Liu. *Computational Geometry*. Academic Press, 1989.

[454] M. Szilvasi-Nagy. Flexible rounding operation for polyhedra. *Computer Aided Design*, 23(9):629–633, 1991.

[455] J. Thompson, Z. Warsi, and C. Mastin. *Numerical Grid Generation: Foundations and Applications*. North-Holland, 1985.

[456] W. Tiller. Rational B-splines for curve and surface representation. *IEEE Computer Graphics and Applications*, 3(6):61–69, 1983.

[457] W. Tiller and E. Hanson. Offsets of two-dimensional profiles. *IEEE Computer Graphics and Applications*, 4:36–46, 1984.

[458] P. Todd and R. McLeod. Numerical estimation of the curvature of surfaces. *Computer Aided Design*, 18(1):33–37, 1986.

[459] C. van Overveld. Family of recursively defined curves, related to the cubic Bézier curve. *Computer Aided Design*, 22(9):591–597, 1990.

[460] J. van Wijk. Bicubic patches for approximating non-rectangular control-point meshes. *Computer Aided Geometric Design*, 3(1):1–13, 1986.

[461] T. Varady. Survey and new results in n-sided patch generation. In R. Martin, editor, *The Mathematics of Surfaces II*, pages 203–236, Oxford University Press, 1987.

[462] T. Varady. Overlap patches: a new scheme for interpolating curve networks with *n*-sided regions. *Computer Aided Geometric Design*, 8(1):7–26, 1991.

[463] T. Varady and M.J. Pratt. Design techniques for the definition of solid objects with free-form geometry. *Computer Aided Geometric Design*, 1(3):207–225, 1984.

[464] D. Vernet. Expression mathématique des formes. *Ingenieurs de l'Automobile,* 10:509–520, 1971.

[465] M. Veron, G. Ris, and J. Musse. Continuity of biparametric surface patches. *Computer Aided Design,* 8(4):267–273, 1976.

[466] K. Vesprille. *Computer Aided Design Applications of the Rational B-spline Approximation Form.* PhD thesis, Syracuse U., 1975.

[467] A. Vinacua and P. Brunet. A construction for VC^1 continuity for rational Bézier patches. In T. Lyche and L. Schumaker, editors, *Mathematical Methods in Computer Aided Geometric Design,* pages 601–611, Academic Press, 1989.

[468] R. Walter. Visibility of surfaces via differential geometry. *Computer Aided Geometric Design,* 7(1-4):353, 1990.

[469] C. Wang. Shape classification of the parametric cubic curve and parametric B-spline cubic curve. *Computer Aided Design,* 13(4):199–206, 1981.

[470] M. Watkins and A. Worsey. Degree reduction for Bézier curves. *Computer Aided Design,* 20(7):398–405, 1988.

[471] U. Wever. Optimal parametrization for cubic splines. *Computer Aided Design,* 23(9):641–644, 1991.

[472] R. Wielinga. Constrained interpolation using Bézier curves as a new tool in computer aided geometric design. In R. Barnhill and R. Riesenfeld, editors, *Computer Aided Geometric Design,* pages 153–172, Academic Press, 1974.

[473] F. Wolter and S. Tuhoy. Curvature computations for degenerate surface patches. *Computer Aided Geometric Design,* 7, 1992.

[474] A. Worsey and G. Farin. An n-dimensional Clough-Tocher element. *Constructive Approximation,* 3:99–110, 1987.

[475] F. Yamaguchi. *Curves and Surfaces in Computer Aided Geometric Design.* Springer, 1988.

[476] C. Yao and J. Rokne. An efficient algorithm for subdividing linear Coons surfaces. *Computer Aided Geometric Design,* 8(4):291–304, 1991.

[477] A. Zenisek. Interpolation polynomials on the triangle. *Numerische Math.,* 15:283–296, 1970.

[478] A. Zenisek. Polynomial approximation on tetrahedrons in the finite element method. *J Approx. Theory,* 7:334–351, 1973.

[479] C.-Z. Zhou. On the convexity of parametric Bézier triangular surfaces. *Computer Aided Geometric Design,* 7(6):459–464, 1990.

[480] J. Zhou. *The Positivity and Convexity of Bézier Polynomials over Triangles.* PhD thesis, Beijing U., 1985.

Index